Unimolecular
Reactions

Unimolecular Reactions

SECOND EDITION

KENNETH A. HOLBROOK
School of Chemistry
University of Hull

MICHAEL J. PILLING
School of Chemistry
University of Leeds

STRUAN H. ROBERTSON
School of Chemistry
University of Leeds

JOHN WILEY & SONS
Chichester · New York · Brisbane · Toronto · Singapore

Other Wiley Editorial Offices

John Wiley & Sons, Inc., 605 Third Avenue,
New York, NY 10158-0012, USA

Jacaranda Wiley Ltd, 33 Park Road, Milton,
Queensland 4064, Australia

John Wiley & Sons (Canada) Ltd, 22 Worcester Road,
Rexdale, Ontario M9W 1L1, Canada

John Wiley & Sons (Asia) Pte Ltd, 2 Clementi Loop #02-01,
Jin Xing Distripark, Singapore 0512

Library of Congress Cataloging-in-Publication Data

Holbrook, Kenneth A.
    Unimolecular reactions / K. A. Holbrook, M. J. Pilling, S. H.
Robertson. – 2nd ed.
      p.   cm.
    Rev. ed. of: Unimolecular reactions / P. J. Robinson. 1972
    Includes bibliographical references (p.   –   ) and index.
    ISBN 0-471-92268-4 (hardcover : alk. paper)
    1. Unimolecular reactions. I. Pilling, M. J. II. Robertson, S. H.
III. Robinson, Peter John. Unimolecular reactions. IV. Title.
QD501.H7137   1996
541.3′ 94–dc20   95-40902
                                                                  CIP

British Library Cataloguing in Publication Data

A catalogue record for this book is available from the British Library

ISBN 0 471 92268 4

Typeset in 10/12pt Times by Keytec Typesetting Ltd, Bridport, Dorset
Printed and bound in Great Britain by Biddles Ltd, Guildford, Surrey
This book is printed on acid-free paper responsibly manufactured from sustainable forestation,
for which at least two trees are planted for each one used for paper production.

To

A. M. H., G. M. P. and T. R.

Preface to the Second Edition

Since the first edition of this book was published in 1972[1], rapid strides have been made in both the theoretical description and the experimental investigation of unimolecular reactions. The nature of vibrational modes at the high energies required for chemical reaction, and the flow of energy between them, have received the attention of both experimentalists and theoreticians[2]. The construction of *ab initio* potential energy surfaces[3] and the understanding of the basis for applying the approach embodied in transition state theory to reactions on such surfaces[4] have advanced considerably. New experimental techniques have been applied to the direct study of the kinetics of unimolecular reactions and of the reverse association (combination or addition) reactions[2,5] and of energy transfer[6]. All of these aspects are covered in this new edition which is, of necessity, more selective and less comprehensive in its coverage of experimental data than was the first. For earlier data the reader is referred to extensive tabulations provided in the first edition[1] and to subsequent review articles[7,8]. Despite these advances, the basis for the description of unimolecular reactions remains unchanged. There is much to be learned and perspective gained from following the historical development of the subject, an approach we have once more adopted. As in the first edition, our aim has been to develop the subject-matter to the point at which the reader is able to apply theoretical results to experimental data and to provide, in effect, a handbook for experimentalists rather than a detailed critique of theory. The structure of the present book is indicated below. Chapter 2 presents a condensed version of the material previously dealt with in the first three chapters of the first edition. The fundamental Lindemann[9] mechanism and its development by Hinshelwood[10], essential for any understanding of modern theories of unimolecular reactions is described and the application to the isomerisation of *cis*-but-2-ene is retained. The *dynamical* theory of Slater[11], which is concerned with detailed treatment of molecular vibrations and the behaviour of particular molecular coordinates as a function of time is given in outline, as is the RRK theory[12], an early statistical theory which formed the basis of the later RRKM theory (Chapter 3). Application of the RRKM theory depends very much on an assessment of the distribution of quantum states of molecular systems, the relevant quantities being the density of quantum states at a given energy and the number of quantum states with energies less

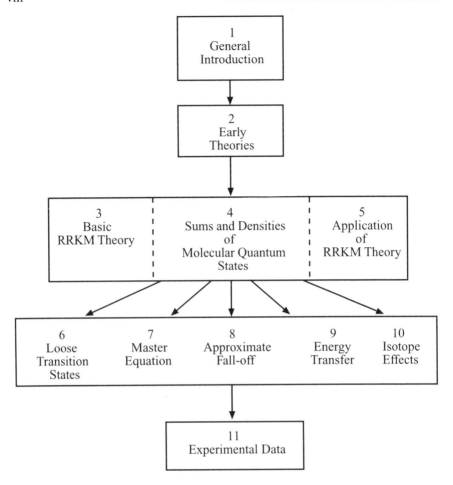

Figure P.1   The logical sequence of chapters in this book.

than this value. The evaluation of these quantities is a common source of difficulty and is dealt with separately in Chapter 4. Chapter 5 discusses the technique of application of the RRKM theory. It includes methods for constructing a model of a given reaction and detailed numerical examples of the application of RRKM theory to typical unimolecular reactions involving 'rigid' and 'loose' transition states. Chapter 6 includes a discussion of models developed to deal with reactions in which there is no well-defined energy maximum along the reaction coordinate, as, for example, in the dissociation of ethane. These models include the variational approach of Bunker and

Pattengill[13], the statistical–adiabatic channel model of Quack and Troe[14] and flexible transition state theory developed by Wardlaw and Marcus[15] (Chapter 6).

The master equation approach is utilised in Chapter 7 in connection with collisional energy transfer and in the treatment of chemically activated species. Chapter 8 discusses approximate energy expressions that have been developed, especially by Troe and his collaborators, to represent and fit fall-off data. The experimental data on inter- and intramolecular energy transfer is presented in Chapter 9 and isotope effects and quantum-mechanical tunnelling are the topics dealt with in Chapter 10. Chapter 11 gives a review of experimental data. Since the first edition, there has been a significant expansion in the number of techniques which have been applied to the study of unimolecular reactions and their reverse association reactions. In addition, unimolecular rate theories are increasingly applied to bimolecular reactions occurring via a bound complex[16]. Finally, ion–molecule reactions continue to be widely studied[17] and there seems, at last, to be some contact between the neutral and ion–molecule communities. The vast amount of experimental data published in these various areas precludes the comprehensive survey which was possible in the first edition, instead, selected examples are presented to illustrate the range of experimental studies undertaken over the last 20 years.

Among the Appendices, the reader's attention is drawn to Appendix 1 which summarises the more important nomenclature used in this book and also to Appendix 2 which contains some relevant statistical mechanical results and derivations.

REFERENCES

1. P. J. Robinson and K. A. Holbrook, *Unimolecular Reactions*, Wiley, New York (1972).
2. Intramolecular kinetics, *Faraday Disc. Chem. Soc.*, **75** (1983).
3. *Potential Energy Surfaces and Dynamics Calculations*, D. G. Truhlar, ed. Plenum, New York (1981).
4. M. S. Child in *Modern Gas Kinetics*, ed. M. J. Pilling and I. W. M. Smith, Blackwell, Oxford (1987).
5. J. E. Baggott and M. J. Pilling, *Ann. Rep. (RSC)*, **79c**, 199 (1982).
6. I. Oref and D. C. Tardy, *Chem. Rev.*, **90**, 1407 (1990).
7. P. J. Robinson in *Reaction Kinetics*, A Specialist Periodical Report, ed. P. G. Ashmore, The Chemical Society, London, Vol. 1, p. 93 (1975).
8. K. A. Holbrook, *Chem. Soc. Rev.*, 163 (1983).
9. F. A. Lindemann, *Trans. Faraday Soc.*, **17**, 598 (1922); see also J. A. Christiansen, Ph.D. thesis, Copenhagen (1921).
10. C. N. Hinshelwood, *Proc. Roy. Soc.* **A113**, 230 (1927).
11. N. B. Slater, *Proc. Camb. Phil. Soc.*, **35**, 56 (1939).

12. L. S. Kassel, *Kinetics of Homogeneous Gas Reactions*, Chemical Catalog Co.,
    New York (1932) and refs. cited therein.
13. D. L. Bunker and M. Pattengill, *J. Chem. Phys.*, **48**, 772 (1968).
14. M. Quack and J. Troe, *Ber. Bunsenges. Phys. Chem.*, **78**, 240 (1974); **79**, 170,
    469 (1975).
15. D. M. Wardlaw and R. A. Marcus, *J. Phys. Chem.*, **90**, 5383 (1986).
16. R. Patrick, J. Barker and D. M. Golden, *J. Phys. Chem.*, **88**, 128 (1994).
17. W. J. Chesnavich and M. T. Bowers, *Gas Phase Ion Chemistry*, ed. M. T.
    Bowers, Academic Press, New York, p. 119 (1979).

Acknowledgements

We wish to acknowledge the help and advice of many colleagues during the preparation of this book. We acknowledge the important contribution of Dr P. J. Robinson to the first edition. We are very grateful to our secretaries especially to Christina Knight, Sandra Walker and Tanya Wainwright for their expertise in dealing patiently with a difficult manuscript, and to Maxine Tyler for her help in preparation of the figures. Our thanks are also due to the editorial staff of John Wiley and Sons Ltd, for their understanding and efficient cooperation.

Contents

1 General Introduction

A unimolecular reaction is in principle the simplest kind of elementary reaction, since it involves the isomerisation or decomposition of a single isolated reactant molecule A through a transition state A^+ which involves no other molecule:

$$A \rightarrow A^+ \rightarrow products$$

This book is concerned almost exclusively with gas-phase reactions, since the so-called 'unimolecular reactions' in condensed phases must necessarily involve participation of the surrounding molecules, and are therefore not unimolecular in the strict sense of the term. Despite their apparent simplicity, gas-phase unimolecular reactions comprise a complex interaction between the collisional processes of energisation and de-energisation of the reactant molecule, and the redistribution of the energy acquired by collision within the molecule.

In order to undergo unimolecular reaction, molecules must first acquire the necessary internal energy; molecules which possess sufficient energy to react are termed *energised molecules* (or sometimes activated molecules). In thermal systems energisation is achieved by intermolecular collisions, although it can also be brought about by other means such as the absorption of radiation. Energised molecules have sufficient energy to react but do not necessarily do so; the energy may need to be redistributed among the internal degrees of freedom of the molecule and even with a suitable energy distribution the molecular vibrations must come correctly into phase before the molecular rearrangement can occur. The molecule then attains a *critical configuration* which is structurally intermediate between reactant and products, and is said to be a *transition state*.

The processes of energy transfer and molecular motion are conveniently discussed in terms of *potential energy surfaces*. The potential energy for a molecule can be expressed in general as a function of n internal coordinates of the molecule. For a non-linear molecule $n = 3N - 6$, and for a linear molecule $n = 3N - 5$, where N is the number of atoms in the molecule. If the potential energy is plotted against these n coordinates, then a hypersurface of dimension $(n + 1)$ results.

The simplest case to consider is that of a non-rotating $(J = 0)$ diatomic molecule which has a single internal coordinate—the internuclear distance. When the potential energy of the system is plotted against this distance a two-dimensional diagram results. For real molecules the result is of the general form shown as a solid curve in Figure 1.1. Various empirical equations have been proposed to represent the shape of this curve, the best known being the Morse equation

$$\text{Potential energy } U(r) = D_e\{1 - \exp[-\beta_M(r - r_e)]\}^2 \qquad (1.1)$$

in which D_e is the electronic binding energy (equal to the ground-state dissociation energy D_0 plus the zero-point energy) and β_M is a constant for a given molecule. Near the equilibrium internuclear distance ($r \approx r_e$), the curve is approximately harmonic. Vibrational energy levels are represented by horizontal lines such as AB in Figure 1.1. After the first few vibrational levels the solid curve diverges from the harmonic curve and tends asymptotically to the energy level representing the two separated atoms at rest at infinite internuclear separation.

In order for unimolecular decomposition of a diatomic molecule to occur, the molecule must acquire the necessary energy, D_0. Since there is only one mode of vibration for a diatomic molecule, every energised molecule will achieve a critical extension (say $r = r_c$) and thus be considered to dissociate during the period of its first vibration. The fraction of molecules with energy

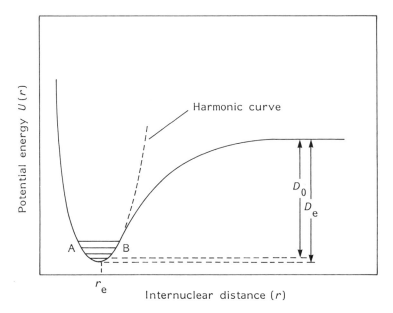

Figure 1.1 The potential energy curve for a diatomic molecule. Reproduced by permission of John Wiley and Sons Ltd.

$\geqslant D_0$ in thermally energised systems is given by the Boltzmann factor $\exp(-D_0/kT)$, and hence the rate constant for the reaction at high pressure is given by (1.2), where v is the vibration frequency of the molecule:

$$k_\infty = v\exp(-D_0/kT) \tag{1.2}$$

For a triatomic molecule restricted to linear configurations, the potential energy can be represented in terms of two internuclear distances. The resulting contour diagram obtained by plotting potential energy for the symmetrical molecule Y_1XY_2 against the distances Y_1X and XY_2 is shown in Figure 1.2. This figure differs from the more familiar contour diagram for the $H + H_2$ system in that the triatomic species exists in a deep potential well and is thus a stable molecule. Normally the molecule vibrates about its mean configuration in its various modes of vibration. The symmetrical vibration corresponding to simultaneous increase or decrease of the two XY distances is represented by a motion along the line AB. Unimolecular dissociation of the molecule into $Y_1 + XY_2$ corresponds more to the asymmetric vibration and is represented by motion along OC, and dissociation into $Y_1X + Y_2$ corresponds to motion along OD. The path OC represents

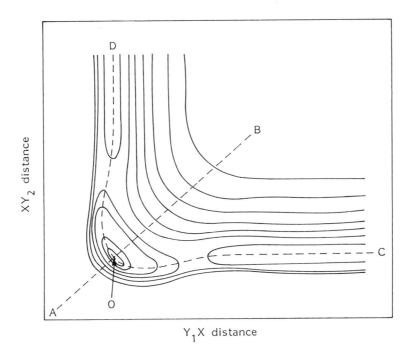

Figure 1.2 The potential energy surface for a symmetrical linear triatomic molecule Y_1XY_2. Reproduced by permission of John Wiley and Sons Ltd.

the lowest energy path on the surface by which the molecule undergoes the reaction Y_1XY_2 and is known as the *reaction coordinate* for this reaction. A section through the surface along the reaction coordinate is the potential energy profile shown in Figure 1.3.

The potential energy profile shows the transition state $(Y_1 \ldots XY_2)^{\ddagger}$ corresponding to the highest energy point on this path. It is customary to regard as transition states those species lying within an arbitrary length δ at the top of the energy barrier, and to consider them as normal molecules except that they have one mode of vibration replaced by a free translational motion along the reaction coordinate (see also sections 3.1 and 3.5). The potential energy profile along the reaction coordinate represents the minimum energy path for the reaction, and the *critical energy* (or threshold energy) E_0 is the difference between the zero-point energies of the reactant and the transition state. In general this critical energy is not exactly equal to the experimental activation energy E_{Arr}, which is defined quite generally by (1.3) in terms of the slope of the Arrhenius plot.

$$E_{\mathrm{Arr}} = -k\frac{\mathrm{d}\ln k}{\mathrm{d}(1/T)} = kT^2\frac{\mathrm{d}\ln k}{\mathrm{d}T} \tag{1.3}$$

The predicted expression for E_{Arr} from any given theoretical treatment is

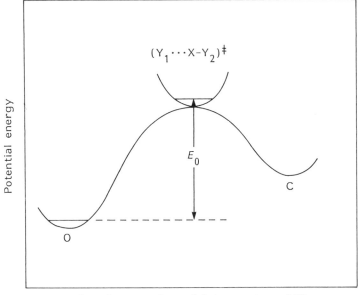

Figure 1.3 Potential energy profile for the reaction $Y_1XY_2 \rightarrow Y_1 + XY_2$. Reproduced by permission of John Wiley and Sons Ltd.

similarly obtained by applying (1.3) to the theoretical expression for k. Tolman[1] showed that E_{Arr} for a unimolecular reaction at its high-pressure limit was equal to the average total energy $\bar{E}^{\ddagger}$ of the molecules which are actually undergoing the reaction (i.e. the transition states) minus the average total energy $\bar{E}$ of all the molecules:

$$E_{Arr} = \bar{E}^{\ddagger} - \bar{E} \qquad (1.4)$$

Since both molecules and transition states have average energies in excess of their zero-point energies, due mainly to vibrational and rotational contributions which will not in general be equal, E_{Arr} will not in general be equal to E_0; usually $E_{Arr} > E_0$. The heat or enthalpy of activation, $\Delta H^{\ddagger}$, is sometimes quoted, and for a unimolecular reaction is equal to $E_{Arr} - kT$. For reactions involving more complicated molecules, the potential energy surfaces are usually too difficult to visualise since too many dimensions are involved. Sometimes it is possible to simplify matters by considering only two coordinates and keeping all others fixed, thus obtaining a section through the complete hypersurface. This is in fact what has been done here with the linear triatomic molecule, since the bending modes of vibration have been ignored.

Accurate calculations of potential energy surfaces by *ab initio* quantum-mechanical methods have been performed in recent years, but only in the simplest cases (e.g. for triatomic systems) can such calculations produce reliable estimates of the activation energy[2]. Approximate calculations such as the 'semiempirical' or 'empirical' methods, on the other hand, have provided much insight into the way in which chemical reactions occur. The treatment of chemical reactions as motions of a point mass over a multi-dimensional surface and the problem of the location of the transition state on such surfaces have been central to the development of chemical reaction rate theory and to unimolecular theory in particular. Such matters are discussed in more detail in later chapters in this book. The most important formulation of these concepts was in the theory put forward by Eyring[3] and by Evans and Polany[4] and now known as canonical transition state theory[5].

1.1 CANONICAL TRANSITION STATE THEORY (CTST)[5]

One of the main assumptions of CTST is that transition states are effectively in equilibrium with reactant molecules. It is argued that the rate of reaction is given by the product of the concentration of transition states passing over the energy barrier located at the col dividing reactants from products on the potential energy surface and the frequency associated with this motion. This frequency is found to be given by kT/h or k and the concentration of transition states is calculated from the statistical-mechanical expression for the equilibrium constant $K^{\ddagger}$ in terms of the appropriate partition functions. For a reaction A + B → products the scheme is

$$A + B \underset{}{\overset{K^{\ddagger}}{\rightleftharpoons}} AB \overset{k^{\ddagger}}{\rightarrow} \text{products} \tag{1.5}$$

and the overall rate constant $k^{\ddagger}$ is then given by the equation

$$k^{\ddagger} = \frac{-\mathrm{d}[\mathrm{A}]/\mathrm{d}t}{[\mathrm{A}][\mathrm{B}]} = \frac{kT}{h} \frac{Q^{\ddagger}}{Q_{\mathrm{A}} Q_{\mathrm{B}}} \exp\left(-E_0/kT\right) \tag{1.6}$$

In this equation E_0 is the critical energy referred to above, Q_{A} and Q_{B} are the complete partition functions per unit volume for the reactants and $Q^{\ddagger}$ is the partition function for all the degrees of freedom of the transition state except the reaction coordinate. The motion in the reaction coordinate has been considered separately and its partition function included in the factor kT/h.

For a unimolecular reaction there is only one reactant and the CTST expression becomes:

$$k_{\infty} = \frac{kT}{h} \frac{Q^{\ddagger}}{Q} \exp\left(-E_0/kT\right) \tag{1.7}$$

it will be seen later that CTST gives the limiting high-pressure rate constant k_{∞} (see section 3.7). The translational contributions to Q and $Q^{\ddagger}$ are identical since the transition state has the same mass as the reactant molecule. Assuming that rotational contributions also cancel and that classical harmonic vibrational partition functions may be inserted ($Q = kT/h\nu$), the approximation (1.8) results in which $\nu_1, \ldots, \nu_n$ are the n frequencies of the molecule and $\nu_1^{\ddagger}, \ldots, \nu_{n-1}^{\ddagger}$ are the $(n-1)$ frequencies of the transition state.

$$k_{\infty} = \frac{\nu_1 \nu_2 \nu_3 \cdots \nu_n}{\nu_1^{\ddagger} \nu_2^{\ddagger} \nu_3^{\ddagger} \cdots \nu_{n-1}^{\ddagger}} \exp\left(-E_0/kT\right) \tag{1.8}$$

The pre-exponential factor at high pressures is therefore generally expected to be of the order of a vibration frequency, i.e. about 10^{13} s^{-1}. Pre-exponential factors of this order of magnitude have, in fact, been found for a large number of unimolecular reactions. If the geometry of the transition state is very different from that of the reactant molecule, the external rotational contributions to Q and $Q^{\ddagger}$ do not cancel and the rate constant is multiplied by a factor $(I_{\mathrm{A}}^{\ddagger} I_{\mathrm{B}}^{\ddagger} I_{\mathrm{C}}^{\ddagger}/I_{\mathrm{A}} I_{\mathrm{B}} I_{\mathrm{C}})^{1/2}$ involving the principal moments of inertia of the reactant molecule and the transition state. When 'loose' transition states are formed, vibrational degrees of freedom in the reactant molecule are replaced by rotational degrees of freedom in the transition state, and this factor contributes to the much higher than normal pre-exponential factors found for such reactions.

1.2 VARIATIONAL TRANSITION STATE THEORY (VTST)

The assumption of CTST that transition states are in equilibrium with reactant molecules is related to the assumption that transition states once formed do not recross the potential barrier at the col on the potential energy surface and re-form reactant molecules.

One of the recent developments of transition state theory is the variational transition state theory put forward by Keck[6] following earlier work by Horiuti[7] and Wigner[8] and developed among others by Truhlar and co-workers[9]. In this theory, a dividing surface or plane perpendicular to the minimum energy path is considered and multiple crossings of this surface by the transition states, some of which return to the reactants, are allowed. The total rate constant is obtained by first calculating microcanonical rate constants at particular fixed energy and angular momentum and then averaging such microcanonical rate constants over a Boltzmann energy distribution.

When various dividing surfaces are considered and the corresponding rate constants calculated, the calculated values are in general higher than the true rate constant which corresponds only to crossings in one direction, from reactants to products. The variational method consists of repeating these calculations until a minimum rate constant is found which may then be regarded as an upper estimate of the true value. This version of transition state theory is of particular importance in classical trajectory studies of unimolecular reactions and also particularly for reactions with no clear maximum in the energy along the reaction path. This and related topics are dealt with in more detail later in Chapter 6.

A microcanonical approach, rather than the canonical approach of CTST, is also required, in theoretical descriptions of unimolecular reactions away from the high-pressure limit, since the competition between reaction and collisional energy transfer is a central aspect of the rate process. The microcanonical rate constant for dissociation is obtained by evaluating the rate of conversion of energised molecules into transition states. The early theories of unimolecular reactions took extreme views of this process, crucial to which is the question of whether all configurations with a given total energy are freely interconvertible by redistribution of the energy between the different degrees of freedom. If they are, then, given sufficient time, every energised molecule may react. This is the case with the 'statistical' theories due to Rice, Ramsperger, Kassel and Marcus (RRKM)[10]. The Slater theory[11], on the other hand, has more stringent requirements concerning energization of the molecule; in the harmonic form of the theory there is no energy flow between different modes of vibration and not all molecules with total energy greater than E_0 are able to react.

Early theories of unimolecular reactions and the more advanced theories such as RRKM theory and the statistical adiabatic channel theory[12,13] have a common mechanism of collisional energisation and de-energisation derived from the original Lindemann mechanism. The historical development of the early theories is outlined in Chapter 2.

REFERENCES

1. R. C. Tolman, *Statistical Mechanics with Applications to Physics and Chemistry*, Chemical Catalog Co., New York (1927).
2. *Potential Energy Surfaces and Dynamics Calculations*, D. G. Truhlar ed. Plenum New York (1981).
3. H. Eyring, *J. Chem. Phys.*, **3**, 107 (1935).
4. M. G. Evans and M. Polanyi, *Trans. Faraday Soc.*, **31**, 875 (1935).
5. K. J. Laidler, *Theories of Chemical Reaction Rates*, McGraw-Hill, London, (1969), Ch. 3.
6. J. C. Keck, *J. Chem. Phys.*, **53**, 2041 (1970).
7. J. Horiuti, *Bull. Chem. Soc. Jpn*, **13**, 210 (1938).
8. E. Wigner, *J. Chem. Phys.*, **5**, 720 (1937).
9. D. G. Truhlar and B. C. Garrett, *Acc. Chem. Res.*, **13**, 440 (1980).
10. D. G. Truhlar W. L. Hase and J. T. Hynes, *J. Phys. Chem.*, **87**, 2664 (1983).
11. N. B. Slater, *Theory of Unimolecular Reactions*, Cornell Univ. Press, Ithaca, (1959).
12. M. Quack and J. Troe, *Ber. Bunsenges. Phys. Chem.*, **78**, 240 (1974); **79**, 170, 469 (1975).
13. D. M. Wardlaw and R. A. Marcus, *J. Phys. Chem.*, **90**, 5383 (1986).

2 Early Theories

2.1 HISTORICAL DEVELOPMENT

At the beginning of the twentieth century, many gas-phase reactions were known to be first-order processes and were assumed to be unimolecular, and unimolecular reactions were thought to be first order under all conditions. Many reactions studied then, such as the thermolyses of simple ketones, aldehydes and ethers, have been subsequently found not to be unimolecular processes according to the modern definition, but to involve free radical chains. Despite this complexity, the early studies of these reactions were important in the development of unimolecular reaction theory as they focused attention upon the central problem of how the reacting molecule acquires the activation energy needed for reaction to take place. It was difficult to see how first-order processes could result if molecules were energised by bimolecular collisions which would be expected to be second-order processes. It was argued (quite wrongly) that, even at very low pressures where molecular collisions were rare, unimolecular reactions would continue to occur with the same first-order rate constant and hence molecular collisions would appear to be unimportant. In 1919, Perrin[1] therefore proposed the radiation hypothesis in which molecules were supposed to acquire energy by the absorption of infrared radiation from the walls of the reaction vessel. The rate constant k for a first-order reaction would then be given by (2.1) in which v is the frequency of the radiation absorbed.

$$k = \text{constant} \times \exp\left(-hv/kT\right) \qquad (2.1)$$

It was soon shown by Langmuir[2] and others that the density of infrared radiation available from the walls at the temperatures concerned was not sufficient to account for the observed reaction rates. In addition, experimental evidence was rapidly accumulated to show that infrared radiation is generally ineffective photochemically, and indeed many molecules do not absorb in the frequency region implied by (2.1) and the observed rate constant. These facts led to the abandonment of the radiation theory and its replacement by theories in which molecular collisions were involved as the means of providing the activation energy.

In the theory of Christiansen and Kramers[3] an overall first-order rate was achieved, despite second-order collisional energisation, by supposing that product molecules were produced with an excess of energy which could be used to re-energise reactant molecules. This theory proved to be unsatisfactory in two major respects. First, most unimolecular reactions are endothermic rather than exothermic processes, and so product molecules are not formed with sufficient internal energy to energise more reactant molecules. Secondly, inert gases would be expected to remove the excess energy of the product molecules and hence to reduce the overall rate of the reaction. In practice it is found that inert gases often increase the rates of unimolecular reactions.

The disadvantages of earlier theories were overcome by the theory of Lindemann, outlined in the next section. The importance of molecular collisions in the energisation process was finally established when it was found that the first-order rate constant for unimolecular reactions is not a true constant but does decline at low pressures. The decline or 'fall-off' in the first-order rate constant with pressure has since become recognized as an important experimental criterion of unimolecular reactions.

A consequence of this decline in the rate constant is that at low pressures the initial rate becomes proportional to the total pressure as well as to the concentration of reactant. The reaction is then second order overall, although for reasons which will be apparent later the time development of a given reaction mixture remains first order. At still lower pressures it is possible that wall effects may become important in energisation processes and there is some evidence that the rate constant then becomes a true first-order constant again.

2.2 THE LINDEMANN THEORY

The theory known as *Lindemann theory* forms the basis for all modern theories of unimolecular reactions and has been developed from ideas published almost simultaneously by Lindemann and Christiansen[4]. The main concepts of the theory can be stated briefly as follows:

1. By collisions, a fraction of the molecules gain energy in excess of a critical quantity E_0. The rate of the energisation process depends upon the rate of bimolecular collisions. In the most general terms this process can be written as

$$A + M \overset{k_1}{\rightarrow} A^* + M \qquad (2.2)$$

where M can represent a product molecule, an added 'inert' gas molecule or a second molecule of reactant. In the simple Lindemann theory k_1 is taken to be energy independent and is calculated from the simple collision theory equation.

2. Energised molecules are de-energised by collision. This is the reverse of process (2.2) and may be written as

$$A^* + M \xrightarrow{k_2} A + M \tag{2.3}$$

The rate constant k_2 is taken to be energy independent, and is equated with the number of collisions experienced by an A^* molecule per second per unit concentration of M, i.e. it is assumed that every collision of A^* leads to de-energisation.

In order to generate first-order kinetics, Lindemann suggested that the energised molecule does not immediately dissociate, but has, at least at high pressures, a lifetime which is sufficiently long that collisional de-energisation is much more rapid than is dissociation. An energy-independent rate constant, k_3, is ascribed to the dissociation/isomerisation process

$$A^* \xrightarrow{k_3} B + C \tag{2.4}$$

where k_3 is much smaller than a vibrational period.

When Lindemann first proposed his theory, he thought of the critical energy acquired by the molecule as rotational energy and wrote of a 'centrifugal bursting' of the molecule. Although the RRKM theory (to be dealt with later) considers rotational energy sometimes to be important, there is no doubt that the major contribution is usually from the vibrational energy of the molecule, and the energisation process is considered largely as one of translational–vibrational or vibrational–vibrational energy transfer.

The consequences of the Lindemann mechanism, expressed by reactions (2.2)–(2.4), may be seen by application of the steady-state hypothesis to the concentration of A^*. The overall rate of reaction (v) is then given by

$$v = k_3[A^*] = \frac{k_1 k_3 [A][M]}{k_2[M] + k_3} = \frac{(k_1 k_3 / k_2)[A]}{1 + k_3 / k_2[M]} \tag{2.5}$$

At high pressures, when $k_2[M] \gg k_3$, (2.5) becomes

$$v = \left(\frac{k_1 k_3}{k_2} \right)[A] = k_\infty[A] \tag{2.6}$$

At low pressures, $k_2[M] \ll k_3$ and (2.5) becomes

$$v_{\text{bim}} = k_1[A][M] = k_{\text{bim}}[A][M] \tag{2.7}$$

i.e. the rate of reaction is then equal to the second-order rate of energisation. In the absence of added gases, if reactant and product molecules differ substantially in their ability to energise molecules of A a more complex treatment is required[5].

It is seen that the Lindemann theory predicts a change in the order of the initial rate of reaction with respect to concentration at low pressures. This is reflected in a decline in the pseudo-first-order rate constant k_{uni} with concentration, where k_{uni} is defined as in

$$k_{uni} = \frac{1}{[A]}\left(\frac{-d[A]}{dt}\right) = \frac{k_1[M]}{1 + k_2[M]/k_3} \tag{2.8}$$

The extent of this 'fall-off' is conveniently measured in terms of the ratio k_{uni}/k_∞, given by

$$k_{uni}/k_\infty = \frac{1}{1 + k_3/k_2[M]} \tag{2.9}$$

and a *transition-pressure* $p_{\frac{1}{2}}$ at which $k_{uni}/k_\infty = \frac{1}{2}$ is then found by putting $k_3 = k_2[M]$. Replacing $[M]$ by the total pressure, we find that $p_{\frac{1}{2}}$ is given by

$$p_{\frac{1}{2}} = k_3/k_2 = k_\infty/k_1 \tag{2.10}$$

Although the 'fall-off curve' of k_{uni} against pressure is not always obtainable over the whole range of pressure, it is sometimes useful to quote the transition pressure for a particular reaction at a particular temperature.

It is easily seen from (2.9) that the predicted effect of adding inert gas below the high-pressure limit of a unimolecular reaction (i.e. in the fall-off region) is to increase the rate constant back to its high-pressure value. The effect is comparable to increasing the pressure of the reactant, although added gases are often less efficient than the reactant itself in the energisation process. Studies of the efficiency of collisional energy transfer and their significance for unimolecular rate theories are treated in more detail in Chapter 9.

The simple conclusion from Lindemann theory concerning the high-pressure rate constant k_∞ is that this is a true constant independent of pressure. No effect of added inert gases would therefore be predicted for a unimolecular reaction in its first-order region, since here the equilibrium proportion of energised molecules is already achieved. It is, however, possible that a small change in k_∞ may occur at very high pressures, due to a non-zero volume of activation $\Delta V^{\ddagger}$; the isomerisation of ethylcyclobutane[6] provides an example of this phenomenon.

2.3 COMPARISON OF LINDEMANN THEORY WITH EXPERIMENT

Early tests of the Lindemann theory were made by comparing the calculated rates of energisation with the observed overall rates of unimolecular reactions. Since many of the examples chosen for such tests are now known not to be simple unimolecular reactions, calculations involving these reactions have now lost their original value. It is instructive instead to apply the simple Lindemann equation to some reliable unimolecular reaction data and to compare the experimental and theoretical results. From equations (2.9) and (2.10) the general first-order rate constant can be written as

$$k_{uni} = \frac{k_\infty}{1 + k_\infty/k_1 p} \qquad (2.11)$$

in which [M] has been replaced by the total pressure p.

The rate constant k_{uni} was calculated from equation (2.11) for the isomerization of cis- to trans-but-2-ene at 469 °C using the experimental data of Rabinovitch and Michel[7]. The experimental high-pressure rate constant at 469 °C is $k_\infty = 1.90 \times 10^{-5}$ s^{-1} and the high-pressure activation energy of $E_\infty = 262.8$ kJ mol^{-1} is simply equated with E_0.[a] The rate constant k_1 was calculated from the collision theory expression

$$k_1 = Z_1 \exp(-E_0/kT) \qquad (2.12)$$

with $Z_1 = 7.66 \times 10^6$ Torr^{-1} s^{-1} calculated from the collision theory expression with the collision diameter σ_d for cis-but-2-ene taken as 5×10^{-10} m.

In Figure 2.1, k_{uni} is plotted logarithmically against reactant pressure p. It will be seen that the transition pressure calculated in this way is about 9×10^6 Torr, whereas the experimental data give a value of 0.04 Torr. Thus although the simple Lindemann theory correctly predicts a fall-off in k_{uni} with pressure, there is a gross discrepancy between the calculated and experimental fall-off curves. This is a general observation for all calculations using the simple Lindemann theory, and the principal reason is the erroneous calculation of k_1 by equation (2.12). This equation gives the frequency of collisions in which the relative kinetic energy along the line of centres is $\geqslant E_0$ and takes no account of the internal energy of the molecules. It will be seen later that a much more realistic model is obtained when this internal energy is taken into account in calculating the rate of energisation. A further consequence of the use of (2.12) to calculate k_1 is seen by rearranging (2.10) to give $k_3 = k_\infty(k_2/k_1)$. Since $k_2 = Z_2 \approx Z_1$ and $E_\infty \approx E_0$ it follows (2.13) that $k_3 \approx A_\infty$.

$$(k_3)_{\text{Lindemann}} = [A_\infty \exp(-E_\infty/kT)]Z_2/[Z_1 \exp(-E_0/kT)] \approx A_\infty \qquad (2.13)$$

The high-pressure A-factor is normally found experimentally to be about $10^{13.5}$ s^{-1}; hence the theory in this form does not require a time-lag since k_3 would similarly be commensurate with a vibration frequency.

---

[a] Values of activation energy and other energies taken from early papers quoted in this book have been converted to kJ mol^{-1} units in accordance with SI convention. The non-SI unit Torr is retained for convenience where 1 Torr $\equiv (101.325/760)$ kN m^{-2}.

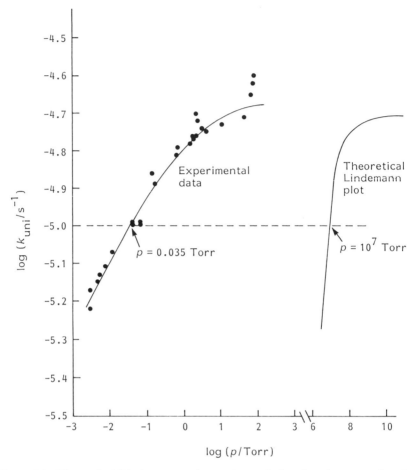

Figure 2.1   Theoretical Lindemann and experimental plots for *cis* → *trans*-but-2-ene at 469 °C.

The 'time-lag' concept therefore requires k_3 to be much less than $10^{13.5}$ s^{-1} and conversely k_1 to be much larger than its value calculated from the collision theory expression. A modification of the theory, due to Hinshelwood, calculates k_1 with allowance for energy in the internal degrees of freedom of the molecule, and is described in section 2.4.

2.4   THE HINSHELWOOD MODIFICATION

The failure of Lindemann theory in its simple form has been illustrated by the calculation of transition pressures $p_{\frac{1}{2}}$ which are much too high to agree with experiment. Alternatively it could be shown that the rate of energisation given by equation (2.12) is too low to account for the observed rate of reaction. Both situations are remedied by an increase in the rate constant k_1. According to simple collision theory, the rate is calculated from the product of the collision number and the chance that two colliding molecules have relative translational energy $\geq E_0$ along the line of centres, i.e. have energy $\geq E_0$ in two classical degrees of freedom. Hinshelwood[8] developed a suggestion of Lindemann's that a more realistic model would be obtained by assuming that the required energy could be drawn in part from the internal degrees of freedom (mainly vibrational) of the reactant molecule. The chance of a molecule containing energy $\geq E_0$ clearly increases with the number of degrees of freedom which contribute, and the rate of energisation is thus increased, as is required. Hinshelwood showed that the probability of a molecule possessing total energy $\geq E_0$ in s classical degrees of freedom is much higher than $\exp(-E_0/kT)$, and is in fact approximately $\exp(-E_0/kT)(E_0/kT)^{s-1}/(s-1)!$ The rate constant k_1 in the modified Hinshelwood–Lindemann theory is therefore given by

$$k_1 = \frac{Z_1}{(s-1)!}\left(\frac{E_0}{kT}\right)^{s-1} \exp(-E_0/kT) \tag{2.14}$$

which even for moderate values of s leads to much bigger values of k_1 than does (2.12).

To derive the Hinshelwood expression for k_1, some results from statistical mechanics are required. The derivation starts from the statement that, in a classical system, the proportion of molecules possessing energy between E and $E + \delta E$ in one degree of vibrational freedom (two square terms) is given by

$$f(E)\delta E = (1/kT)\exp(-E/kT)\delta E \tag{2.15}$$

Now suppose there are s degrees of freedom in a molecule and the energy is distributed in such a way that the amount of energy in the first degree of freedom is between E_1 and $E_1 + \delta E_1$, the amount in the second is between E_2 and $E_2 + \delta E_2$ and so on. The probability of this happening is found by the multiplication law of probabilities—the probability of the simultaneous occurrence of two or more events is the product of the probabilities of the individual events. If E is the total energy (per molecule) in all degrees of freedom, i.e. $E = E_1 + E_2 + \ldots + E_s$, then the fraction of molecules having energy E distributed in this way among s degrees of freedom is given by

$$f(E)\delta E = (1/kT)^s \exp(-E_1/kT)\exp(-E_2/kT) \ldots \delta E_1 \ldots \delta E_s$$
$$= (1/kT)^s \exp(-E/kT)\delta E_1 \ldots \delta E_s \tag{2.16}$$

To obtain the fraction of molecules containing total energy E in a specified range E to $E + \delta E$, this expression has to be integrated for all possible distributions of the energy among the degrees of freedom, and the required result is thus given as equation (2.17).

$$f(E)\delta E = \left(\frac{1}{kT}\right)^s \exp\left(-\frac{E}{kT}\right)\!\int \ldots \int dE_1 \ldots dE_s \qquad (2.17)$$

The multiple integral (2.17) may be evaluated by first expressing the restrictions on E in terms of the Dirac δ function, i.e.

$$\delta(E - \Sigma_i E_i) = \begin{cases} 1 \text{ if } E = \Sigma_i E_i \\ 0 \text{ otherwise} \end{cases}$$

This allows (2.17) to be rewritten in the form

$$f(E)\delta E = \left(\frac{1}{kT}\right)^s \exp\left(-E/kT\right)\int_0^\infty \ldots \int_0^\infty \delta(E - \Sigma_i E_i)\,dE_1 \ldots dE_s \quad (2.18)$$

The multiple integral in (2.18) is of a form that appears regularly in unimolecular theory. Such integrals can be tackled using either Laplace transforms or the general result due to Dirichlet. Section 4.4 gives details of such integrals. Using the results of Section 4.4, it can be shown that the fraction of molecules with energies in the range E to $E + \delta E$, distributed in any way among s degrees of freedom, is then given by

$$f(E)\delta E = \left(\frac{E}{kT}\right)^{s-1} \exp\left(-E/kT\right)\frac{1}{(s-1)!}\left(\frac{\delta E}{kT}\right) \qquad (2.19)$$

In the Hinshelwood–Lindemann mechanism, k_1 is a function of energy, and attention is first concentrated upon the rate constant $k_{1(E \rightarrow E + \delta E)}$ for energisation to a small energy range E to $E + \delta E$. The rate constant k_2 for de-energisation is still assumed to be independent of energy. The Hinshelwood–Lindemann mechanism is then as follows:

$$A + M \xrightarrow{\delta k_{1(E \rightarrow E + \delta E)}} A^*_{(E \rightarrow E + \delta E)} + M$$

$$A^*_{(E \rightarrow E + \delta E)} + M \xrightarrow{k_2} A + M$$

$$A^*_{(E)} \xrightarrow{k_3} \text{products}$$

As in the simple Lindemann treatment, the high-pressure limit corresponds to a situation in which the energised molecules are present at their equilibrium proportion, given now by (2.19) for any specified small energy range. The equilibrium proportion of molecules energised in this range is also given by $\delta k_{1(E \rightarrow E + \delta E)}/k_2$ and if this is equated to the right-hand side of (2.19) an equation for $\delta k_1/k_2$ is obtained. By integrating between the limits $E = E_0$ and ∞, where E_0 is the critical energy per molecule required to bring about unimolecular reaction, equation (2.20) for k_1/k_2 is then obtained. Changing the variable to $\exp(-E/kT)$ and integrating by parts, one obtains equation (2.21).

$$\frac{k_1}{k_2} = \int_{E=E_0}^{\infty} \frac{dk_1}{k_2} = \int_{E=E_0}^{\infty} \frac{1}{(s-1)!}\left(\frac{E}{kT}\right)^{s-1} \exp(-E/kT)\, d(E/kT) \quad (2.20)$$

$$\frac{k_1}{k_2} = \left[\frac{1}{(s-1)!}\left(\frac{E_0}{kT}\right)^{s-1} + \frac{1}{(s-2)!}\left(\frac{E_0}{kT}\right)^{s-2} + \ldots\right]\exp(-E_0/kT) \quad (2.21)$$

Some calculations of the relative magnitudes of the first and second terms in equation (2.21) for some unimolecular reactions are given in Table 2.1, and it can be seen that since usually $E_0 \gg (s-1)kT$, the second term in (2.21) can normally be neglected compared with the first. Neglect of the second term may cause an error in k_1 by a factor of 1.2, but a much larger error in k_1 actually occurs due to the uncertainty in the value of s which is required by the theory. If the second term is neglected, then k_1/k_2 is given by

$$\frac{k_1}{k_2} = \frac{1}{(s-1)!}\left(\frac{E_0}{kT}\right)^{s-1} \exp(-E_0/kT) \quad (2.22)$$

and, if $k_2 = Z_2 \approx Z_1$, one obtains the Hinshelwood expression for k_1 given previously (equation 2.14). The high-pressure rate constant according to the Hinshelwood–Lindemann theory is given by

$$k_\infty = \frac{k_1 k_3}{k_2} = \frac{k_3}{(s-1)!}\left(\frac{E_0}{kT}\right)^{s-1} \exp(-E_0/kT) \quad (2.23)$$

Equations (2.14) for k_1 and (2.23) for k_∞ are thus the basic Hinshelwood–Lindemann equations for the rate constants at low and high pressures, and the essential features of the results will be illustrated in section 2.5.

From equations (2.14) and (2.23) and utilising (1.3) we can derive the Arrhenius activation energies E_{bim} and E_∞ predicted by the theory for the low- and high-pressure regions. The activation energy at low pressures,

Table 2.1 Comparison of terms in equation (2.21)

Unimolecular reaction	$N_A E_0$ (kJ mol⁻¹)	RT (kJ mol⁻¹)	$N_A E_0/RT$	s	2nd term in (2.21) / 1st term in (2.21)
$N_2O \rightarrow N_2 + O$	251.0	7.45	34	2^a	0.03
$O_3 \rightarrow O_2 + O$	104.6	3.14	34	3^a	0.06
$H_2O_2 \rightarrow 2OH$	200.8	6.07	33	6^a	0.15
$N_2O_5 \rightarrow NO_3 + NO_2$	104.6	2.51	41	7^a	0.15
Cyclopropane $\rightarrow$ propene	271.9	6.61	41	7^b	0.15
$C_2H_5\,Cl \rightarrow C_2H_4 + HCl$	251.0	6.49	39	8^b	0.18

[a]Value for which HL theory predicts approximately the correct value of k_{bim}.
[b]Value for which HL theory predicts approximately the correct value of $p_{1/2}$.

associated with k_1 in concentration units ($Z_1 \propto T^{1/2}$) is thus given by

$$E_{\text{bim}} = E_0 - (s - \tfrac{3}{2})kT \tag{2.24}$$

and that at high pressures is given by

$$E_\infty = E_0 - (s - 1)kT \tag{2.25}$$

The activation energies predicted by (2.24) and (2.25) are very similar, and thus virtually no change in activation energy with pressure is predicted. Unimolecular reactions do in practice show a substantial decline in activation energy with pressure (section 3.8), and this behaviour is predicted by the later theories.

It was pointed out previously that an increase in k_1 at constant k_∞ requires, all other things being equal, a corresponding decrease in the rate constant k_3. According to the Hinshelwood–Lindemann theory, the rate constant k_3 is independent of the energy of the energised molecule but does depend upon the molecular complexity since it is a function of the number of vibrational modes, s. From equation (2.10), $k_3 = k_\infty k_2/k_1$. Calculations using equations (2.22) and (2.25) with the typical parameters $A_\infty = 10^{13}$ s^{-1}, $E_\infty/kT = 40$ yield the following results[9]:

$$s = \quad 1 \qquad 5 \qquad 10$$

$$k_3/\text{s}^{-1} = 10^{13} \quad 10^{9.5} \quad 10^{6.5}$$

It is therefore clear that the lifetimes of the energised molecules (which are given by $\tau \approx 1/k_3$) become greater when the molecule can store energy among a greater number of degrees of freedom. If only one degree of freedom is involved, e.g. for a diatomic molecule, the energised molecule has a very short lifetime and disappears during the course of a single vibration.

2.5 COMPARISON OF HINSHELWOOD–LINDEMANN THEORY WITH EXPERIMENT

Calculations of a similar kind to those described in section 2.3 can now be carried out for the general rate constant k_{uni} at various pressures, using equation (2.11), but with k_1 values calculated from equation (2.14):

$$k_{\text{uni}} = \frac{k_\infty}{1 + k_\infty/k_1 p} \tag{2.11}$$

Figure 2.2 illustrates such calculations for the isomerisation of *cis*- to *trans*-but-2-ene at 469 °C, using the experimental data of Rabinovitch and Michel as before. It can be seen that the inclusion of moderate values of s in equation (2.14) has the effect of reducing dramatically the calculated transi-

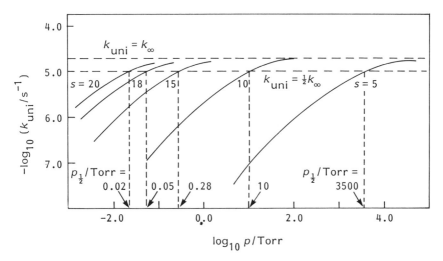

Figure 2.2   Hishelwood–Lindemann plots for *cis* → *trans*-but-2-ene at 469 °C. Reproduced by permission of John Wiley and Sons Ltd.

tion pressure. For $s = 18$, the calculated and observed transition pressures are in good agreement. It may be noted that the number of degrees of vibrational freedom for this molecule is given by $3N - 6 = 30$, which gives a maximum possible value of $s = 30$.

The example given here is again typical, and it is generally possible to obtain agreement between theory and experiment at the transition pressure by choosing a suitable value of s. One unsatisfactory feature which remains, however, is that there is no *a priori* method of calculating s for a particular reaction on this theory. Although the values of s found are feasible in the sense that they are less than (usually about one-half) the number of modes of vibration, no significance can be attached to their actual values. In addition, although the theoretical curve may be made to give approximate agreement with experiment at the transition pressure, it is a poor fit to the experimental curve over the whole pressure range. This is well shown by plotting $1/k_{uni}$ versus $1/p$, and such a plot for *cis*-but-2-ene isomerisation is shown in Figure 2.3. Whereas the Hinshelwood–Lindemann equation (2.11) can be rearranged into the linear form (2.26), the experimental plot is strongly curved and is only poorly approximated by the Hinshelwood–Lindemann line irrespective of whether the parameters are chosen to fit the results at high or at low pressures.

$$\frac{1}{k_{uni}} = \frac{1}{k_1} \cdot \frac{1}{p} + \frac{1}{k_\infty} \qquad (2.26)$$

For this reason the experimental value of k_∞ was derived for this reaction using instead an empirical plot of $1/k_{uni}$ versus $1/p^{1/2}$.[7]

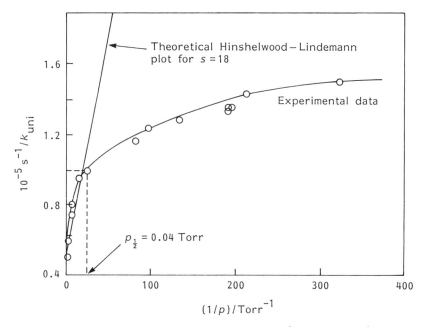

Figure 2.3 Theoretical and experimental[7] plots of $1/k_{uni}$ against $1/$pressure for *cis* → *trans*-but-2-ene at 469 °C. Reproduced by permission of John Wiley and Sons Ltd.

The reason for the curvature in this plot lies in the fact that the constant k_3 as well as k_1 is really a function of the energy possessed by the molecule whereas, in the simple treatment given so far, it has been assumed to be a true constant for molecules of all energies above the critical energy. It is intuitively more plausible that k_3 should increase with the energy possessed by the molecule in excess of the critical energy. Now that molecules can be made to undergo unimolecular reaction at different energy levels, for example by chemical activation (see Chapters 7 and 9) or laser pumping (Chapter 9), there is clear experimental evidence for the energy dependence of k_3. The major subsequent theoretical developments are concerned to a large extent with the calculation of the rate of unimolecular reaction of an energised molecule in terms of its energy content.

2.6   THE FURTHER DEVELOPMENT OF UNIMOLECULAR REACTION RATE THEORIES

In subsequent unimolecular reaction rate theories, both the rate constants k_1 and k_3 are made energy dependent. The rate constant k_1 is considered, as in Hinshelwood–Lindemann theory, by the contribution $\delta k_{1(E \to E + \delta E)}$ for energisation to the small energy range E to $E + \delta E$. The rate constant k_3 is replaced by the energy-dependent rate constant $k_a(E)$ (or the related L in Slater's theory, see below), and the reaction scheme becomes

$$A + M \xrightarrow{\delta k_1} A^*_{(E \to E + \delta E)} \tag{2.27}$$

$$A^*_{(E \to E + \delta E)} + M \xrightarrow{k_2} A + M \tag{2.28}$$

$$A^*_{(E)} \xrightarrow{k_a(E)} products \tag{2.29}$$

Application of the steady-state treatment to the concentration of $A^*_{(E \to E + \delta E)}$ leads to the expression (2.30) for the contribution to the unimolecular rate constant k_{uni} from energised molecules in this energy range, and the total rate constant k_{uni} is then obtained by integration as in (2.31):

$$\delta k_{uni(E \to E + \delta E)} = \frac{k_a(E)(\delta k_1/k_2)_{(E \to E + \delta E)}}{1 + k_a(E)/k_2[M]} \tag{2.30}$$

$$k_{uni} = \int_{E=E_0}^{\infty} \frac{k_a(E)\, dk_{1(E \to E + \delta E)}/k_2}{1 + k_a(E)/k_2[M]} \tag{2.31}$$

Two important early theories are discussed in the remainder of this chapter. They represent separate and complementary approaches to the theory of unimolecular reaction rates—the dynamical approach of Slater theory and the statistical Rice, Ramsperger, Kassel (RRK) theory.

In the Slater theory the rate constant k_a is related to a 'specific dissociation probability' L; this is the frequency with which a chosen critical coordinate in the molecule reaches a critical value, and can be calculated for the case in which the vibrations of the molecule are assumed to be harmonic. The specific dissociation probability L for a system of independent harmonic oscillators is actually a function of the energies in the individual oscillators [denoted $L(E_1, \ldots, E_n)$, cf. equation (2.41)], and not simply of the total energy E of the molecule. The single integration in (2.31) is therefore replaced in Slater theory by a multiple integral over all the possible energy ranges of the individual oscillators.

In RRK theory the assumption is made that k_a (microcanonical rate constant) is related to the chance that the critical energy E_0 is concentrated in one part of the molecule, e.g. in one oscillator (Kassel theory) or in one squared term (Rice–Ramsperger theory). This probability is clearly a function of the total energy E of the energised molecule.

2.7 SLATER'S HARMONIC THEORY

The theory of Slater, first put forward in 1939[10] and explained in detail in his book[11], was the first serious attempt to relate the kinetics of unimolecular reactions to our knowledge of molecular vibrations. The theory accepts the basic Hinshelwood–Lindemann mechanism of collisional energisation with, however, a more restricted definition of the energised molecule. The molecule undergoing reaction is pictured as an assembly of harmonic oscillators of particular amplitudes and phases. Reaction is said to occur when a chosen coordinate in the molecule (such as a combination of bond distances and/or bond angles) attains a critical extension. The rate constant k_a of the scheme given previously is replaced by a 'specific dissociation probability' L which is the frequency with which the 'critical coordinate' attains the critical extension. By calculating L in this way and integrating the necessary equations, Slater was able to derive expressions for the rate constant k_{uni} at high, low and intermediate pressures.

In order to understand more fully the derivation of these equations and the assumptions involved it is necessary to consider the vibrational analysis of polyatomic molecules. The reader unfamiliar with this topic is referred to Appendix 3, where the method of vibrational analysis is discussed and applied to some simple examples. It is shown there how it is possible to derive the form of normal mode vibrations described by the normal coordinates Q_k from solutions of the equations of motion. The motions of the atoms are first described in terms of Cartesian displacement coordinates (x), then internal coordinates (r) relating to displacements of the bonds within the molecule and finally it is shown how, in some simple cases, internal coordinates can be combined to give normal coordinates.

For most molecules it is not possible to guess the relationship between the internal coordinates and normal coordinates which permits direct solution of the then independent equations of motion. Instead, use is made of the known symmetry of the molecule to construct symmetry coordinates from internal coordinates and these reduce the order of the secular equation which has to be solved. The method is described in detail in the book by Wilson, Decius and Cross[12].

For the application of Slater theory, we are interested in the form of the normal mode vibrations and the way in which they influence the behaviour of a particular internal coordinate with time. We need, therefore, to know both the normal mode frequencies ν_k and the force constants which occur in the secular equation. Substitution of any particular value of $\lambda_k = 4\pi^2 \nu_k^2$ in the secular equation allows the ratios of the displacement coordinates to each other in the corresponding normal mode to be found.

If the secular equation has been written in terms of $3N - 6$ coordinates q_i (which are chosen so that the transformation to normal coordinates is orthogonal) then for mode 1, for example, the ratios of the q_i will be obtained from the equation

$$q_1:q_2:q_3: \dots :q_{3N-6} = a_{11}:a_{12}:a_{13}: \dots :a_{1,3N-6}$$

where the quantities a_{1i} are related to the normal coordinate Q_1 by

$$Q_1 = a_{11}q_1 + a_{12}q_2 + a_{13}q_3 + \ldots + a_{1,3N-6}q_{3N-6} \qquad (2.32)$$

In terms of symmetry coordinates S_i, the result of vibrational analysis similarly yields ratios of symmetry coordinates to each other in a given normal mode, so that Q_1 for example is given by

$$Q_1 = L_{11}S_1 + L_{12}S_2 + \ldots + L_{1,3N-6}S_{3N-6} \qquad (2.33)$$

In general there will be fewer than $3N - 6$ symmetry coordinates involved in a particular normal coordinate Q_1, since only those of the same symmetry as Q_1 have non-zero L_{1j} values, and this leads to considerable simplification.

For the purposes of Slater theory, we need the contribution of each normal coordinate to a particular internal coordinate or to a particular symmetry coordinate. These contributions are found from the inverse relations corresponding to (2.32) and (2.33). Thus if (2.32) is written in matrix notation as $\mathbf{Q} = \mathbf{a}\mathbf{q}$ the inverse relation is given by

$$\mathbf{q} = \mathbf{a}^{-1}\mathbf{Q} = \alpha\mathbf{Q} \qquad (2.34)$$

That is,

$$\left. \begin{array}{l} q_1 = \alpha_{11}Q_1 + \alpha_{12}Q_2 + \ldots + \alpha_{1,3N-6}Q_{3N-6} \\ \quad \cdot \qquad \cdot \qquad \cdot \qquad \cdot \qquad \cdot \qquad \cdot \\ q_{3N-6} = \alpha_{3N-6,1}Q_1 + \ldots + \alpha_{3N-6,3N-6}Q_{3N-6} \end{array} \right\} \qquad (2.35)$$

Similarly, if there are m symmetry coordinates, which are related to m normal coordinates of a given symmetry type by equations of the type (2.33), or in matrix notation $\mathbf{Q} = \mathbf{L}\mathbf{S}$, then

$$\mathbf{S} = \mathbf{L}^{-1}\mathbf{Q} = \mathbf{L}'\mathbf{Q}$$

That is,

$$S_1 = L'_{11}Q_1 + L'_{12}Q_2 + \ldots + L'_{1m}Q_m$$

$$\cdot \qquad \cdot \qquad \cdot \qquad \cdot \qquad \cdot$$

$$S_m = L'_{m1}Q_1 + L'_{m2}Q_2 + \ldots + L'_{mm}Q_m$$

When direct solution of the secular equation yields the $\mathbf{a}$ matrix or the $\mathbf{L}$ matrix, then the inverse matrices α or $\mathbf{L}'$ must be found by the standard techniques of matrix inversion.

For the example of a linear triatomic molecule discussed in Appendix 3, an equation is derived for the time dependence of internal coordinate r_1 when both normal coordinates Q_1 and Q_2 are non-zero. This equation is of the form

$$r_1 = \alpha_{11}A_1 \cos(2\pi v_1 t + \psi_1) + \alpha_{12}A_2 \cos(2\pi v_2 t + \psi_2)$$

where the coefficients α relate r_1 to Q_1 and Q_2 by the equation

$$r_1 = \alpha_{11}Q_1 + \alpha_{12}Q_2$$

For any general internal coordinate q_i related to $n = 3N - 6$ normal coordinates by the equation

$$q_i = \sum_{k=1}^{n} \alpha_{ik}Q_k$$

it can be similarly shown that the time dependence of q_i is given by

$$q_i = \sum_{k=1}^{n} \alpha_{ik}A_k \cos(2\pi v_k t + \psi_k) \tag{2.36}$$

Slater's theory regards the behaviour of coordinate q_i as a superposition of n independent contributions from the normal modes of vibration. The function q_i varies in a complicated way with time, only occasionally attaining high values (Figure 2.4). An internal coordinate q_1 (or possibly a combination of internal coordinates) is chosen as the critical coordinate. Unimolecular reaction is then said to occur if the chosen coordinate exceeds a critical value q_0, harmonic motion being assumed up to the point of rupture. From

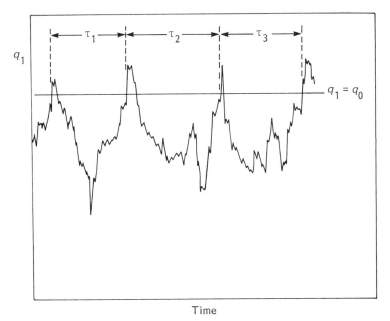

Time

Figure 2.4 Variation of a typical internal coordinate q_1 with time showing gaps between successive attainments of a critical value q_0. Reproduced by permission of John Wiley and Sons Ltd.

(2.36) the maximum value of a particular coordinate q_1 is given by (2.37), where E_k is the total energy in normal mode k and $A_k = \sqrt{E_k}$ by virtue of the special properties of normal coordinates

$$(q_1)_{max} = \sum_{k=1}^{n} |\alpha_{1k}| A_k = \sum_{k=1}^{n} |\alpha_{1k}| \sqrt{E_k} \qquad (2.37)$$

A molecule is said to be energised if the last collision has given it sufficient energy to satisfy the inequality

$$\sum_{k=1}^{n} |\alpha_{1k}| \sqrt{E_k} \geq q_0 \qquad (2.38)$$

It can then be shown that the minimum total energy E_0 for which this is possible is given by

$$E_0 = q_0^2/\alpha^2 \qquad (2.39)$$

where

$$\alpha = \left(\sum_{k=1}^{n} \alpha_{1k}^2 \right)^{1/2}$$

This energy is the critical energy for reaction and will be seen later to be equal to the high-pressure activation energy. Equation (2.39) represents the minimum energy which the molecule must possess in order to undergo unimolecular reaction, although it will not do so unless the energy is distributed among the normal modes in such a way that equation (2.38) is also satisfied. Whether or not the energised molecule reacts is then dependent upon the relative chance of the critical coordinate exceeding the value q_0 and the chance of deactivation by collision. The frequency with which the former occurs is called by Slater the specific dissociation rate or probability L.

The Slater calculation of L is based upon the behaviour of a sum of harmonic vibrations. If the average behaviour of a coordinate over a long time interval is required, the initial phase is irrelevant and (2.36) becomes, for a particular coordinate q_1,

$$q_1 = \sum_{k=1}^{n} \alpha_{1k} \sqrt{E_k} \cos(2\pi v_k t) \qquad (2.40)$$

Slater makes use of a formula due to Kac[13] for the frequency of 'up-zeros' of the function

$$\sum_{k=1}^{n} \alpha_{1k} A_k \cos(2\pi v_k t) - q_0$$

where there are n independent frequencies v_k.

Unfortunately, the precise equation for L is too cumbersome to deal with in Slater theory, except at the high-pressure limit, and an approximate expression is used instead. Slater has shown that when the sum of the amplitudes $\Sigma \alpha_{1k} A_k$ is not much greater than q_0, then the equation for L becomes

$$L(E_1, E_2, \ldots, E_n) = \frac{1}{\Gamma(\frac{1}{2}n + \frac{1}{2})} \left(\frac{\sum\limits_{k=1}^{n} a_k v_k^2}{a_1 a_2 \ldots a_n} \right)^{\frac{1}{2}} \left(\frac{a - q_0}{2\pi} \right)^{\frac{1}{2}(n-1)} \quad (2.41)$$

where

$$a_k = \alpha_{1k} \sqrt{E_k} \quad \text{and} \quad a = \sum_{k=1}^{n} a_k$$

It will be noted that (2.41) gives L for a specified distribution of energy among the normal modes and not simply for a given total energy as in most other theories. The corresponding rate constant $k_a(E)$ for a given total energy E, which is the term used in most other theories, may be calculated by averaging L over all the possible distributions of the energy (see ref. 11, p. 92).

THE GENERAL-PRESSURE RATE CONSTANT

The Slater theory makes use of earlier ideas concerning the overall mechanism of unimolecular reactions in order to express the rate constant at any pressure. These may be conveniently discussed in terms of the basic mechanism given earlier by equations (2.27)–(2.29):

$$A + M \xrightarrow{\delta k_1} A^*_{(E \to E + \delta E)} + M \quad (2.27)$$

$$A^*_{(E \to E + \delta E)} + M \xrightarrow{k_1} A + M \quad (2.28)$$

$$A^*(E) \xrightarrow{k_a(E)} \text{products} \quad (2.29)$$

and this leads to the expression (2.31) for the first-order rate constant k_{uni}.

$$k_{uni} = \int_{E=E_0}^{\infty} \frac{k_a(E) \, dk_{1(E \to E + \delta E)}/k_2}{1 + k_a(E)/k_2[M]} \quad (2.31)$$

In the Hinshelwood–Lindemann treatment $\delta k_1/k_2$ is equated to the equilibrium fraction of molecules having total energy in the range E to $E + \delta E$, and this is obtained by integrating the fraction of molecules having energy E distributed in a specified way among the s degrees of freedom (2.16). Because of the different meaning of the term 'energisation' in Slater theory a different approach is required here; it is necessary to integrate over the relevant ranges of energy for each degree of freedom rather than the corresponding range of total energy. Thus the expression for k_{uni} is evaluated by replacing $k_a(E)$ in (2.31) by $L(E_1, \ldots, E_n)$, and replacing dk_1/k_2 by the expression $(1/kT)^n \exp(-E/kT) dE_1 \ldots dE_n$ (2.16). This expression gives the equilibrium fraction of molecules having E_i to $E_i + dE_i$ in the ith mode of vibration, where $E = E_1 + E_2 + \ldots + E_n$ is the total energy in the n contributing modes of vibration. Integration over all the relevant energy distributions then gives the following expression for k_{uni}:

$$k_{uni} = \int\int \cdots \int \frac{L(E_1, \ldots, E_n) \exp(-E/kT)}{1 + L(E_1, \ldots, E_n)/k_2[M]} \prod_{k=1}^{n} \left(\frac{dE_k}{kT}\right) \qquad (2.42)$$

The integrations with respect to $dE_1 \ldots dE_n$ in (2.42) are now over the restrictive energy ranges governed by (2.38) and (2.39), which reduce to the condition

$$\sum_{k=1}^{n} \mu_k \sqrt{E_k} \geq \sqrt{E_0} \qquad (E_k \geq 0) \qquad (2.43)$$

where

$$\mu_k = |\alpha_{1k}|/\alpha = |\alpha_{1k}|/\left(\sum_{k=1}^{n} \alpha_{1k}^2\right)^{1/2} \qquad (2.44)$$

The integral (2.42) reduces to a manageable form if the approximate expression (2.41) is inserted for L and the result of the integration is

$$k_{uni} = v \exp(-b) I_n(\theta) \qquad (2.45)$$

$$I_n(\theta) = \frac{1}{\Gamma(\frac{1}{2}n + \frac{1}{2})} \int_{x=0}^{\infty} \frac{x^{1/2(n-1)} \exp(-x) dx}{1 + x^{1/2(n-1)} \theta^{-1}} \qquad (2.46)$$

$$v^2 = \sum_{k=1}^{n} \alpha_k^2 v_k^2/\alpha^2 = \sum_{k=1}^{n} \mu_k^2 v_k^2 \qquad (2.47)$$

$$b = E_0/kT \qquad (2.48)$$

$$\theta = (k_2 p/v) b^{1/2(n-1)} f_n \qquad (2.49)$$

$$f_n = (4\pi)^{1/2(n-1)} \Gamma(\frac{1}{2}n + \frac{1}{2}) \prod_{k=1}^{n} \mu_k \qquad (2.50)$$

In (2.49), p is the total pressure and k_2, the rate constant for collisional de-energisation, is given by the collision-theory expression

$$k_2 = 4\sigma_d^2 (\pi R T/M)^{1/2} \qquad (2.51)$$

THE LIMITING FORMS AT HIGH AND LOW PRESSURES

When the pressure p tends to infinity, so does θ, and $I_n(\theta)$ tends to unity, so that k_∞ is given by

$$k_\infty = v \exp(-b) = \left(\sum_{k=1}^{n} \mu_k^2 v_k^2 \right)^{1/2} \exp(-E_0/kT) \qquad (2.52)$$

Slater's theory thus predicts that the high-pressure activation energy is equal to the critical energy E_0 and that the high-pressure A-factor is a weighted mean of the vibration frequencies in the molecule.

At low pressures, θ tends to zero and k_{uni} becomes proportional to the first power of the pressure, i.e. the rate is second order. The rate constant k_{uni} then becomes

$$k_{uni} = k_{bim}p = k_2 p (4\pi E_0/kT)^{1/2(n-1)} \exp(-E_0/kT) \prod_{k=1}^{n} \mu_k$$

and the second-order rate constant k_{bim} (also referred to as k_0) is given by

$$k_{bim} = k_2 (4\pi E_0/kT)^{1/2(n-1)} \exp(-E_0/kT) \prod_{k=1}^{n} \mu_k \qquad (2.53)$$

THE THEORETICAL FALL-OFF CURVE

From (2.45) and (2.52), the fall-off curve is defined by

$$k_{uni}/k_\infty = I_n(\theta)$$

The function $I_n(\theta)$, the 'Slater integral' defined in (2.46) depends solely upon n (the number of contributing modes of vibration) and upon θ, which is proportional to p at a constant temperature T. At a given temperature there is thus a single curve of $\log(k_{uni}/k_\infty)$ against $\log p$ for each value of n. The effect on this curve of increasing the temperature from T_1 to T_2, for a given value of n, is to translate the curve to higher pressures at the higher temperature by an amount

$$\Delta \log p = \tfrac{1}{2} n \log(T_2/T_1) \qquad (2.54)$$

From (2.49) and (2.50), it can be seen that the parameter θ which determines k_{uni}/k_∞ depends upon the value of n. For a given value of E_0/kT, θ is lower and k_{uni}/k_∞ is therefore lower the smaller the value of n. For two molecules having similar values of n, k_{uni}/k_∞ will be lower for the one having the lower value of E_0/kT. These considerations are helpful in deciding which of several possible unimolecular rate constants is likely to decline first from its high-pressure limit in a decomposition occurring by a complex mechanism. For example, in the generally accepted mechanism for the thermolysis of ethane, there are two unimolecular steps, equations (2.55) and (2.56).

$$C_2H_6 \rightarrow 2CH_3 \qquad (2.55)$$

$$C_2H_5 \rightarrow C_2H_4 + H \qquad (2.56)$$

From the above considerations, one would expect reaction (2.56), which has a lower activation energy than (2.55), to depart from first-order behaviour before (2.55) on lowering the total pressure. This is confirmed by experimental evidence[14].

THE CHANGE IN ACTIVATION ENERGY WITH PRESSURE

Since v is a temperature-independent constant for a given reaction (see 2.47), the activation energy at high pressure, E_∞, is predicted from (2.52) to be equal to the critical energy E_0. At low pressure, Slater theory predicts a decline in activation energy. The experimental activation energy is defined by

$$E_{Arr} = kT^2(d \ln k/dT) \qquad (1.3)$$

so that at low pressure, from (2.53),

$$E_{bim} = kT^2(d \ln k_{bim}/dT)$$
$$= E_0 - \tfrac{1}{2}(n - 1)kT + kT^2(d \ln k_2/dT)$$

The evaluation of E_{bim} depends on whether the rate constants k_2 and k_{bim} are expressed in concentration units or in pressure units. Equation (2.51) defines k_2 in concentration units (e.g. $cm^3 \, mol^{-1} s^{-1}$), from which it is seen that in these units, $k_2 \propto T^{1/2}$. Similarly, in pressure units, $k_2 \propto T^{-1/2}$. Thus if E_{bim} is determined from experimental values of k_{bim} measured in concentration units, it is related to E_∞ by

$$E_{bim} = E_\infty - (\tfrac{1}{2}n - 1)kT \text{ (concentration units)} \qquad (2.57)$$

but if E_{bim} is determined from k_{bim} values in terms of pressure units, the relationship is

$$E_{bim} = E_\infty - \tfrac{1}{2}nkT \text{ (pressure units)} \qquad (2.58)$$

In both cases it is predicted that $E_{bim} < E_\infty$ by an amount of 20–30 kJ mol^{-1} for moderately complex molecules ($n \approx 10$–15) at temperatures around 400–500 °C.

THE ASSUMPTIONS OF SLATER THEORY

The description of Slater theory given above follows the early classical harmonic form of the theory. It has long been realised, however, that molecular vibrations are neither classical nor purely harmonic under the conditions of unimolecular reactions. Considerable work has been done to produce modifications to the theory avoiding these assumptions, but, as Slater noted, it is difficult to abandon all the assumptions simultaneously.

THE HARMONIC ASSUMPTION

The assumption of harmonicity has the useful corollary of allowing the molecular vibrations to be treated in terms of normal modes of vibration. Qualitatively, the inclusion of anharmonicity provides a means whereby energy can flow between normal modes. Thus at sufficiently low pressures, the restrictive Slater condition (2.38) on the normal mode energies no longer applies, and every molecule with energy $\geq E_0$ is capable of reacting. The low-pressure second-order rate constant is then found by integrating equation (2.42) subject only to the condition

$$\sum_{k=1}^{n} E_k = E \geq E_0$$

The resulting form of the second-order rate constant is then

$$k_{\mathrm{bim}} = k_2 (E_0/kT)^{(n-1)} \exp(-E_0/kT)/(n-1)!$$

instead of (2.53), and this is closely similar to Hinshelwood's expression (2.14) for the rate of energisation.

It is difficult to evaluate quantitatively the effect on the general rate constant of using specific expressions for the potential energy of the molecule involving anharmonic cross-terms. The difficulty arises partly from lack of knowledge of the precise potential functions applicable and partly from the difficulty of handling the resultant equations of motion when these are not amenable to the normal mode approximation. Some progress has been made by numerical integration of the equations for model systems such as linear triatomic[15] and tetratomic[16] molecules with Morse potentials.

A quantum-mechanical version of the harmonic theory was worked out by Slater[17], and results in the same form of the results but with b and the μ_k redefined. The main difference is that the new b has a higher value than that from (2.48). The fall-off curve is given by the same function and therefore has the same shape for both the quantum and classical theories, but a given extent of fall-off occurs at lower pressures according to the quantum theory. A major difficulty of the theory is that it predicts an A-factor which is lower than the mean of the vibration frequencies in the molecule, and it has not been tested in any detail.

THE RANDOM-GAP ASSUMPTION

A further assumption of Slater theory is the 'random-gap' assumption. The 'gap length' is the time between successive crossings by the critical coordinate q_1 of the critical boundary $q_1 = q_0$ (see Figure 2.4). It is sometimes convenient to formulate unimolecular reaction as the movement of a representative point on a hypersurface defined by the coordinates and momenta of the molecule (see Appendix 2, section A2.7). Under these conditions the gap length becomes the time (τ) between successive crossings of a critical surface. If the incidence of this event is random, the probability of a particular value of τ is given by

$$P(\tau) = L \exp(-L\tau)$$

The validity of this assumption has been discussed by Bunker[18] and Thiele[19] and shown to be inherent in many theories of unimolecular reactions; it will be discussed in more detail in connection with RRKM theory in section 3.12. Slater's 'new approach' to rate theory[20] examines the consequences of adopting different gap distributions and concludes that despite differences at intermediate pressures, the limiting high- and low-pressure rate constants are the same for all gap distributions.

THE STRONG-COLLISION ASSUMPTION

The harmonic theory of Slater, the RRK theory and the basic formulation of RRKM theory all assume implicitly that every collision between an energised and a non-energised molecule results in de-energisation. While this assumption is often true for thermally energised polyatomic molecules of moderate complexity, it has been found increasingly necessary in recent years to abandon this assumption, particularly for small molecules with a high degree of energisation such as those produced in shock tubes or chemical activation experiments. The detailed treatment of weak collisions is considered in Chapter 7.

2.8 THE RICE–RAMSPERGER–KASSEL THEORIES

The theories due to Rice and Ramsperger[21,22] and Kassel[23,24], were developed at about the same time, are very similar in approach and depend on the development of expressions for the energy dependence of $k_a(E)$ in the scheme below:

$$A + M \underset{k_2}{\overset{\delta k_{1(E \rightarrow E + \delta E)}}{\rightleftharpoons}} A^*_{(E \rightarrow E + \delta E)} + M \qquad (2.59)$$

$$A^*_{(E)} \underset{}{\overset{k_a(E)}{\rightleftharpoons}} \text{products} \qquad (2.60)$$

Following the basic Hinshelwood process of energisation and de-energisation (2.59) the rate of conversion of an energised molecule to products is taken to be a function of its energy content. Both theories evaluate $k_a(E)$ from the chance that a critical amount of energy E_0 is concentrated in one particular part of the molecule given that all contributing molecular oscillators are statistically equally probable when a given molecular energy is distributed among them. The differences between the theory of Rice and Ramsperger and that of Kassel are relatively minor and the description which follows resembles more closely the treatment of Kassel although it is commonly referred to nowadays as the RRK theory to indicate the contributions made by all these workers.

CLASSICAL RRK THEORY

Classical RRK theory is based on a calculation of the probability that a system of s classical oscillators with total energy E should have energy $\geq E_0$ in one chosen oscillator; the derivation is clearly documented elsewhere[23] and the result is

$$\text{probability (energy} \geq E_0 \text{ in one chosen oscillator)} = \left(\frac{E - E_0}{E}\right)^{s-1} \qquad (2.61)$$

It is then assumed that the rate constant $k_a(E)$ for conversion of energised molecules to products is proportional to this probability, as in

$$k_a(E) = A\left(\frac{E - E_0}{E}\right)^{s-1} = A(1 - E_0/E)^{s-1} \qquad (2.62)$$

At this stage A is no more than a proportionality constant, but it acquires significance when (2.62) is inserted into (2.31), using Hinshelwood's expression for δk_1. The high-pressure limit is then given by (2.63), which on substitution of $x = (E - E_0)/kT$ and use of the defining integral for $\Gamma(s) = (s - 1)!$ (Appendix 4) gives

$$k_\infty = \int_{E = E_0}^{\infty} [A(1 - E_0/E)^{s-1}]\left[\frac{(E/kT)^{s-1}\exp(-E/kT)}{(s-1)!}\right]d\left(\frac{E}{kT}\right) \qquad (2.63)$$

$$= A\exp(-E_0/kT) \qquad (2.64)$$

Thus the theory predicts strict adherence to the Arrhenius equation, and the constant A is now identified with the high-pressure A-factor for the reaction. Slater[11] has shown that if Hinshelwood's δk_1 is used then (2.62) is the only form of $k_a(E)$ which will give a temperature dependence of the Arrhenius form. At present most unimolecular reactions which have been tested over a wide temperature range do appear to conform to Arrhenius behaviour, but there would be expected to be a small temperature dependence of both A and E on theoretical grounds[25].

The corresponding equation for k_{uni} in the fall-off region is easily written down, and division by (2.64) results in

$$\frac{k_{uni}}{k_\infty} = \frac{1}{(s-1)!} \int_{x=0}^{\infty} \frac{x^{s-1} \exp(x)\, dx}{1 + (A/k_2[M])[x/(x + E_0/kT)]^{s-1}} \qquad (2.65)$$

This last equation is probably the most widely used result of Kassel's work, although this is unfortunate since the quantum version, to be described below, is substantially more realistic. In applying (2.65) $k_2[M]$ is generally replaced by Zp (the total collision frequency at unit pressure of A^*), and the experimental high-pressure A-factor and activation energy are used for A and E_0. The integral is evaluated numerically for various values of s, using methods such as the Gauss or Gauss–Laguerre procedures[26], or possibly more elaborate techniques if manual work is required[27].

QUANTUM RRK THEORY

The *quantum version* of RRK theory is in essence very similar to the classical theory outlined above. It assumes, in its simplest form, that there are s identical oscillators in the molecule, all having frequency v. The energies are expressed in quanta, the critical number of quanta being $m = E_0/hv$. The expression corresponding to (2.61) is now the probability that if s oscillators contain a total of n quanta (where $n = E/hv$), one chosen oscillator will contain at least m quanta; the result is

$$\text{probability (energy} \geq m \text{ quanta in chosen oscillator)} = \frac{n!(n - m + s - 1)!}{(n - m)!(n + s - 1)!} \qquad (2.66)$$

and the corresponding expression for k_a is thus

$$k_a(nhv) = A\frac{n!(n - m + s - 1)!}{(n - m)!(n + s - 1)!} \qquad (2.67)$$

The expression used for $k_1(E)$ is a similar development of Hinshelwood's expression, namely (2.68)[23,24]:

$$k_1(nhv) = k_2 \alpha^n (1 - \alpha)^s \frac{(n + s - 1)!}{n!(s - 1)!} \quad \text{where } \alpha = \exp(-hv/kT) \qquad (2.68)$$

This is no longer a differential quantity, since it refers to energisation into a specific quantum state rather than into an energy range E to $E + \delta E$. The overall rate constant is obtained by summing over the discrete energy levels; the high-pressure rate constant is given by equation (2.69) and this can be shown to result once more in the Arrhenius form equation (2.64):

$$k_\infty = \sum_{n=m}^{\infty} \frac{k_1(n h v) k_a(n h v)}{k_2} \qquad (2.69)$$

$$= A \exp(-E_0/kT) \qquad (2.64)$$

The interpretation of the constant A is thus the same as in the classical case. The rather cumbersome expression for k_{uni}/k_∞ is similarly derived as,

$$\frac{k_{uni}}{k_\infty} = (1-\alpha)^s \sum_{p=0}^{\infty} \frac{\alpha^p (p+s-1)! [p! (s-1)!]}{1 + (A/k_2[M])(p+m)!(p+s-1)!/[(p+m+s-1)!p!]} \qquad (2.70)$$

in which $n - m$ has been replaced by p. Kassel further developed this type of model to deal with more realistic cases where the s oscillators were not all of the same frequency. Detailed equations were given for the case where there are two frequencies, one an exact multiple of the other, and the extensions to other cases were hinted at.

It will be noted that the limiting case of classical behaviour may be obtained from the quantum theory by letting the quantum interval $h v$ become very small, so that n and m are both much greater than s. If (2.67) and (2.68) are written as in (2.71) and (2.72) the approximations involved in replacing them by the corresponding classical results will be more readily appreciated. In (2.72) $k_1(n h v)$ is divided by the quantum energy $h v$ for comparison with $dk_1(E)/dE$, the classical rate of energisation into a unit energy range.

$$k_a(n h v) = A \frac{\overline{(n-m+1)(n-m+2) \ldots (n-m+s-1)}}{(n+1)(n+2) \ldots (n+s-1)}$$

$$\approx A \frac{(n-m)^{s-1}}{n^{s-1}}$$

$$[\equiv A(1 - E_0/E)^{s-1}] \text{ for } n - m \gg s \qquad (2.71)$$

$$\frac{1}{h v} \frac{k_1(n h v)}{k_2} = \frac{\exp(-E/kT)}{(s-1)!} \frac{(E/kT)^{s-1}}{kT}$$

$$\times \left\{ \frac{(n+1)(n+2) \ldots (n+s-1)}{n^{s-1}} \left[\frac{1 - \exp(-h v/kT)}{h v/kT} \right]^s \right\}$$

$$\approx \frac{\exp(-E/kT)}{(s-1)!} \frac{(E/kT)^{s-1}}{kT} \quad \text{for } n \gg s \text{ and } h v/kT \ll 1$$

$$\qquad (2.72)$$

2.9 DEVELOPMENT AND APPLICATION OF THE RRK THEORIES

The use of the classical RRK equations (2.62), (2.64) and (2.65) does represent an improvement over the Hinshelwood–Lindemann treatment. An early application due to Kassel was to the decomposition of azomethane. Although this is now known not to be a simple unimolecular reaction and the fall-off observed is probably attributable to the declining rate of a unimolecular initiation process[28], it is seen from Figure 2.5 that the Kassel curves reproduce much more closely the relatively shallow curvature of the experimental fall-off curve.

An improvement in shape also occurs when the calculated and experimental data[29] are compared in the form of reciprocal plots such as $1/k_{uni}$ versus $1/p$ which are no longer linear as found in the case of the Hinshelwood–Lindemann (HL) theory (see Fig. 2.3).

One unsatisfactory feature of the classical Kassel theory is that in order to obtain agreement with experiment using a reasonable value of σ_d, the required value of s is usually about half the number of oscillators in the

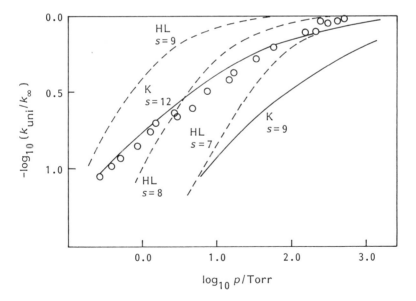

Figure 2.5 Fall-off curves for azomethane decomposition at 603 K. Circles are Ramsperger's experimental results[29]. Broken lines are HL curves with $s = 7, 8$ or 9; solid lines are classical Kassel curves[23] with $s = 9$ or 12; $\sigma_d^2 = 150$ Å^2 throughout. (1 Å = 10^{-10} m). Reproduced by permission of John Wiley and Sons Ltd.

molecule. Admittedly this fraction can vary over a wide range[30], but about one-half is the norm for a large number of reactions. Although it is *a priori* possible that not all the vibrational modes of an energised molecule can contribute their energy to the reaction, it would be a curious coincidence if for most molecules about half did so, and later theoretical and experimental work has in fact indicated that probably all modes can contribute in most cases. The above anomaly arises from the serious inadequacy of the classical statistical mechanics used, and the quantum Kassel theory gives satisfactory results assuming all the modes to be 'active'. Figure 2.6 shows two curves calculated[23] from the quantum version, which fit the experimental data just as well as the classical $s = 12$ curve in Figure 2.5. The curve for $s = 24$, $m = 15$, $\sigma_d^2 = 6$ Å^2 (1 Å $= 10^{-10}$ m) assumes all vibrations to be active and is almost identical to the classical curves for $s = 12$, $\sigma_d^2 = 150$ Å^2 or $s = 14$–15, $\sigma_d^2 = 6$ Å^2. This value of m corresponds to an oscillator frequency of 1220 cm^{-1}, which must be close to the geometric mean frequency for the molecule, so the model is reasonably realistic.

The use of a single oscillator frequency in the quantum RRK treatment was soon recognised as an oversimplification. Eyring and Giddings[31] treated

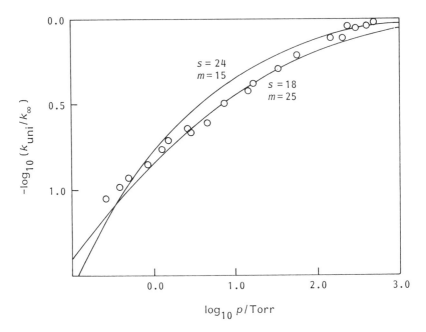

Figure 2.6 Kassel quantum curves[23] for azomethane decomposition at 603 K ($\sigma_d^2 = 6$ Å^2). Circles are Ramsperger's experimental results[29] (1 Å $= 10^{-10}$ m). Reproduced by permission of John Wiley and Sons Ltd.

the energised molecule as a set of s oscillators with frequencies v_j ($j = 1$ to s) and the transition state as a set of $s - 1$ oscillators with potentially different frequencies v_i ($i = 1$ to $s - 1$).

Using classical statistical mechanics, $k_a(E)$ for this model is given by

$$k_a(E) = \left(\prod_{j=1}^{s} v_j \Big/ \prod_{i=1}^{s-1} v_i\right)\left(\frac{E - E_0}{E}\right)^{s-1} \qquad (2.73)$$

which is identical with Kassel's $k_a(E)$ with the constant A (and therefore the theoretical high-pressure A-factor) now being equated to $\Pi^s v_j / \Pi^{s-1} v_i$. Thus A should be of the order of magnitude of a vibration frequency ($\approx 10^{13}$ s^{-1}), as is found experimentally to be the case for many unimolecular reactions.

The quantum s-fold degenerate oscillator RRK theory has found application to the theory of multiphoton dissociation processes[32] where it has been preferred to the more sophisticated RRKM theory (to be discussed in Chapter 3) on the grounds of simplicity of application and avoidance of ambiguity in selection of transition state parameters. Arguments have been presented (Chapter 3) for identification of the frequency v with an arithmetic rather than a geometric mean of the molecular frequencies for these purposes.

Realising that the major inadequacy of the classical RRK theory lies in the classical treatment of density of states, Benson[33] proposed an independent calculation of s from the molar vibrational heat capacity C_v by the equation $s = C_v/R$. (Troe and Wagner[34] have similarly taken s to be given by $s = U/RT$ where U is internal energy).

Golden, Solly and Benson[35] compared RRK and RRKM theories taking the RRK s value $= C_v/R$ for a variety of thermally energised unimolecular reactions with a range of A and E values and concluded that the RRK method is adequate for predicting the unimolecular fall-off to within a factor of 2 or 3, provided that $k_{uni} > 10^{-4}$ s^{-1}. Skinner and Rabinovitch[36] subsequently found that no unique value of s exists which successfully predicts both the position of fall-off curves with respect to pressure and their detailed shape. The arguments have been taken a step further by Schranz, Nordholm and Hamer[37] who have analysed the reasons for the necessity of two s values for a given molecule in order correctly to calculate the unimolecular rate constant. By comparing the quantum RRK theory using two s values appropriate to the different energy ranges accessible to energised molecules and transition states with RRKM calculations they have improved the quantum RRK theory at the expense of some added computational complexity.

These authors recognise that at this level the extra demands of the more realistic RRKM treatment are not much greater and that further refinement of the quantum RRK treatment is not justified.

REFERENCES

1. J. Perrin, *Ann. Phys.*, **11**, 1 (1919).
2. I. Langmuir, *J. Amer. Chem. Soc.*, **42**, 2190 (1920).
3. J. A. Christiansen and H. A. Kramers, *Zeit. Phys. Chem.*, **104**, 451 (1923).
4. D. C. Truhlar and B. C. Garrett, *Acc. Chem. Res.* **13**, 440 (1980).
5. M. Volpe and H. S. Johnston, *J. Amer. Chem. Soc.*, **78**, 3910 (1956); see also J. M. Alvarino and J. E. Figuerelo, *J. Chem. Ed.*, **54**, 675 (1977).
6. J. Aspden, N. A. Khawaja, J. Reardon and D. J. Wilson, *J. Amer. Chem. Soc.*, **91**, 7580 (1969).
7. B. S. Rabinovitch and K. W. Michel, *J. Amer. Chem. Soc.*, **81**, 5065 (1959).
8. C. N. Hinshelwood, *Proc. Roy. Soc.*, **A113**, 230 (1927).
9. A. F. Trotman-Dickenson, *Gas Kinetics*, Butterworths, London, 1955 p. 55.
10. N. B. Slater, *Proc. Camb. Phil. Soc.*, **35**, 56 (1939).
11. N. B. Slater, *Theory of Unimolecular Reactions*, Methuen, London, 1959.
12. E. B. Wilson, J. C. Decius and P. C. Cross, *Molecular Vibrations*, McGraw-Hill, New York, 1955, p. 21.
13. M. Kac, *Amer. J. Math*, **65**, 609 (1943).
14. C. P. Quinn, *Proc. Roy. Soc.*, **A275**, 190 (1963).
15. E. Thiele and D. J. Wilson, *J. Chem. Phys.*, **35**, 1256 (1961).
16. R. J. Harter, E. B. Altermann and D. J. Wilson, *J. Chem. Phys.*, **40**, 2137 (1964).
17. J. H. Current and B. S. Rabinovitch, *J. Chem. Phys.*, **38**, 783 (1963) and references therein.
18. D. L. Bunker, *Theory of Elementary Gas Reaction Rates*, Pergamon, Oxford (1966).
19. E. Thiele, *J. Chem. Phys.*, **36**, 1466 (1962); **38**, 1959 (1963).
20. J. R. Barker, *J. Phys. Chem.*, **88**, 11 (1984).
21. O. K. Rice and H. C. Ramsperger, *J. Amer. Chem. Soc.*, **49**, 1617 (1927).
22. O. K. Rice and H. C. Ramsperger, *J. Amer. Chem. Soc.*, **50**, 617 (1928).
23. L. S. Kassel, *J. Phys. Chem.*, **32**, 225, 1065 (1928).
24. L. S. Kassel, *Kinetics of Homogeneous Gas Reactions*, Chemical Catalog Co., New York, (1932).
25. J. H. Gibbs, *J. Chem. Phys.*, **57**, 4473 (1972).
26. H. E. Salzer and R. Zucker, *Bull. Amer. Math. Soc.*, **55**, 1004 (1949).
27. N. B. Slater, *Theory of Unimolecular Reactions*, Methuen, London, 1959 p. 168.
28. A. F. Trotman-Dickenson, *Gas Kinetics*, Butterworths, London, p. 70.
29. H. C. Ramsperger, *J. Amer. Chem. Soc.*, **49**, 1495 (1927).
30. D. W. Placzek, B. S. Rabinovitch and G. Z. Whitten, *J. Chem. Phys.*, **43**, 4071 (1965).
31. J. C. Giddings and H. Eyring, *J. Chem. Phys.*, **22**, 538 (1954).
32. E. Thiele, J. Stone and M. F. Goodman, *Chem. Phys. Lett.*, **76**, 579 (1980) and refs. cited therein.
33. S. W. Benson, *Thermochemical Kinetics*, Wiley, New York (1968).
34. J. Troe and H. G. Wagner, *Ber. Bunsenges, Phys. Chem.*, **71**, 937 (1967).
35. D. M. Golden, R. K. Solly and S. W. Benson, *J. Phys. Chem.*, **75**, 1333 (1971).
36. G. B. Skinner and B. S. Rabinovitch, *J. Phys. Chem.*, **76**, 2418 (1972).
37. H. W. Schranz, S. Nordholm and N. D. Hamer, *Int. J. Chem. Kinet.*, **14**, 543 (1982).

3 Basic RRKM Theory

The theory presented in this chapter is known generally as the RRKM (Rice–Ramsperger–Kassel–Marcus) theory and is the logical successor to the RRK theory described in Chapter 2. It is due to the work of R. A. Marcus[1] following an earlier paper by R. A. Marcus and O. K. Rice[2] and further developed in subsequent papers[3]. The treatment which follows retains largely that of the first edition of this book, although accounts have subsequently appeared in other places[4]. The reader is referred to Appendix 2 where some relevant statistical-mechanical results are given. A continuing problem is that of consistency of nomenclature, and the currently most widely used symbols for the various quantities involved are given in Appendix 1.

3.1 THE RRKM REACTION SCHEME

As indicated in Chapter 2, the reaction scheme used in the RRKM theory comprises the reactions shown in equations (3.1) and (3.2):

$$A + M \underset{k_2}{\overset{\delta k_1(E^* \to E^* + \delta E^*)}{\rightleftharpoons}} A^*_{(E^* \to E^* + \delta E^*)} + M \tag{3.1}$$

$$A^*_{(E^*)} \xrightarrow{k_a(E^*)} A^+ \xrightarrow{k^+} \text{products} \tag{3.2}$$

This scheme is a more detailed version of the RRK mechanism, in which (3.2) was condensed to (3.3) (cf. (2.60), (2.61)).

$$A^*_{(E^*)} \xrightarrow{k_a(E^*)} \text{products} \tag{3.3}$$

The superscript * in E^* has been introduced for reasons that will emerge later (section 3.3). As in any theory of this type the overall first-order rate constant k_{uni} is given by (2.31), reproduced here as

$$k_{uni} = -\frac{1}{[A]}\frac{d[A]}{dt} = \int_{E^*=E_0}^{\infty} \frac{k_a(E^*)\,dk_1(E^* \to E^* + dE^*)/k_2}{1 + k_a(E^*)/k_2[M]} \tag{3.4}$$

There are essentially two new principles involved in the RRKM treatment[1,2]. Firstly, the energisation rate constant k_1 in (3.1) is evaluated as a function of energy by a quantum-statistical-mechanical treatment instead of the classical treatment used in the classical RRK and Slater theories. The de-energisation rate constant k_2 is considered as in those theories to be independent of energy, and is often equated to the collision number Z or to $\beta_c Z$ where β_c is a collisional deactivation efficiency. The assumption of constant k_2 can be relaxed eventually, but this type of calculation really requires a quite different approach (see Chapter 7). The second major feature of RRKM theory is the application of canonical transition state theory (CTST) to the calculation of $k_a(E^*)$. For this purpose the overall reaction (3.3) is written in terms of two steps, (3.2), in which a careful distinction has been made between the *energised molecule* A* (sometimes called the active molecule) and the *transition state* A$^+$ (occasionally called the activated complex).

The *energised molecule* A* is basically an A molecule, but is characterised loosely by having enough energy to react; a more precise definition will be possible in section 3.3. The energy distribution, however, will not usually be such that reaction occurs immediately, there will be numerous quantum states of the energised molecule in a given small energy range, and only a few of these will correspond to energy distributions with which the molecule can actually undergo conversion to products. The energised molecules thus have lifetimes to decomposition which are much greater than the periods of their vibrations ($\approx 10^{-13}$ s). The actual lifetimes to de-energisation or decomposition depend on the values of $k_2[M]$ and $k_a(E^*)$ respectively, but are typically in the range 10^{-9}–10^{-4} s.

The *transition state* A$^+$ is basically a species which is recognisable as being intermediate between reactant and products, and is characterised by having a configuration corresponding to the top of an energy barrier between reactants and products. Bunker[5] originally criticised the use of the term *transition state* in this context and he and Pattengill[6] proposed an alternative *critical molecular configuration* corresponding to a minimum in the density of internal quantum states along the reaction coordinate. More recently it has been recognised that a minimum in the sum of states is a more appropriate criterion for locating the transition state[7]. The precise location of transition states, particularly in relation to *loose transition states*, is considered in more detail in Chapter 6.

In RRKM theory as in CTST, the conversion of A* into products is considered in terms of translational motion in the reaction coordinate. The energy profile along the reaction coordinate often involves a potential energy barrier between reactant and products, of height E_0 (the critical energy requirement), and this barrier must be surmounted for reaction to occur. The transition state is a molecule for which the extension of the reaction coordinate lies in an arbitrarily small range δ at the top of the barrier (see, however, the footnote in section 3.5). The transition state is thus unstable to movement in either direction along the reaction coordinate and, in contrast to an energised molecule, has no measurable life. There will usually be more than one quantum state of A^+ which can be formed from a given A^*, because of the different possible distributions of the energy between the reaction coordinate and the vibrational and rotational degrees of freedom of the transition state. Thus the rate constant $k_a(E^*)$ (equation 3.2) will be evaluated, as in CTST, as the sum of a set of contributions from the various possible transition states.

3.2 CLASSIFICATION OF ENERGIES AND DEGREES OF FREEDOM

The RRKM theory uses statistical mechanics for calculating the equilibrium concentrations of A^* and A^+, and is thus concerned very much with evaluating the number of ways of distributing a given amount of energy between the various degrees of freedom of a molecule. Any *fixed energy*, which cannot be redistributed, is clearly of no interest from this point of view; for example, the zero-point energy of the molecular vibrations is always present and always the same, and hence is said to be fixed. Overall translation of a molecule has no effect on the rate of its unimolecular reactions, and translational energy of the molecule as a whole is therefore fixed. In contrast, some of the energy content of the molecule (the *non-fixed energy*) is not fixed by any basic principle and is considered to be free to move around the molecule. In particular, the vibrational energy of the molecule (apart from the fixed zero-point energy) is assumed to be subject to rapid statistical redistribution. This assumption is the same as that of the RRK theories and is diametrically opposed to that of the Slater theory.

Rotational energy poses a more difficult problem; it might well contribute to a reaction (for example, centrifugal force might assist a bond-breaking reaction), but there might be limitations due to the requirement of conservation of the angular momentum. If, as a result, a rotational degree of freedom stays in the same quantum state during the reaction, the mode is said to be *adiabatic*. There is no *random* exchange of energy between an adiabatic mode and the other degrees of freedom, but if the moment of inertia changes, the rotational energy must also change, and this will affect the rate of reaction (see section 3.10).

The RRKM theory as normally applied assumes that all the non-adiabatic degrees of freedom of the molecule are active, i.e. can contribute their non-fixed energy to the reaction. Marcus[1] has also made provision for some of the modes to be inactive although not adiabatic; such degrees of freedom exchange energy between themselves but cannot contribute it towards the energy required for surmounting the potential energy barrier.

In this chapter we treat only the case where all the modes are either active or adiabatic. The concept of inactive modes implying that not all the modes rapidly exchange energy intramolecularly is related to so-called non-RRKM behaviour. This is considered further in section (3.12) which deals with the assumptions of basic RRKM theory and in Chapter 9 on intramolecular vibrational relaxation.

3.3 TERMINOLOGY FOR ENERGIES

In order to handle the statistical mechanics used in the RRKM theory it is important to understand clearly the significance of the various energy terms used, and we are now in a position to discuss these in detail. As an aid to the discussion, Figure 3.1 illustrates the energetics of a simple unimolecular reaction and some of the relationships between the terms to be used.

An energised molecule, A^*, can now be defined precisely as a molecule which contains in its active degrees of freedom a non-fixed energy E^* greater than a critical value E_0 below which classical reaction cannot occur.

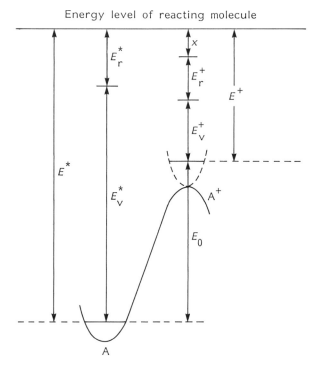

Energy level of reacting molecule

Figure 3.1   Illustration of energy terminology for a unimolecular reaction; adiabatic and inactive degrees of freedom are excluded. Reproduced by permission of John Wiley and Sons Ltd.

This critical energy may alternatively be defined as the difference between the ground-state energies of A^+ and A; it is closely related to the Arrhenius activation energy of the reaction, but may differ from it by a few multiples of kT. The energy E^* may comprise both vibrational and rotational non-fixed energy, denoted by E_v^* and E_r^* respectively, and the sum of these is clearly E^*, which will sometimes be subscripted as follows for emphasis:

$$E^* \equiv E_{vr}^* = E_v^* + E_r^*$$

The total non-fixed energy of a given transition state A^+ is denoted by E^+. This quantity is closely related to the non-fixed energy E^* of the energised molecule A^* from which the transition state was formed. In this process an amount of energy E_0 has been used to surmount the energy barrier and has thus become fixed; any redistribution of this energy would correspond to motion in the reaction coordinate and the molecule would then no longer be at the transition state.

Thus E^+ and E^* for a corresponding transition state and energised molecule are related simply by

$$E^+ = E^* - E_0$$

The energy E^+ may, like E^*, contain both vibrational and rotational non-fixed energy, denoted by E_v^+ and E_r^+ respectively, the sum of these being E_{vr}^+. It is also necessary to separate for special treatment the energy x associated with the translational motion of A^+ in the reaction coordinate, and we therefore have

$$E^+ = E_{vr}^+ + x = E_v^+ + E_r^+ + x$$

The different possible distributions of the energy between vibration, rotation and the translation along the reaction coordinate are an important factor in determining the number of quantum states corresponding to a given E^* or E^+, and hence the equilibrium concentrations of A^* and A^+.

3.4   EXPRESSION FOR $\delta k_{1(E^* \to E^* + \delta E^*)}/k_2$

The quantity $\delta k_{1(E^* \to E^* + \delta E^*)}/k_2$, which is required for the evaluation of (3.4), is the equilibrium constant for the reaction (3.1) in which the A molecules are thermally energised into the small range E^* to $E^* + \delta E^*$ (cf. section 2.4). This equilibrium constant is given by statistical mechanics as the partition function ratio $Q(A^*_{(E^* \to E^* + \delta E^*)})/Q(A)$, both partition functions being calculated from the common energy zero, that of A (see Appendix 2, section A2.5). The function $Q(A)$, henceforth denoted Q_2, is simply the ordinary molecular partition function for all the active modes of A; these usually comprise all the vibrations and internal rotations of the molecule (see section 3.12). The function $Q(A^*_{(E^* \to E^* + \delta E^*)})$ is the partition function for those A^* having non-fixed energy specifically in the range E^* to $E^* + \delta E^*$. If δE^* is small the exponential terms in the expression $Q = \Sigma g_i \exp(-E_i/kT)$ are all the same, simply $\exp(-E^*/kT)$, and the partition function can be expressed as follows:

$$Q(A^*_{(E^* \to E^* + \delta E^*)}) = \left(\sum_{E^* \to E^* + \delta E^*} g_i \right) \exp(-E^*/kT)$$

$$= \rho(E^*) \delta E^* \exp(-E^*/kT)$$

Division by $Q(A) \equiv Q_2$ then gives the required result (equation 3.5):

$$\frac{\delta k_1(E^* \to E^* + \delta E^*)}{k_2} = \frac{\rho(E^*) \exp(-E^*/kT)}{Q_2} \delta E^* \qquad (3.5)$$

this expression is in fact the quantum Boltzmann distribution function $K(E)\delta E$ giving the thermal equilibrium proportion of molecules in the given energy range.

In these equations, Σg_i, the number of quantum states of A^* in the energy range E^* to $E^* + \delta E^*$, has been replaced by $\rho(E^*)\delta E^*$. Thus $\rho(E^*)$ is the *density of quantum states* or the *number of quantum states per unit range of energy* at energies close to E^*, although the physical significance of $\rho(E^*)\delta E^*$ is probably easier to grasp than that of $\rho(E^*)$ itself. A further discussion of state densities and related quantities is given in section 3.11. The replacement of Σg_i by $\rho(E^*)\delta E^*$ involves the implicit assumption that the energy levels are closely spaced relative to the width δE^* of any energy range of interest. Only under these conditions is it adequate to replace the essentially stepwise increase in the number of energy levels by a continuous distribution function. The density of energy levels at the energies of interest for the energised molecule is, in fact, usually very high. For example, cyclopropane at energies E^* near the critical energy ($E_0 \equiv 264 \text{ kJ mol}^{-1}$) has a vibrational state density of about 8×10^9 states per kJ energy range, so that a continuous distribution is quite adequate even for energy ranges δE^* as low as $4 \times 10^{-4} \text{ J mol}^{-1}$. Incidentally, this treatment does not necessarily imply that the vibrations have to be treated on a classical basis; $\rho(E^*)$ can be evaluated on a quantum-mechanical basis, giving values which can be quite different from those obtained from a classical approximation (see section 4.4). Rice[8] has argued further in favour of the continuous distribution treatment, on the basis that the energised molecules have a limited mean life $\tau \approx 1/k_a(E^*)$ and that the Heisenberg Uncertainty Principle thus requires the energy levels themselves to be broadened by an amount $E \approx h/\tau$, where h is Planck's constant. It so happens that this broadening is, as a matter of principle, of the same order of magnitude as the spacing of the energy levels, so that the broadened levels overlap to some extent and the stepwise variation of energy approaches a continuous variation. For example, for cyclopropane at $2\,kJ\,mol^{-1}$ above the threshold, $k_a(E^*) \approx 10^4\,s^{-1}$. Since $h \approx 4 \times 10^{-13}\,kJ\,mol^{-1}\,s$, the energy levels are blurred to the extent of $\Delta E \approx 4 \times 10^{-9}\,kJ\,mol^{-1}$, compared with the energy-level spacing of $1/\rho(E^*) \approx 4 \times 10^{-9}\,kJ\,mol^{-1}$ at this energy.

3.5 EXPRESSION FOR $k_a(E^*)$

An expression for $k_a(E^*)$, the rate constant for the formation of transition states from energised molecules with non-fixed energy E^*, is obtained by a detailed application of the steady-state treatment to A^+ in reaction (3.2). For any particular energy levels these are of the form:

$$A^* \xrightarrow{k_a(E^*)} A^+ \xrightarrow{k^+} \text{products}$$

and the steady-state treatment gives $k_a[A^*] = k^+[\overrightarrow{A^+}]$ and hence

$$k_a(E^*) = k^+\left(\frac{[\overrightarrow{A^+}]}{[A^*]}\right)_{\text{steady state}} \tag{3.6}$$

$$= \tfrac{1}{2}k^+\left(\frac{[A^+]}{[A^*]}\right)_{\text{eqm}}$$

In deducing (3.6) it has been assumed, as in CTST, that the steady-state concentration $[\overrightarrow{A^+}]$, of transition states which are crossing the barrier in the forward direction during the reaction is the same as it would be if A^* and A^+ were in equilibrium. At equilibrium the concentration of forward-crossing transition states is half the total concentration of transition states, and since $[A^+]$ here denotes the total concentration this accounts for the factor $\tfrac{1}{2}$ in (3.6). The validity of this 'equilibrium assumption' is discussed in section 3.12.

The non-fixed energy of a transition state A^+ formed from an energised molecule A^* of energy E^* is $E^+ = E^* - E_0$ (see section 3.3 and Figure 3.1) and this energy E^+ can be divided in different ways into energy of vibration and rotation (E_{vr}^+) and energy of translational motion in the reaction coordinate (x). Accordingly, the overall reaction (3.2) is broken down into contributions from the different transition states as shown in (3.7), and $k_a(E^*)$ is evaluated as the sum (3.8) of corresponding contributions of the form (3.6).

$$A^*(E^*) \xrightarrow{k_a(E_{vr}^+,x)} A^+_{(E_{vr}^+,x)} \xrightarrow{k^+(x)} \text{products} \tag{3.7}$$

$$k_a(E^*) = \sum_{E_{vr}^+=0}^{E^+} \tfrac{1}{2}k^+(x)\left(\frac{[A^+_{(E_{vr}^+,x)}]}{[A^*_{(E^*)}]}\right)_{\text{eqm}} \tag{3.8}$$

where

$$E^* = E_0 + E^+ \quad \text{and} \quad x = E^+ - E_{vr}^+$$

The sum covers all possible distributions of E^+ between E_{vr}^+ and x, from $E_{vr}^+ = 0$ (in which case all the non-fixed energy is translational energy in the reaction coordinate and $x = E^+$), up to the maximum possible value of E_{vr}^+ below E^+. The latter case corresponds to transition states having the minimum possible energy in the reaction coordinate (for the given E^+), but since E_{vr}^+ is quantised it may well not have a value corresponding precisely to E^+. The rate constant k^+ depends only on x and is denoted $k^+(x)$, but the contribution to $[A^+]/[A^*]$ depends on E_{vr}^*, x and E^*, any two of which, together with E_0, serve to determine the third. The treatments of $k^+(x)$ and $[A^+]/[A^*]$ are detailed separately in the following sections.

EVALUATION OF $k^+(x)$

Marcus treated decomposition of the transition state into products as the translation of a particle of mass μ in a one-dimensional box of length δ, the small region at the top of the energy barrier which is considered to define the transition state.[a] The mass μ is often considered to be a reduced mass for the atoms involved in the reaction coordinate, but its precise significance, like that of δ, need not be considered in detail because both cancel out in the final expression. If the energy in the reaction coordinate is x, the speed of translation is $(2x/\mu)^{1/2}$ and the time taken for the mass μ to cross the box is $\delta/(2x/\mu)^{1/2}$. The rate constant for crossing of the barrier is therefore

$$k^+(x) = (2x/\mu)^{1/2}/\delta = (2x/\mu\delta^2)^{1/2} \tag{3.9}$$

EVALUATION OF $([A^+]/[A^*])_{eqm}$

As previously indicated, the RRKM theory uses the equilibrium ratio of concentrations of A^+ and A^*. This is calculated from statistical mechanics as the ratio $Q(A^+)/Q(A^*)$ of the partition functions of the transition state and energised molecule, using energies reckoned from a common energy zero, that of the A molecule. Since the two species under consideration both have a total energy in the small range $E^* \to E^* + \delta E^*$, each partition function is

---

[a]This aspect of the CTST formulation has been criticised and an alternative derivation of the same result suggested by Slater[9]. The number of complexes crossing the barrier in unit time is given by their speed in the reaction coordinate divided by their 'distance apart along the reaction coordinate', thus avoiding the concept of an arbitrary length δ which has no physical significance. The 'trap' described by Slater is avoided in the type of derivation used here, which does not make the simplification of giving all complexes the mean velocity or mean life. There are also other treatments, again leading to the same result[10].

of the form $(\Sigma g_i) \exp(-E^*/kT)$, where Σg_i is the number of quantum states in this small energy range, and $Q(A^+)/Q(A^*)$ reduces simply to $\Sigma g_i^+/\Sigma g_i^*$. Although the A^+ and the A^* have the same total energy, the former has much less *non-fixed* energy, and there are thus many fewer quantum states for A^+ in the given energy range and $[A^+]/[A^*]$ will be small, as is physically reasonable. As previously (section 3.4), Σg_i^* can be replaced by a continuous distribution function $\rho(E^*)\delta E^*$, and a similar treatment is valid at this stage for the transition state, since it contains a translational degree of freedom (the reaction coordinate). Translational motion usually has an extremely close spacing of energy levels (see Appendix 2, section A2.2) and the energy can reasonably be treated as continuous rather than quantised. The number of quantum states of the transition state in the range of total energy $E^* \rightarrow E^* + \delta E^*$ is commonly written as $\rho(E^+)$, where $E^+ = E^* - E_0$, thus the concentration ratio for the small energy range of interest reduces to

$$([A^+]/[A^*])_{eqm} = \Sigma g_i/\Sigma_{g_i}^* = \rho(E^+)/\rho(E^*) \tag{3.10}$$

In order to evaluate the concentration ratio in (3.8) we need the density of states $\rho(E^+)$ for a specified division of E^+ into the vibrational–rotational energy E_{vr}^+ and the translational energy x in the reaction coordinate. This will be denoted by $\rho(E_{vr}^+, x)$ to avoid confusion with the total state density $\rho(E^+)$. Although the translational energy x can be treated in terms of a continuous distribution, the vibrational and rotational energies of the transition state will both be treated as quantised for the time being. The quantum treatment is certainly essential for the vibrational degrees of freedom of the transition state, because the non-fixed energy of A^+ is in general relatively small, so that the vibrations of A^+ are much less highly excited than those of A^*. A continuous treatment is only reasonable at high energies where the density of quantum states is high; see sections 3.11 and 4.4. For example, for cyclopropane with $E^* = 295 \text{ kJ mol}^{-1}$ and $E^+ = 21 \text{ kJ mol}^{-1}$, A^* has $\approx 10^9$ vibrational quantum states per kJ energy range, while A^+ has only ≈ 24; a continuous function is much less valid for A^+ than for A^*.

Thus in order to split $\rho(E_{vr}^+, x)$ into contributions from E_{vr}^+ and x we define:

$P(E_{vr}^+) =$ number of vibrational–rotational quantum states of A^+ with vibrational–rotational non-fixed energy equal to E_{vr}^+ (precisely);

$\rho_{rc}(x)\delta x =$ number of translational quantum states of A^+ with energy in the reaction coordinate in the range $x \rightarrow x + \delta x$.

The overall number of states in an energy range x is given by the product of these degeneracies, and is $\rho(E_{vr}^+, x)\delta x$, whence

$$\rho(E_{vr}^+, x) = P(E_{vr}^+)\rho_{rc}(x) \tag{3.11}$$

The required concentration ratio is thus obtained by substituting $\rho(E_{vr}^+, x)$ from (3.11) for $\rho(E^+)$ in (3.10), the result being

$$\left(\frac{[A_{(E)_{vr}^+ x)}^+]}{[A_{(E^*)}^*]}\right)_{eqm} = \frac{P(E_{vr}^+)\rho_{rc}(x)}{\rho(E^*)} \tag{3.12}$$

EXPRESSION FOR $\rho_{rc}(x)$

This is derived in a manner similar to that used in the more sophisticated versions of CTST[11]. The wave-mechanical treatment[12] of the translation of a particle of mass μ in a box of length δ produces the result that the energy is quantised, the energy x of the nth level being

$$x = n^2 h^2/8\mu\delta^2$$

where h is Planck's constant. Thus the number of quantum states with energy up to and including x is $n = (8\mu\delta^2 x/h^2)^{1/2}$. The number of states in the energy range $x \rightarrow x + \delta x$ (where δx, although small, is sufficiently large compared with the energy level spacing) is $\delta n = (dn/dx)\delta x$, and this number of states is equal to $\rho_{rc}(x)\delta x$, whence

$$\rho_{rc}(x) = dn/dx = (2\mu\delta^2/h^2x)^{1/2} \tag{3.13}$$

RESULT FOR $k_a(E^*)$

The expression (3.8) for $k_a(E^*)$ can now be evaluated using (3.9), (3.12) and (3.13) to form the summand for a given energy distribution, and the result obtained is

$$k_a(E^*) = \sum_{E_{vr}^+=0}^{E^+} \frac{1}{2}\left(\frac{2x}{\mu\delta^2}\right)^{1/2} \frac{P(E_{vr}^+)(2\mu\delta^2/h^2x)^{1/2}}{\rho(E^*)}$$

$$= \sum_{E_{vr}^+=0}^{E^+} \frac{P(E_{vr}^+)}{h\rho(E^*)} \tag{3.14}$$

hence

$$k_a(E^*) = \frac{1}{h\rho(E^*)}\sum_{E_{vr}^+=0}^{E^+} P(E_{vr}^+) = \frac{W(E_{vr}^+)}{h\rho(E^*)} \tag{3.15}$$

where

$$W(E_{vr}^+) = \sum_{E_{vr}^+=0}^{E^+} P(E_{vr}^+)$$

It will be noted that when the contribution to k_a from transition states with a specified distribution of E^+ between E_{vr}^+ and x is evaluated (i.e. the summand in (3.14)) the explicit characteristics of the reaction coordinate (μ and δ) disappear. The value of the contribution thus depends only on the properties of the energy levels in the energised molecule and in the ordinary degrees of freedom of the transition state (i.e. excluding the reaction coordinate).

The sum in (3.15), often written less explicitly as $\Sigma P(E_{vr}^+)$, is the sum of the numbers of vibrational–rotational quantum states at all the quantized energy levels of energy less than or equal to E^+; i.e. it is simply the total number of vibrational–rotational quantum states of the transition state with energies $\leqslant E^+$. If all the energy levels of the transition state were plotted on an energy diagram, $\Sigma P(E_{vr}^+)$ could be evaluated by simply counting the number of levels at energies up to and including E^+. The physical significance of this quantity is better appreciated when expressed as $\Sigma P(E_{vr}^+)$ rather than the abbreviation $W(E_{vr}^+)$ and hence the former notation is retained here and also in section 3.11 and Chapter 4.

TWO MODIFICATIONS, AND FINAL RESULT FOR $k_a(E^*)$

It will be profitable to introduce at this stage two minor modifications, concerned with adiabatic rotations and statistical factors respectively.

It has already been pointed out that adiabatic rotations, which stay in the same quantum state while the energised molecule becomes a transition state, nevertheless suffer an energy change because of the different moments of inertia of A^* and A^+. Usually energy is released into the other degrees of freedom of the molecule (see section 3.10), and this obviously has an effect on the rate constant $k_a(E^*)$. The CTST treatment of the reaction at high pressures (section 3.7) shows that in this limiting case the correct k_∞ is obtained if the expression (3.15) for $k_a(E^*)$ is multiplied by the factor Q_1^+/Q_1, where Q_1^+ and Q_1 are the partition functions for the adiabatic rotations in the transition state and the A molecule respectively. The appearance of this factor may be traced to the calculation of $[A^+]/[A^*]$. Marcus suggested originally[1] that the same correction is reasonably accurate even at low pressures, although later work by Marcus[3,13] and others has shown that a more sophisticated treatment is really required. Such treatments are discussed in section 3.10, but the crude correction by the factor Q_1^+/Q_1 will be included for the time being.

The second modification concerns the possibility that a reaction can proceed by several distinct paths which are kinetically equivalent, i.e. equivalent with regard to the energetics and rate of reaction. The number of such paths is termed the *statistical factor* $L^{\ddagger}$. A simple case would be the dissociation of H_2O to $OH + H$, for which $L^{\ddagger} = 2$ since either of the two identical OH bonds can be broken. The problem is discussed in detail in section 3.9, from which it emerges that the correct rate constant $k_a(E^*)$ for disappearance of energised molecules by all the paths together is obtained by including a factor $L^{\ddagger}$ in (3.15). When $L^{\ddagger}$ is defined as in section 3.9, symmetry numbers must be omitted from the rotational partition functions.

The final equation for $k_a(E^*)$ is therefore (3.16), and unless a better treatment of adiabatic rotations is required, (3.16) is the expression generally used in the RRKM formulation.

$$
\begin{aligned}
k_a(E^*) &= L^{\ddagger}\frac{Q_1^+}{Q_1}\frac{1}{h\rho(E^*)}\sum_{E_{vr}^+=0}^{E^+} P(E_{vr}^+) \\[2mm]
&\equiv L^{\ddagger}\frac{Q_1^+}{Q_1}\frac{W(E_{vr}^+)}{h\rho(E^*)}
\end{aligned}
\tag{3.16}
$$

3.6   RRKM EXPRESSION FOR k_{uni}

As indicated in section 3.1 the overall first-order rate constant k_{uni} is obtained by substituting the expressions (3.5) and (3.16) into (3.4), whence

$$
k_{uni} = \frac{L^{\ddagger}Q_1^+}{hQ_1Q_2}\int_{E^*=E_0}^{\infty}\frac{\left\{\sum P(E_{vr}^+)\right\}\exp(-E^*/kT)\,dE^*}{1 + k_a(E^*)/k_2[M]}
$$

Or, since $E^* = E_0 + E^+$ and $dE^* = dE^+$, we obtain the final result:

$$
k_{uni} = \frac{L^{\ddagger}Q_1^+\exp(-E_0/kT)}{hQ_1Q_2}\int_{E^+=0}^{\infty}\frac{\left\{\displaystyle\sum_{E_{vr}^+=0}^{E^+} P(E_{vr}^+)\right\}\exp(-E^+/kT)\,dE^+}{1 + k_a(E_0 + E^+)/k_2[M]}
$$

$$\tag{3.17}$$

in which $k_a(E_0 + E^+)$ is given by (3.16), rewritten in the equivalent form:

$$
k_a(E_0 + E^+) = \frac{L^{\ddagger}Q_1^+}{hQ_1\rho(E_0 + E^+)}\sum_{E_{vr}^+=0}^{E^+} P(E_{vr}^+)
\tag{3.18}
$$

Since the result of the summation in (3.17) and (3.18) is simply a function of E^+ (see section 3.11) the whole integrand is a function of E^+ and can be integrated numerically to give k_{uni} if $\rho(E_0 + E^+)$ and $\sum P(E_{vr}^+)$ are known as functions of energy, i.e. if the distributions of the vibrational–rotational energy levels of the reactant and the transition state are known. The physical significance of these quantities is discussed in detail and further illustrated in section 3.11. Their exact evaluation is usually laborious in the extreme, and application of the RRKM theory depends very much on the availability of simple but reasonably accurate approximations which will be discussed in Chapter 4. The basic assumptions which have been made in the derivation of (3.17) are reviewed and their validity discussed in section 3.12.

Apart from the approximate treatment of adiabatic rotations, (3.17) is to be regarded as the fundamental RRKM result for k_{uni}. If adiabatic rotations are to be ignored (by putting $Q_1^+/Q_1 = 1$) the result is then accurate within the framework of the basic assumptions discussed in section 3.12. Reference to section 3.10 is desirable if adiabatic rotations are to be included; it emerges that (3.17) and (3.18) give erroneous results in the fall-off and low-pressure regions. For low values of Q_1^+/Q_1 a better result is obtained by including this factor in (3.17) but not in (3.18)[3a], but in cases where the effect is really worth considering, further modifications are necessary.

It will be noted that in (3.17) vibrations and rotations have both been treated as quantised, in contrast to the original formulations. In Chapter 4 a classical treatment will be applied to the active rotations, following Marcus and leading to his result, but this simplification is not an essential part of the basic theory. The integration over a continuously varying E^+ arises, despite the quantum treatment and the consequent stepwise variation of $\Sigma P(E_{vr}^+)$ and $k_a(E^*)$, from the essentially continuous energies of the translational degree of freedom in A^+ (the reaction coordinate) and the very high density of energy levels of A^* at the energies of interest, already referred to in section 3.4.

3.7   THE HIGH-PRESSURE LIMIT

It is interesting to consider the result obtained from the RRKM theory for a unimolecular reaction at high pressures, and to compare this with the predictions of other theories. In particular the RRKM theory should give a similar result to CTST since both are based on a statistical-mechanical approach. The high-pressure limit is easily obtained from (3.17) by putting $[M] \to \infty$, when the pseudo-first-order rate constant k_{uni} becomes the genuine pressure-independent first-order rate constant k_∞ given by

$$k_\infty = \frac{L^\ddagger Q_1^+}{hQ_1Q_2} \exp\left(-E_0/kT\right) \int_{E^+=0}^\infty \left\{\exp\left(-E^+/kT\right) \sum_{E_{vr}^+=0}^{E^+} P(E_{vr}^+)\right\} dE^+ \quad (3.19)$$

It will be noted that $\rho(E^*)$ does not appear in this result, except implicitly in the calculation of Q_2, the partition function for the active degrees of freedom of the reactant molecule A. The integral in (3.19) can be evaluated by reversing the order of summation and integration, with careful attention to the limits involved. It is necessary to consider all possible transition states, and this is achieved in (3.19) by integrating from $E^+ = 0$ to ∞ (thus covering all possible total energies of the transition state), and for each value of E^+ summing the contributions from the different transition states having the different possible divisions of this energy E^+ between E_{vr}^+ and x, i.e. all the transition states with $0 \leqslant E_{vr}^+ \leqslant E^+$.

$$
k_\infty = \frac{L^\ddagger Q_1^+}{hQ_1Q_2} \exp\left(-E_0/kT\right) \sum_{E_{vr}^+=0}^{\infty} \left\{ P(E_{vr}^+) \int_{E^+=E_{vr}^+}^{\infty} \exp\left(-E^+/kT\right) dE^+ \right\}
$$

(3.20)

In the reversed expression (3.20) the same effect is achieved by summing over all possible values of E_{vr}^+ (from 0 to ∞), and for each value of E_{vr}^+ considering all possible x (from 0 to ∞) so that E^+ varies continuously from E_{vr}^+ to ∞, these being the limits for the inner integration. In (3.20) $P(E_{vr}^+)$ stays constant while the integral is evaluated as a simple function of E_{vr}^+, as follows:

$$
\int_{E^+=E_{vr}^+}^{\infty} \exp\left(-E^+/kT\right) dE^+ = kT \exp\left(-E_{vr}^+/kT\right)
$$

Summation over all possible values of E_{vr}^+ then gives

$$
k_\infty = \frac{L^\ddagger Q_1^+}{hQ_1Q_2} kT \exp\left(-E_0/kT\right) \sum_{E_{vr}^+=0}^{\infty} \left[P(E_{vr}^+)\exp\left(-E_{vr}^+/kT\right)\right]
$$

A little consideration shows that the sum in this expression is simply the partition function Q_2^+ for the active vibrations and rotations in the transition state (cf. Appendix 2, section A2.1). This is very similar to the partition function Q_2 for the active vibrations and rotations in the ordinary molecule A, except that the frequencies and moments of inertia will have changed and that the transition state has one degree of freedom fewer, since the motion in the reaction coordinate has been singled out for special treatment. The high-pressure limit from the RRKM theory is thus

$$
k_\infty = L^\ddagger \frac{kT}{h} \frac{Q^+}{Q} \exp\left(-E_0/kT\right)
$$

(3.21)

in which Q and Q^+ are the complete vibrational–rotational partition functions for the reactant and the transition state ($Q = Q_1Q_2$ and $Q^+ = Q_1^+Q_2^+$) respectively. Except for the omission of a transmission coefficient this result is identical with that obtained from CTST for a unimolecular reaction, which is reasonable in view of the similarity of the treatments involved. Canonical transition state theory[14] calculates the rate on the assumption that the transition states A$^+$ are in thermal equilibrium with the reactant molecules A, i.e. that the thermal Boltzmann distribution is maintained at all energies, which is true at sufficiently high pressures. The CTST scheme is thus

$$M + A \xrightleftharpoons{eqm} A^+ + M$$

$$A^+ \rightarrow products$$

(3.22)

The RRKM theory, in the general case, admits equilibrium between A^+ and A^*, but not between A^* and A in the reactions (3.1) and (3.2). At high pressures, however, A^* and A are also in equilibrium so that the model becomes the same as that treated in CTST and the results naturally coincide.

It may be noted that the agreement between the RRKM and CTST results for k_∞ is not complete unless the RRKM formulation includes a correction which reduces to Q_1^+/Q_1 at high pressures. The idea that a constant factor of Q_1^+/Q_1 will give approximately the correct result for k_{uni} at lower pressures has only limited validity—see section 3.10.

3.8   THE LOW-PRESSURE LIMIT

In the limit of very low pressures the first-order rate constant from (3.17) becomes proportional to the pressure; the second-order rate constant k_{bim} is then given by

$$
\begin{aligned}
k_{bim} &= \lim_{[M]\rightarrow 0}\left(\frac{k_{uni}}{[M]}\right) \\
&= \frac{k_2}{Q_2}\exp\left(-E_0/kT\right)\int_{E^+=0}^{\infty}\rho(E_0 + E^+)\exp\left(-E^+/kT\right)dE^+ \\
&= \frac{k_2}{Q_2}\int_{E^*=E_0}^{\infty}\rho(E^*)\exp\left(-E^*/kT\right)dE^* = \frac{k_2 Q_2^*}{Q_2}
\end{aligned}
$$

(3.23)

where Q_2^* is the partition function for *energised molecules* (i.e. specifically those A molecules which have non-fixed energy greater than E_0) *using the ground state of A for the zero of energy*. Thus Q_2^* is a part of the series

$$g_1 \exp\left(-E_1/kT\right) + g_2 \exp\left(-E_2/kT\right) + \ldots$$

which defines Q_2 (cf. Appendix 2), but includes only the terms for which $E_i \geqslant E_0$.

An alternative derivation of (3.23) starts by taking the low-pressure limit of (3.4), whence

$$k_{bim} = \int_{E^*=E_0}^{\infty} dk_1$$

Insertion of the RRKM expression for δk_1 from (3.5) leads as above to

(3.23). This is equivalent to noting that at sufficiently low pressures all the energised molecules react; the rate of reaction is thus simply the rate of energisation, and the rate of formation of A^+ from A^* is no longer relevant. An expression for the rate of energisation may be obtained as before by considering the equilibrium $M + A \rightleftharpoons M + A^*$; at equilibrium the rate of energisation is equal to the rate of de-energisation, $k_2[M][A^*]$, and the concentration of energised molecules is given by $[A^*]/[A] = Q_2^*/Q_2$. Thus the rate of energisation is $(k_2 Q_2^*/Q_2)[M][A]$, and this is assumed to be correct even when A^* are removed by reaction (see section 3.12). There are no statistical factors to be included; this is because they occur in the rate constant k_a for formation of A^+ from A^*, and this no longer affects the rate of reaction. Similarly there is no factor involving the adiabatic rotations in (3.23); this is a fault of the original Marcus treatment which is corrected in the later versions discussed in section 3.10. Apart from this reservation, however, we have the interesting result that the low-pressure rate constant depends only on the properties of the reactant molecules and the height of the energy barrier, and in no way depends on the detailed properties of the transition state.

Equation (3.23) may alternatively be written in terms of the partition function $Q_2^{*'}$ for energised molecules *relative to the ground state of A^+ as the zero of energy*; this level is in fact the 'ground state' for energised molecules, being the lowest energy that an A^* can have, and in some ways forms a more natural choice of energy zero. The two partition functions are related simply by the equation $Q_2^{*'} = Q_2^* \exp(E_0/kT)$(see Appendix 2, section A2.1), and k_{bim} is therefore given by

$$k_{bim} = \frac{Q_2^{*'}}{Q_2} k_2 \exp(-E_0/kT) \qquad (3.24)$$

The terms $k_2 \exp(-E_0/kT)$ in this equation correspond to Lindemann's k_1, and the ratio $Q_2^{*'}/Q_2$ is thus the quantum statistical-mechanical equivalent of the term $(E_0/kT)^{s-1}/(s-1)!$ in Hinshelwood's k_1 (equation (2.14)). The density of quantum states increases rapidly with energy (see section 3.11), and $Q_2^{*'}$ is therefore much greater than Q_2. Thus (3.24), like the Hinshelwood equation, can give rates of energisation which are many times greater than those predicted on the simple collision theory picture.

It has already been seen that the Arrhenius activation energy of a unimolecular reaction varies with pressure. The RRKM theory does not lead to any simple equation for this variation, but the theoretical activation energy E_{Arr} at any pressure may be obtained in the usual way (equation 1.3) from the first-order rate constants calculated at a series of temperatures. The marked variation which can occur is illustrated by some calculated results for the isomerisation of 1,1-dichlorocyclopropane (see section 5.4); with a critical energy of $E_0 = 232\ \text{kJ mol}^{-1}$, E_{Arr} is calculated to be

241 kJ mol^{-1} at 1000 Torr and 203 kJ mol^{-1} at 10^{-5} Torr, these being close
to the high- and low-pressure limits of E_{Arr}. The extent of the variation
increases with the complexity of the molecule; for isomerisation of
the relatively small methyl isocyanide molecule $E_0 = 159$ kJ mol^{-1}, $E_\infty =$
161 kJ mol^{-1} and $E_{bim} = 151$ kJ mol^{-1}, the variation being in good agreement with experiment.

The basic reason for the variation of E_{Arr} with pressure is the change in
the energy distribution of the reacting molecules, illustrated in Figure 3.2 for
the isomerization of 1,1-dichlorocyclopropane. This figure is constructed
from the data summarised in Table 5.7. At sufficiently low pressures, all the
molecules which become energised react. The energy distribution is that
associated with k_1, which decreases rapidly as the energy increases, thus
favouring the reaction of molecules with energies near the critical energy. At
high pressures there is a competition between reaction and collisional
de-energisation. The energised molecules with energies near the critical
energy have long lifetimes before reaction, and are thus more likely to be

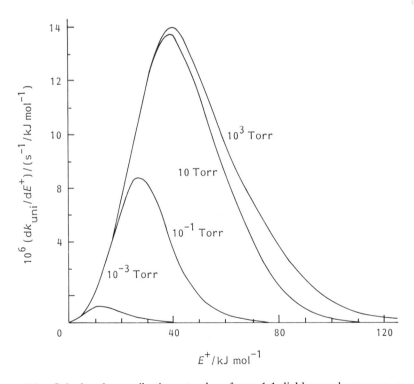

Figure 3.2 Calculated contributions to k_{uni} from 1,1-dichlorocyclopropane complexes with various energies as a function of pressure.

de-energised than the rapidly reacting molecules with higher energies. The more highly energised molecules thus contribute more heavily to the overall rate of reaction, and the activation energy is correspondingly higher.

3.9   STATISTICAL FACTORS

The straightforward RRKM formulation (3.15) of $k_a(E^*)$ refers basically to the rate of reaction by a single reaction path from the reactant molecule to the products. It often happens that there are several paths which are physically distinct but nevertheless completely equivalent so far as the rate calculation is concerned. Such reaction paths involve transition states which are geometrical or optical isomers of each other, and if diagrams are drawn showing the movements of the various atoms during the reaction, the diagrams for the different paths will also be geometrically or optically isomeric. In this case the calculated rate must be increased by an appropriate factor known as the *statistical factor* or *reaction path degeneracy*, denoted here by $L^{\ddagger}$. The terms are not always used with exactly the same meaning as here, but any difference will be clear from the context. The factor $L^{\ddagger}$ appears basically in the rate constant $k_a(E^*)$ for formation of transition states from energised molecules, (e.g. equation (3.16)), and hence appears in the final equation for k_{uni} both as part of k_a in the denominator and in front of the whole expression, (e.g. equation (3.17)). In CTST the factor is generally derived from the ratio σ/σ^+ of the symmetry numbers of the rotations (both internal and overall) of A and A^+. Thus the usual formulation can be written as

$$k_{\infty} = \frac{\sigma}{\sigma^+} \frac{kT}{h} \frac{Q^+}{Q} \exp\left(-E_0/kT\right)$$

(in which the partition functions do not include the symmetry numbers), and the statistical factor is found to be σ/σ^+.

It has been recognised for some time that this procedure can be in error, however, in connection with both equilibrium[15,16] and rate[16-18] studies, and a direct count of the number of reaction paths has generally been preferred. In what follows we have particularised the treatments of various workers to refer specifically to unimolecular reactions. Bishop and Laidler[18,19] defined a statistical factor $l^{\ddagger}$ as the number of different transition states that can be formed if all identical atoms in A are numbered. With this definition the correct rate of reaction was shown to be obtained by omitting the symmetry numbers from the partition functions and multiplying the rate expression by $l^{\ddagger}$. This remains a simple and accurate way of handling the problem, the $l^{\ddagger}$ of Bishop and Laidler being identical with our statistical factor $L^{\ddagger}$. Elliott and Frey[20] similarly defined the reaction path degeneracy $L^{\ddagger}$ as 'the number of different operations producing all equivalent transition states and comprising equivalent motion of different sets of atoms or different motion of equivalent sets of atoms'.

As an example, consider the isomerisation of 1,1-dichlorocyclopropane through a transition state of the type shown in Figure 3.3. We number the atoms in the molecule A and count four equivalent but distinct transition states I–IV; the statistical factor for this calculation is thus 4. Note that I is distinct from IV (and II from III) because of the numbering, although physically superimposable by rotation, and that I and IV are enantiomorphic with II and III, i.e. they are non-superimposable mirror-images of II and III.

Schlag and Haller[21] have given an alternative and carefully defined direct-count procedure for determining the statistical factor. The atoms are again numbered, and a count is made of the number of transition states *which would be superimposable if the numbering were removed* (and are not related by simple rotation with the numbering present). This number is then multiplied by two if the reactant is symmetric and the transition state asymmetric, and divided by two if the reverse is true. In the above example there are only two superimposable transition states (I and IV or II and III), but this number is multiplied by two since the reactant is symmetric and the transition states asymmetric, producing a statistical factor of 4 as before.

It is also possible to formulate a correct answer in terms of symmetry numbers, but an additional factor may be necessary. Bishop and Laidler[18] showed that $l^{\ddagger} = r^{\ddagger}\sigma/\sigma^{+}$, where $r^{\ddagger}$ is the statistical factor for the return of transition states to reactant molecules, defined in a similar way to $l^{\ddagger}$, and σ

Molecule A

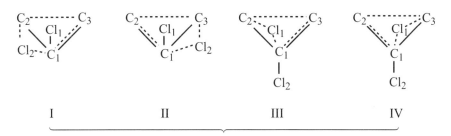

Figure 3.3   Transition states for isomerisation of 1,1-dichlorocyclopropane. Reproduced by permission of John Wiley and Sons Ltd.

and σ^+ are the symmetry numbers, taken as $\frac{1}{2}$ for asymmetric molecules[16]. In the above example $\sigma = 2$, $\sigma^+ = \frac{1}{2}$ and $r^{\ddagger} = 1$ (since each transition state I–IV can return to reactant by only one path); hence the statistical factor is again 4. The application of this approach has been simplified by the realisation[22] that a transition state can only have $r^{\ddagger} = 1$. It is not possible to have a species in a configuration corresponding to the top of an energy barrier in more than one degree of freedom at a time, so that a transition state can have only one route for returning to reactant and only one route for going on to product. If this appears not to be the case then the species under consideration is not a transition state according to the normal defini- tion. For example, a plausible-looking transition state (V) for cyclopropane isomerisation is symmetrical with a C—C bond stretched to a critical length; H migration will occur only after the barrier is passed and will occur by four distinct paths. In fact[22] there must be four equivalent routes of lower potential energy, proceeding via four distinct transition states through each of which there is only one path from reactant to products; transition states similar to those in Figure 3.3 would be satisfactory in this respect. Complica- tions have been discussed[23] for the case where reaction proceeds via a single symmetric transition state but the product 'valley' on the potential energy surface branches to give more than one possible set of identical products. The real difficulty in these cases seems to be that the systems require a more detailed mechanistic scheme than the simple CTST formulation ($A^+ \rightarrow$ products).

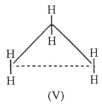

(V)

Provided that these complications are avoided, however, and the transi- tion state is chosen so that it can form products or return to reactants by one route only, then the statistical factor is given by $L^{\ddagger} = \sigma/\sigma^+$ if the symmetry numbers are taken as $\frac{1}{2}$ for asymmetric molecules or $L^{\ddagger} = \alpha\sigma/\sigma^+$, where α is the number (1 or 2) of optical isomers of the transition state and the symmetry numbers are now taken as 1 for asymmetric species.

This last equation is effectively the formulation used by Marcus[24,25], the symmetry numbers there being included in the partition functions. In that formulation the rate is also summed for a number of 'geometrical isomers' of the reaction path, but these are really different reactions with different potential energy barriers and different properties for the transition state, and would not normally be dealt with together as parts of one calculation. For example, Wieder and Marcus[3a] drew eight transition states for the

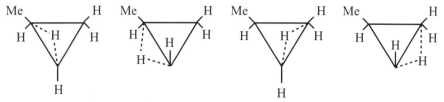

Figure 3.4 Transition states[3a] for isomerisation of methylcyclopropane. Reproduced by permission of John Wiley and Sons Ltd.

isomerisation of methylcyclopropane to but-1-ene and but-2-ene, and obtained a statistical factor of 8 since $\sigma = 1$, $\sigma^+ = 1$, and the number of 'optical isomeric forms' of the transition state $= 8$. The transition states are those shown in Figure 3.4 together with their mirror images (not shown). On careful inspection it is clear that the structures in Figure 3.4 are not *equivalent* transition states, but involve migration of two differently situated hydrogen atoms (*cis-* or *trans-* to CH_3) to two different sites on the ring ($CH_3.CH <$ or $CH_2 <$). The transition states will have different properties and the potential energy curves through them will be different. For each of the four distinct reactions the statistical factor is 2. For example, for the first transition state in Figure 3.4 there are $L^{\ddagger} = 2$ *cis*-hydrogen atoms which can migrate to the $CH_3.CH$ group; alternatively $L^{\ddagger} = r^{\ddagger}\sigma/\sigma^+ = 1 \times 1\frac{1}{2} = 2$; or finally $L^{\ddagger} = \alpha\sigma/\sigma^+ = 2 \times 1/1 = 2$. For computational purposes it may be useful to assume that the disappearance of reactant proceeds with the same rate constant k by each of the four reactions, but this should be taken into account by putting

$$k_{\text{uni}} = \sum_{g=1}^{4}(L^{\ddagger}k_g) = 4(2k) = 8k$$

rather than by calling the eight transition states 'optical isomers'. There can only be one or two geometrically equivalent optical isomers of a structure, since reflection of a mirror image in a mirror is bound to regenerate the original structure and cannot possibly generate a third structure.

We shall return to the question of symmetry numbers in Chapter 4 where we discuss the calculation of sums and densities of states.

3.10 IMPROVED TREATMENT OF ADIABATIC ROTATIONS

In previous sections the adiabatic degrees of freedom (usually the overall rotations of the molecule) have been largely ignored and taken into account only by the semi-empirical correction factor Q_1^+/Q_1. In the present section more sophisticated treatments are discussed. The whole subject has been reviewed by Waage and Rabinovitch.[26]

An adiabatic rotation is one for which the angular momentum stays constant during the conversion of the energised molecule to a transition state, i.e. the rotation stays in the same quantum state throughout this process. Since the energy of the rotation is given by $E_J = (h^2/8\pi^2 I)J(J + 1)$, the energy will change with change in the geometry of the molecule and hence in the moment of inertia I. In most cases where such effects are worth considering, $I^+ > I$ so that $E_J > E_J^+$ and the adiabatic rotations release energy into the other (active) degrees of freedom of the molecule, thus increasing the multiplicity of available quantum states of the transition state and increasing the specific rate constant k_a. An alternative interpretation is that this 'centrifugal effect' allows part of the adiabatic rotational energy to be used for overcoming the potential energy barrier, thus effectively reducing E_0. In bond fission reactions (such as the dissociation of ethane into methyl radicals) the moments of inertia can change substantially and the effect can be quite marked, amounting to an effective reduction of E_0 by more than kT and an increase in k_∞ by more than a factor of e. The effect was first treated by Rice and Gershinowitz,[27] who considered only the high-pressure (equilibrium) limit and derived the correction factor Q_1^+/Q_1 for a simple bond fission reaction. Bunker and Pattengill[6] have extended a similar treatment to the fall-off region for the specific case of a classical triatomic molecule and found improved agreement with the results of Monte Carlo 'experiments' on the same models.

BASIC TREATMENT

The general treatment described below is basically that due to Marcus[24,28], but reaction path degeneracy has been accounted for by the inclusion of a statistical factor $L^\ddagger$ rather than a symmetry number ratio (see section 3.9). In addition the case dealt with is specifically the usual one where there is a single doubly degenerate adiabatic rotational degree of freedom.[a] This corresponds to the model often used for a unimolecular bond fission reaction, as discussed further in section 5.2.

The basic effect of the change in I is best seen in the modified energy diagram of Figure 3.5. This is based on Figure 3.1, but now shows in addition the energies E_J and E_J^+ of the adiabatic rotations in A^* and A^+ respectively in their Jth energy level.[b] The diagram shows that the energy in

---

[a]Marcus has suggested[24] that an extension to other cases can be made by multiplying E_J, etc. in what follows by $l/2$, where l is the number of adiabatic rotational degrees of freedom. In particular, the expression (3.29) for $\langle \Delta E_J \rangle$ is multiplied by $l/2$ and the same factor thus appears in the resulting expressions (3.30) and (3.31).

[b]In ref. 24, J was used to denote 'the totality of quantum numbers that are approximately conserved', i.e. in the present case it would include K quantum number (see section 5.2). Thus where Marcus had $\sum_J(\ldots)$ we must have, for a doubly degenerate rotation $\sum_J(2J + 1)(\ldots)$.

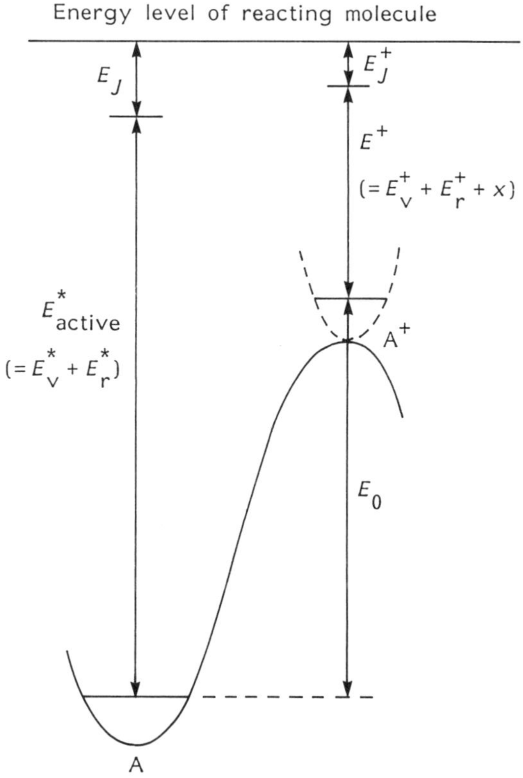

Energy level of reacting molecule

Figure 3.5 Energy diagram for a unimolecular reaction with adiabatic rotations included. Reproduced by permission of John Wiley and Sons Ltd.

the active degrees of freedom of A^*, which was previously $E_0 + E^+$, is now

$$E^*_{\text{active}} = E_0 + E^+ + E_J^+ - E_J = E_0 + E^+ + \Delta E_J$$

where

$$\Delta E_J = E_J^+ - E_J$$

The symbol E^*_{active} is used here to avoid confusion with the use[26,31] of E^* for the quantity $E_0 + E^+$; thus $E^*_{\text{active}} = E^* + \Delta E_J$. The integral corresponding to (3.17) can now be reformulated, in the first place with reference to a particular rotational state with quantum number J. The basic equation is still (3.4), [i.e. equation (2.31)], but a consideration of sections 3.4 and 3.5 shows that the equations (3.5) and (3.15) for $\delta k_1/k_2$ and k_a should be modified as follows:

$$\left(\frac{\delta k_1}{k_2}\right)_J = \frac{\rho(E^*_{\text{active}})\exp[-(E_0 + E^+ + E_J^+)/kT]\delta E^*_{\text{active}}}{Q}$$

$$k_{EJ}(E^*_{\text{active}}) = \frac{L^{\ddagger}}{h\rho(E^*_{\text{active}})} \sum_{E^+_{\text{vr}}=0}^{E^+} P(E^+_{\text{vr}})$$

In the equation for $(\delta k_1/k_2)_J$, Q is the partition function for all the degrees of freedom (both active and adiabatic) of A, and $(E_0 + E^+ + E^+_J)$ is the total energy of the A^* in question relative to the ground state of A (cf. Appendix 2, sections A2.3 and A2.5). The rate constant for conversion of energised molecules to transition states is now a function of J as well as the total energy, and is therefore denoted by k_{EJ}. The integration corresponding to (3.17) is still from $E^+ = 0$ to ∞, and the contribution θ to k_{uni} from each of the $(2J + 1)$ states with quantum number J is therefore given by

$$\theta = \frac{L^{\ddagger} \exp(-E_0/kT) \exp(-E^+_J/kT)}{hQ} \int_{E^+=0}^{\infty} \frac{\left\{\sum P(E^+_{\text{vr}})\right\} \exp(-E^+/kT)\, dE^+}{1 + k_{EJ}(E^*_{\text{active}})/k_2[\text{M}]}$$

(3.25)

The term $\exp(-E^+_J/kT)$ appears outside the integral sign, being a constant for this particular state, and from the above considerations, $dE^*_{\text{active}} = dE^+$ for a specified state. The total k_{uni} is obtained by summing the contributions from all the adiabatic rotational quantum states. In the present case there are $(2J + 1)$ states for each value of J, and k_{uni} is therefore given by

$$k_{\text{uni}} = \frac{L^{\ddagger} \exp(-E_0/kT)}{hQ}$$

$$\times \sum_{J=0}^{\infty} (2J + 1) \exp(-E^+_J/kT) \int_{E^+=0}^{\infty} \frac{\left\{\sum P(E^+_{\text{vr}})\right\} \exp(-E^+/kT)\, dE^+}{1 + k_{EJ}(E_0 + E^+ + \Delta E_J)/k_2[\text{M}]}$$

(3.26)

APPROXIMATE METHOD

Since the first edition of this book appeared, there has, of course, been a breathtaking advance in both the power and availability of computers. As a consequence, integrals such as in (3.26) are very straightforward to evaluate directly. This has rendered obsolete many of the elegant approximations developed in the past to assist integration. Given that this is the case, we present here only a brief description of one of the important approximations developed. The expression (3.26) can be greatly simplified by replacing ΔE_J by its mean value $\langle \Delta E_J \rangle$ for all the rotational states of interest. This step is justified by the statement that k_{EJ} is insensitive to fluctuations of ΔE_J about this mean, and leads to the replacement of k_{EJ} by $k_a(E_0 + E^+ + \langle \Delta E_J \rangle)$, which is no longer a function of J. The appropriate expression for $\langle \Delta E_J \rangle$ is obtained by noting that the rotational energies of A^* and A^+ in a given level J are related by $E_J/E_J^+ = I^+/I$, whence

$$\Delta E_J = E_J^+ - E_J = (1 - I^+/I)E_J^+ \quad \text{and} \quad \langle \Delta E_J \rangle$$

is given by

$$\langle \Delta E_J \rangle = (1 - I^+/I)\langle E_J^+ \rangle \tag{3.27}$$

An expression for $\langle E_J^+ \rangle$ is obtained by averaging E_J^+ with a weighting factor which is proportional to the number of molecules in the given quantum state which undergo reaction, i.e. to the right-hand side of (3.25). At high pressures this is simply $\exp(-E_J^+/kT)$ multiplied by a term independent of J, and at low pressures it is effectively the same since k_{EJ} varies much less rapidly with J than does $\exp(-E_J^+/kT)$. Since in the present case there are $(2J + 1)$ rotational states for each value of J, the expression for $\langle E_J^+ \rangle$ becomes

$$\langle E_J^+ \rangle = \int_{J=0}^{\infty} E_J^+(2J + 1)\exp(-E_J^+/kT)\,\mathrm{d}J \bigg/ \int_{J=0}^{\infty} (2J + 1)\exp(-E_J^+/kT)\,\mathrm{d}J$$

Putting $x = (E_J^+/kT) = h^2 J(J + 1)/8\pi^2 I^+ kT$, whence

$$\mathrm{d}x = (h^2/8\pi^2 I^+ kT)(2J + 1)\,\mathrm{d}J$$

the simple result (3.28) is obtained (the integrals in (3.28) being standard forms and both equal to unity—the first is $\Gamma(2)$, see Appendix 4).

$$\langle E_J^+ \rangle = kT \int_{x=0}^{\infty} x \exp(-x)\,\mathrm{d}x \bigg/ \int_{x=0}^{\infty} \exp(-x)\,\mathrm{d}x = kT \tag{3.28}$$

Thus from (3.27) and (3.28) the value of $\langle \Delta E_J \rangle$ to be inserted into (3.26) is given by

$$\langle \Delta E_J \rangle = (1 - I^+/I)kT \tag{3.29}$$

Alternatively it may be noted that the average rotational energy of the energised molecules (corresponding to (3.27) for the transition states) is

$$\langle E_J \rangle = (I^+/I)\langle E_J^+ \rangle = (I^+/I)kT$$

and therefore

$$\langle \Delta E_J \rangle = \langle E_J^+ \rangle - \langle E_J \rangle = (1 - I^+/I)kT$$

It is interesting to note that $\langle E_J \rangle$ can be considerably in excess of the average rotational energy of all the molecules, which is of course kT. Thus the molecules undergoing reaction are rotationally 'hot'; the centrifugal effect can increase the rate so much that transition states are formed with high preference from the energised molecules having rotational energy considerably above the average of the thermal distribution[29].

When ΔE_J in (3.26) has thus been replaced by $\langle \Delta E_J \rangle$ the only function of

J remaining is $(2J + 1)\exp(-E_J^+/kT)$. The equation can therefore be rewritten as

$$k_{uni} = C\sum_{J=0}^{\infty}[(2J + 1)\exp(-E_J^+/kT)]$$

$$= C\int_{J=0}^{\infty}(2J + 1)\exp(-E_J^+/kT)\,dJ = CQ_1^+$$

in which C comprises the remaining terms in (3.26) and is independent of J, and Q_1^+ is the classical partition function for this two-dimensional rotation.

The final equations for k_{uni} are therefore

$$k_{uni} = \frac{L^{\ddagger}Q_1^+\exp(-E_0/kT)}{hQ}\int_{E^+=0}^{\infty}\frac{\left\{\sum_{E_{vr}^+=0}^{E^+}P(E_{vr}^+)\right\}\exp(-E^+/kT)\,dE^+}{1 + k_a(E_0 + E^+ + \langle\Delta E_J\rangle)/k_2[M]}$$

(3.30)

$$k_a(E_0 + E^+ + \langle\Delta E_J\rangle) = \frac{L^{\ddagger}}{h\rho(E_0 + E^+ - (I^+/I - 1)kT)}\sum_{E_{vr}^+=0}^{E^+}P(E_{vr}^+)$$

(3.31)

These equations differ from (3.17) and (3.18) only in the omission of the factor Q_1^+/Q_1 in the expression for k_a and in the modified energy at which k_a is evaluated. The high-pressure limit is easily derived by the technique of section 3.7; the result is (3.32), which now includes the factor Q_1^+/Q_1 automatically and therefore agrees identically with the CTST expression

$$k_{\infty} = L^{\ddagger}\frac{kT}{h}\frac{Q^+}{Q}\exp(-E_0/kT)$$

(3.32)

where

$$Q^+ = Q_1^+Q_2^+ \quad\text{and}\quad Q = Q_1Q_2$$

3.11 THE QUANTITIES $W(E_{vr}^+) \equiv \sum_{E_{vr}^+=0}^{E^+}P(E_{vr}^+)$ AND $\rho(E^*)$

The sums and densities of states are fundamental to any application of RRKM theory and it is therefore worth while at this point to illustrate by means of simple examples the precise significance of these terms. For this purpose the superscripts $^+$ and * will be dropped and the sum and density of states considered for any general molecule. To simplify the illustration further, active rotations will be excluded and the quantities $\sum_{E_v=0}^{E}P(E_v)$ and $\rho(E)$ relating only to vibrational energy will be discussed. In general it is appropriate to consider the sum of states often written less explicitly as

$\Sigma P(E_v)$ when the quantised nature of molecular vibrations must be recognised, i.e. at relatively low energies. The function $\rho(E)$ relates to a continuously varying energy content and is more appropriate at high energies.

SUM OF STATES $\sum_{E_v=0}^{E} P(E_v)$

Consider first the sum $\Sigma P(E_v)$, which is particularly relevant to the quantised model of the transition state employed in the RRKM theory. The degeneracy $P(E_v)$ is the number of vibrational quantum states with a vibrational energy of E_v, and $\sum_{E_v=0}^{E} P(E_v)$ is thus the total number of states with energy not exceeding E. Consider first the case of a single simple harmonic oscillator of frequency v, representing perhaps a diatomic molecule. Such an oscillator has one quantum state at each energy level 0, $hv, 2hv, \ldots, vhv, \ldots$ relative to the ground state as zero energy. The fixed zero-point energy $\frac{1}{2}hv$ is not included here. Thus $P(E_v) = 1$ at each level, and a plot of $P(E_v)$ against energy appears as shown in the lower part of Figure 3.6(a); this is the canonical energy level diagram plotted on its side. The sum $\Sigma P(E_v)$ is then obtained by counting the number of quantum states up to and including energy E; from $E/hv = 0$ to $0.\dot{9}$ there is only one such

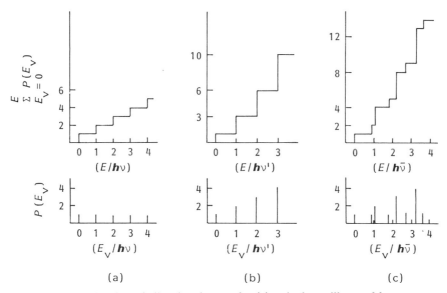

Figure 3.6 Distribution of vibrational states for (a) a single oscillator of frequency v; (b) a doubly degenerate oscillator of frequency v', and (c) a system of three oscillators comprising (a) and (b). Reproduced by permission of John Wiley and Sons Ltd.

state, $v = 0$; from $E/hv = 1$ to $1.\overset{.}{9}$ the total is two, $v = 0$ and $v = 1$ and so on. Thus $\Sigma P(E_v)$ is built up as a stepwise function which has a constant value throughout the energy range between any two of the quantised energy levels; this is shown in the upper part of Figure 3.6(a).

Next consider the case of a doubly degenerate oscillator, i.e. two vibrations A and B with the same frequency v' ($= 1.2v$ for illustration). The energy levels are simply vhv', since the quanta of energy hv' may be fed into either or both of the vibrations, but now there may be more than one quantum state at a given energy level. There are in fact $v + 1$ distinct quantum states of energy vhv', as may readily be verified by inspection in a few simple cases. For $v = 1$ the quantum can be in either oscillator, giving two physically distinct states of the system. For $v = 2$, A and B can contain 2 and 0, 1 and 1, or 0 and 2 quanta respectively, giving three quantum states and so on. The corresponding graphs in Figure 3.6(b) show the resulting upward curvature of the plots of $\Sigma P(E_v)$ against E.

Finally, Figure 3.6(c) shows the irregular type of behaviour which arises when the two preceeding cases are combined, so that the molecule under consideration consists of three oscillators, two of which have the same frequency. It is clear that diagrams of this type become rapidly more complex as the number of atoms in the molecule increases, and this is well illustrated in the typical curves shown in Figures 3.7 and 3.8. Other notable features are the rapid increase in $\Sigma P(E_v)$ with energy (a logarithmic plot is needed to display the results over a reasonable range of energies in Figure 3.8), and the greater numbers of states available to molecules with a greater number of vibrations.

DENSITY OF QUANTUM STATES, $\rho(E)$

It is clear from Figures 3.7 and 3.8 that if attention is focused on highly excited molecules, i.e. those with a large non-fixed energy, the stepwise variation in the number of states becomes much less apparent. An extension of Figure 3.6 to higher energies, shown in Figure 3.9, similarly yields plots of $W(E_v)$ which are much better approximated by a smooth curve than the same plots at low energies. For the single oscillator in Figure 3.9(a), for example, the plot is well approximated by the straight line $\Sigma P(E_v) = (E/hv)$. For the doubly degenerate oscillator of Figure 3.9(b) the plot is approximately represented by the smooth curve

$$\Sigma P(E_v) = \tfrac{1}{2}(E/hv')^2$$

and so on. These approximate representations are likely to be perfectly adequate at the high energies relevant to the energised molecules considered in RRKM theory. These molecules will generally contain at least 20–30 quanta of vibrational energy, and the curves also tend to be smoother for

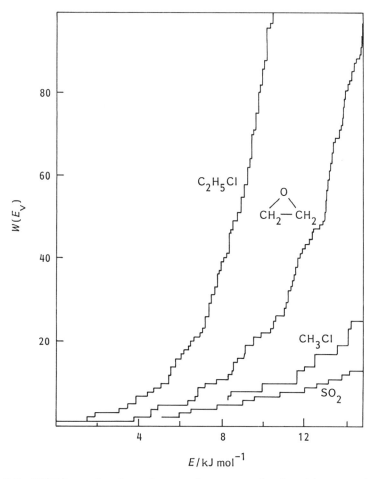

Figure 3.7 $W(E_v)$ as a function of energy for some molecules at low energies; the frequency assignments are detailed in ref. 30 for ethyl chloride and in Chapter 4, ref. 29, for the other molecules.

realistic models having several different vibration frequencies (see, e.g., Figures 3.7 and 3.8).

Under these circumstances it may be useful to formulate results in terms of $\rho(E)$, the *density of quantum states* or the *number of quantum states per unit energy range* at energy E, which is found to be simply the gradient of the plot of $W(E_v)$ against E. This important result will be required when the numerical evaluation of state densities is considered in Chapter 4, and may be verified for the general case as follows. The number of quantum states of any system in a small range E to $E + \delta E$ is the number of states at

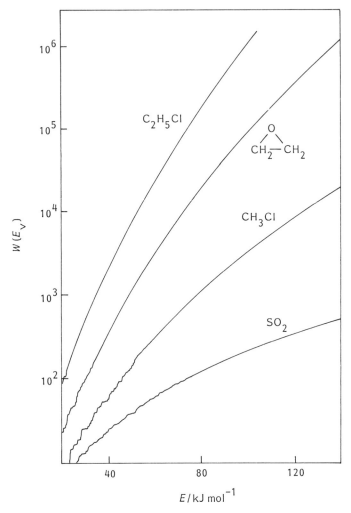

Figure 3.8 Extension of Figure 3.7 to higher energies; note the logarithmic plot required. Reproduced by permission of John Wiley and Sons Ltd.

all energies up to $E + \delta E$ minus the number of states at all energies up to E, i.e. $W(E + \delta E) - W(E)$.

Provided δE is sufficiently small this number of states is also given by $\rho(E)\delta E$, and $\rho(E)$ is therefore given by

$$\rho(E) = \lim_{\delta E \to 0} \left(\frac{W(E + \delta E) - W(E)}{\delta E} \right) = \frac{\mathrm{d}}{\mathrm{d}E} W(E) \tag{3.33}$$

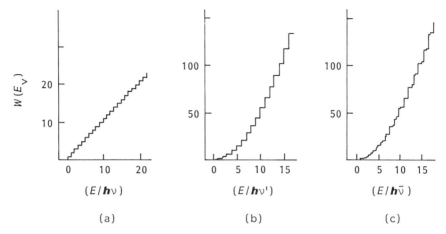

Figure 3.9 Distribution of vibrational quantum states for the systems of Figure 3.6 at higher energies. Reproduced by permission of John Wiley and Sons Ltd.

This result is most useful under conditions where the density of states is high and the $W(E_n)$ plot is essentially a smooth curve; the density of states is then a continuous function and it is normally only under these conditions that a result is formulated in terms of $\rho(E)$. The result is still valid when $W(E_n)$ is a coarsely stepped function, however, in this case $\rho(E)$ is infinite at energies corresponding to the quantised energy levels and zero at intermediate energies, and may be formulated in terms of the Dirac δ function[31].

Logarithmic plots of $\rho(E)$ against energy for some typical vibrational systems are shown in Figure 3.10. A striking feature of these plots is the very high density of quantum states which can be found at quite moderate energies, and this amply justifies the use of a continuous energy distribution function for the energised molecules considered in the RRKM theory. The broken lines at low energy are smoothed versions of the actual spiked variations discussed above; the behaviour at low energies is discussed in more detail in Chapter 4.

3.12 ASSUMPTIONS OF THE BASIC RRKM THEORY

Before proceeding to discuss the application of the RRKM theory it is appropriate to list and examine the fundamental assumptions and approximations which were essential to the derivation of (3.17). Five topics are now discussed, although it will soon become apparent that these are not altogether separable.

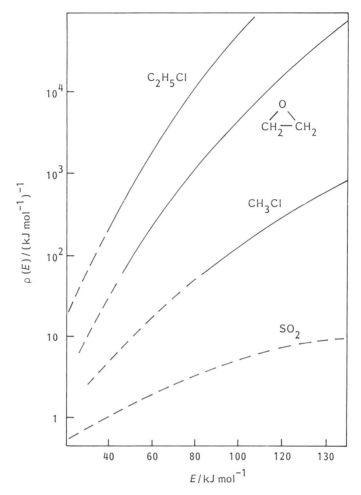

Figure 3.10 Vibrational state density as a function of energy for the molecules referred to in Figures 3.7 and 3.8 (note logarithmic plot). Reproduced by permission of John Wiley and Sons Ltd.

FREE EXCHANGE OF ENERGY BETWEEN OSCILLATORS — RAPID INTRAMOLECULAR ENERGY TRANSFER

The assumption that the non-fixed energy of the active vibrations and rotations is subject to rapid statistical redistribution means that every sufficiently energetic molecule will eventually be converted into products unless deactivated by collision. This is a much less restrictive assumption than that made in Slater's theory, in which there must be not only sufficient energy but also a suitable distribution of the energy before a molecule is able to react. In phase-space terminology, the RRKM theory assumes that the phase space of the molecules concerned cannot be subdivided into smaller regions such that the representative point cannot move from one region to another by any feasible trajectory[5].

If the phase space were decomposable this would imply that not all states of the same energy were freely interchangeable, and purely statistical considerations would not be adequate to describe the model.

Marcus has in fact made provision for some of the degrees of freedom to be completely inactive[1]. This means that the energy in these degrees of freedom cannot flow into the reaction coordinate; an energised molecule A^* is one with non-fixed energy greater than E_0 in the active modes, irrespective of that in the inactive modes. The energy in the inactive modes can, however, be redistributed at random among these modes, and thus does contribute to the number of quantum states at a given energy and hence affects $\rho(E^*)$ and $k_a(E^*)$.

Early attempts were made to formulate a theory in which energy exchange between some of the oscillators occurs at only a limited rate[31]. More recently work on the slow exchange of energy or intramolecular vibrational energy redistribution (IVR) between active and inactive vibrational states has been carried out with a view to understanding the criteria for 'non-RRKM behaviour'. Hase and Bunker[32] have defined non-RRKM behaviour as either apparent or intrinsic. The latter is shown by a molecule possessing weak coupling between certain oscillators or more precisely between certain vibrational states.

One example of a theoretical approach to the problem is the work of Snider[33] who has treated the slow randomisation as a kinetic problem and solved the kinetic equations for a model allowing for thermal unimolecular reaction with slow interconversion of two groups of states of the reactive molecule. It was found that a meaningful rate constant can only be derived if the threshold energy for reaction exceeds that for randomisation by several kT.

Apparent non-RRKM behaviour can arise when the reaction rate is much faster than the rate of intramolecular energy transfer. Much experimental work has now been done in order to discover whether this can occur in a given reaction. Chemical activation experiments largely in the hands of Rabinovitch and co-workers[34] in which energy was initially deposited in a particular part of the reacting molecule have enabled the rate constants for randomisation to be determined and led to values of the order of 10^{12} s^{-1}. In most of the cases studied randomisation has been sufficiently rapid for the RRKM theory to hold.

Some examples of non-randomisation from experiments of this kind have been reported, for example for the 1–5 H atom shift in the isomerisation of *cis*-3 methyl-[1,2-^{2}H$_2$]penta-1-3 diene at high pressures[35] and in the experiments of Rowland and co-workers[36] on chemically activated radicals produced by addition of fluorine atoms to tetrallyl tin and tetrallyl germanium. In Rowland's experiments non-RRKM behaviour is attributed to the blockage to energy transfer between the allyl groups caused by the presence of the heavy metal atom.

Intramolecular energy transfer has recently been studied by a variety of techniques including photochemical activation[37], infrared multiphoton dissociation[38], molecular beams[39], overtone excitation[40] and picosecond vibronic excitation[41]. In some of these techniques it is possible to infer the adherence to statistical redistribution of intramolecular energy from the observed translational energies and angular distributions of product molecules.

Several excellent reviews covering intramolecular energy transfer and the related question of mode specificity have recently appeared[42–44] and energy distributions in the unimolecular decompositions of ions have also been reviewed[45,46]. In view of the importance of these topics they are dealt with in more detail in Chapter 9.

The major conclusion to be reached here is that in the majority of cases studied the assumption of rapid intramolecular energy transfer following excitation in a unimolecular reaction is valid. Under special conditions likely to produce slow randomisation or high rates of decomposition commensurate with the rate of randomisation, it is, however, possible that non-statistical behaviour may be observed.

THE STRONG-COLLISION ASSUMPTION—
INTERMOLECULAR ENERGY TRANSFER

The original formulation of RRKM theory outlined in section 3.1 assumes that energisation–de-energisation occur as essentially single-step processes involving strong collisions in which relatively large amounts of energy ($\gg kT$) are transferred between molecules. While there is a good deal of evidence that the strong-collision assumption is reasonably realistic for thermally energised molecules of moderate size and at moderate temperatures, the advent of newer methods of energisation such as photoactivation and laser irradiation has extended the energy range of the energised species and created a need for the incorporation of weak collision treatments into the theory. A crude allowance for limited energy transfer on collision may be made by replacing $k_2[M]$ in (3.17) and similar equations by λZp, the collision rate constant Zp multiplied by a collisional de-energisation efficiency λ. This assumes that the number of collisions required to de-energise an energised molecule is a constant (λ^{-1}) and leads to a simple shift of the fall-off curve ($\log k_{uni}$ versus $\log p$) to higher pressures by an increment $\log \lambda$.

Here λ is formally equivalent to the collision efficiency per unit collision β_c which relates to both energisation and de-energisation, i.e. reaction (3.1):

$$A + M \rightleftharpoons A^* + M \qquad (3.1)$$

If k_{bim}^{sc} is the strong-collision rate constant for the forward reaction then the limiting low-pressure rate constant k_{bim} (see section 3.8) is given by

$$k_{bim} = \beta_c k_{bim}^{sc}$$

For strong collisions $\beta_c = 1$ and for weak collisions $\beta_c < 1$.

Values of β_c can be obtained from experimental measurements on unimolecular reactions in the presence of various inert collider bath gases, and the related amounts of energy transferred per collision can also be derived. Experiments of this kind are described more fully in Chapter 9 and the incorporation of weak collisions in RRKM theory which is usually now done in the master-equation formulation of the theory is described in Chapter 7.

THE 'EQUILIBRIUM HYPOTHESIS'

The assumption is made in the RRKM theory, as in the CTST, that the concentration of forward-crossing transition states is the same in the steady state as it would be at total equilibrium where no net reaction was occurring. This assumption has been defended on various grounds[47,48], the main argument being that energised molecules about to become transition states do not know that there are no molecules travelling in the opposite direction, and hence enter the critical section of the reaction coordinate at the same rate as they would in a true equilibrium situation. Perhaps more convincingly, non-equilibrium calculations show[49,50] that the equilibrium hypothesis provides a good approximation except again for cases where E_0/kT is small (less than about 10) in which case the reaction is so rapid as to disturb the Boltzmann distribution of molecules with energies near the critical energy. The effect has been calculated[49] to give a maximum error of about 8%, this occurring at $E_0/kT \approx 5$.

The 'quasi-equilibrium' between reactants and transition states assumed in CTST is related to the assumption that no multiple crossings of the dividing surface between reactants and products occur.

It has been shown that in the event of multiple crossings the CTST could give too high an estimate of the rate constant. The VTST and further refinements are considered to be improvements in locating the position of the transition state[51], although the calculated rate constant still represents an upper bound.

RANDOM LIFETIMES

The RRKM theory assumes[52] that the energised molecules A* have random lifetimes before their unimolecular dissociation, and this implies that the process $A^* \rightarrow A^+$ is governed by purely statistical considerations, and that there is no particular tendency for all A* to decompose soon after their formation, or to exist for some particular length of time before decomposing. Molecules with a particular energy have a distribution $P(\tau)$ of lifetimes τ which follows the exponential equation

$$P(\tau) = k_a \exp(-k_a\tau)$$

In this distribution the proportionality constant and the exponential decay constant are both equal to the first-order rate constant for conversion of the energised molecules into transition states. This random distribution is illustrated in Figure 3.11, curve A; it is very similar to the distribution of the number of times a coin can be tossed without a head appearing. It has been noted[52] that even if the distribution of lifetimes is non-random the zero-time value of $P(\tau)$ is still k_a, defined a little more specifically as the high-pressure rate constant for conversion of A* into A^+. Non-random lifetimes may thus correspond to $P(\tau)$ curves which descend at the wrong rate from the initial value, although they may still have the general appearance of exponential

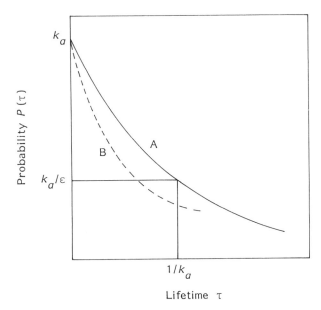

Figure 3.11 Distributions of lifetimes of energised molecules; curve A, random lifetimes $[P(\tau) = k_a \exp(-k_a\tau)]$; curve B, a non-random distribution. Reproduced by permission of John Wiley and Sons Ltd.

decay. Bunker[52] has illustrated some cases where there is abnormally high probability of dissociation in the period immediately following an energising collision, and this behaviour is illustrated in Figure 3.11, curve B.

The random-lifetime assumption is closely related to the assumptions described in the previous two sections; since the collisional activation and deactivation are treated as random statistical processes, the dissociation must clearly be treated in the same way if a consistent theory is to emerge. The validity of this assumption has been discussed at length[5,6,8,52–54] and it appears to be accurate except perhaps[6,52] for small molecules (diatomics and some triatomics) or for cases where E_0/kT is exceptionally low.

The occurrence of non-random lifetimes can be taken as evidence of non-RRKM behaviour and as pointed out by Hase[55] can either be due to non-statistical preparation of the internal vibrational states of the energised molecule (apparent non-RRKM) or due to a bottleneck in the flow of energy between vibrational states which may be present even if statistical redistribution between other states occurs (intrinsic non-RRKM).

CONTINUOUS DISTRIBUTION FUNCTION $\rho(E^*)$

The treatment of the energised molecule in terms of continuously variable energy rather than of quantised energy levels has already been discussed (sections 3.4 and 3.11) and seems to be well justified, at least in the majority of cases. It is perhaps worth noting that a very different treatment would be required for the case where the energy levels of A were significantly separated.

REFERENCES

1. R. A. Marcus, *J. Chem. Phys.*, **20**, 359 (1952).
2. R. A. Marcus and O. K. Rice, *J. Phys. and Colloid Chem.*, **55**, 894 (1951).
3. (a). G. M. Wieder and R. A. Marcus, *J. Chem. Phys.*, **37**, 1835 (1962); (b). R. A. Marcus, *J. Chem. Phys.*, **43**, 2658 (1965).
4. For example, W. Forst, *Theory of Unimolecular Reactions*, Academic Press, New York (1973); M. Quack and J. Troe in *Gas Kinetics and Energy Transfer*, a specialist periodical report ed. P. G. Ashmore and R. J. Donovan), vol. 2, Ch. 5, Chemical Soc., London (1977); I. W. M. Smith, *Kinetics and Dynamics of Elementary Gas Reactions*, Ch. 4, Butterworths, London (1980); A. B. Callear in *Comprehensive Chemical Kinetics*, ed. C. H. Bamford and C. F. H. Tipper, Vol. 24, Ch. 4 (1983).
5. D. L. Bunker, *Theory of Elementary Gas Reaction Rates*, Pergamon, Oxford, p. 54 (1966).
6. D. L. Bunker and M. Pattengill, *J. Chem. Phys.*, **48**, 772 (1968).
7. D. M. Wardlaw and R. A. Marcus, *Chem. Phys. Lett.*, **110**, 230 (1984).
8. O. K. Rice, *J. Phys. Chem.*, **65**, 1588 (1961).

9. N. B. Slater, *Theory of Unimolecular Reactions*, Methuen, London, p. 108 (1959).
10. (a). K. J. Laidler, *Theories of Chemical Reaction Rates*, McGraw-Hill, New York, Ch. 3 (1969); (b). J. D. Macomber and C. Colvin, *Internat, J. Chem. Kinetics*, **1**, 483 (1969); (c). R. P. Bell, *Trans. Faraday Soc.*, **66**, 2770 (1970).
11. R. H. Fowler and E. A. Guggenheim, *Statistical Thermodynamics*, Cambridge University Press § 1208 (1939), Y. N. Kondratiev, *Chemical Kinetics of Gas Reactions*, Pergamon, Oxford, p. 174 (1964).
12. L. D. Landau and E. M. Lifshitz, *Quantum Mechanics*, Pergamon, London, 2nd edn (1965), L. I. Schiff, *Quantum Mechanics*, McGraw-Hill, New York, 2nd edn. (1955).
13. R. A. Marcus, *J. Chem. Phys.*, **52**, 1018 (1970).
14. S. Glasstone, K. J. Laidler and H. Eyring, *The Theory of Rate Processes*, McGraw-Hill, New York and London (1941); see also K. J. Laidler, *Theories of Chemical Reaction Rates*, McGraw-Hill, New York, Ch. 3 (1969).
15. R. P. Bell and E. Gelles, *Proc. Roy. Soc.*, **A210**, 310 (1952); D. R. Augood, *Nature*, **178**, 754 (1956).
16. V. Gold, *Trans. Faraday Soc.*, **60**, 739 (1964).
17. E. A. Guggenheim, *Trans. Faraday Soc.*, **50**, 574 (1954); D. Rapp and R. E. Weston, *J. Chem. Phys.*, **36**, 2807 (1962); S. W. Benson and P. S. Nangia, *J. Chem. Phys.*, **38**, 18 (1963); E. W. Schlag, *J. Chem. Phys.*, **38**, 2480 (1963).
18. D. M. Bishop and K. J. Laidler, *J. Chem. Phys.*, **42**, 1688 (1965); see also *Trans. Faraday Soc.*, **66**, 1685 (1970).
19. K. J. Laidler, *Theories of Chemical Reaction Rates*, McGraw-Hill, New York, p. 65 (1969).
20. C. S. Elliott and H. M. Frey, *Trans. Faraday Soc.*, **64**, 2352 (1968).
21. E. W. Schlag and G. L. Haller, *J. Chem. Phys.*, **42**, 584 (1965).
22. J. N. Murrell and K. J. Laidler, *Trans. Faraday Soc.*, **64**, 371 (1968).
23. J. N. Murrell and G. L. Pratt, *Trans. Faraday Soc.*, **66**, 1680 (1970); see also M. R. Wright and P. G. Wright, *J. Phys. Chem.*, **74**, 4394, 4398 (1970).
24. R. A. Marcus, *J. Chem. Phys.*, **43**, 2658 (1965); **52**, 1018 (1970).
25. R. A. Marcus, *J. Chem. Phys.*, **43**, 1598 (1965).
26. E. V. Waage and B. S. Rabinovitch, *Chem. Rev.*, **70**, 377 (1970).
27. O. K. Rice and H. Gershinowitz, *J. Chem. Phys.*, **2**, 853 (1934).
28. R. A. Marcus, rewritten by H. Heydtmann, Unimolecular reaction rate theory in *Chemische Elementarprozesse*, ed. H. Hartmann, Springer-Verlag, Berlin, Heidelberg and New York, p. 109 (1968).
29. S. W. Benson, *J. Amer. Chem. Soc.*, **91**, 2152 (1969).
30. L. W. Daasch, C. Y. Liang and J. R. Nielsen, *J. Chem. Phys.*, **22**, 1293 (1954).
31. E. K. Gill and K. J. Laidler, *Proc. Roy. Soc.*, **A250**, 121 (1959); M. Solc, *Mol. Phys.*, **11**, 579 (1969); *Chem. Phys. Lett.*, **1**, 160 (1967); N. B. Slater, *Mol. Phys.*, **12**, 107 (1967).
32. W. L. Hase and D. L. Bunker, *J. Chem. Phys.*, **59**, 4621 (1973).
33. N. Snider, *J. Phys. Chem.*, **93**, 5789 (1989).
34. B. S. Rabinovitch, R. F. Kubin and R. E. Harrington, *J. Chem. Phys.*, **88**, 405 (1963); J. D. Rynbrandt and B. S. Rabinovitch, *J. Phys. Chem.*, **75**, 2164 (1971).
35. T. Ibuki and S. Sugita, *J. Chem. Phys.*, **70**, 3989 (1979).
36. P. J. Rogers, J. I. Selco and F. S. Rowland, *Chem. Phys. Lett.*, **97**, 313 (1983).
37. H. Hippler, K. Luther, J. Troe and R. Walsh, *J. Chem. Phys.* **68**, 323 (1979).
38. S. Ruhman, O. Anner, S. Berjhumi and Y. Haas, *Chem. Phys. Lett.*, **99**, 281 (1983); A. J. Grimley and J. C. Stephenson, *J. Chem. Phys.*, **74**, 447 (1981).

39. P. A. Schulz, Aa. S. Sudbø, D. J. Krajnovich, H. S. Kwok, Y. R. Shen and Y. T. Lee, *Ann. Rev. Phys. Chem.*, **30**, 379 (1979).
40. K. V. Reddy and M. J. Berry, *Chem. Phys. Lett.*, **52**, 111 (1977); D. W. Chandler, W. E. Farneth and R. N. Zare, *J. Chem. Phys.*, **77**, 4447 (1982).
41. L. R. Khundkar, R. A. Marcus and A. H. Zewail, *J. Phys. Chem.*, **87**, 2473 (1983).
42. I. Oref and B. S. Rabinovitch, *Acc. Chem. Res.*, **12**, 166 (1979).
43. F. F. Crim, *Ann. Rev. Phys. Chem.*, **35**, 657 (1984).
44. V. E. Bondybey, *Ann. Rev. Phys. Chem.*, **35**, 591 (1984).
45. J. L. Franklin in *Gas Phase Ion Chemistry*, ed. M. T. Bowers, Academic Press, New York, Vol. 1, p. 273 (1979).
46. C. Lifshitz, *J. Phys. Chem.*, **87**, 2304 (1983).
47. (a) K. J. Laidler, *Chemical Kinetics*, 3rd edn., Harper and Row, New York (1987); (b) K. J. Laidler, *Theories of Chemical Reaction Rates*, McGraw-Hill, New York, Ch. 3 (1969); (c) D. M. Bishop and K. J. Laidler, *J. Chem. Phys.*, **42**, 1688 (1965); (d) K. J. Laidler and J. C. Polanyi, *Prog. Reac. Kinetics*, **3**, 1 (1965); (e) K. J. Laidler and A. Tweedale, *Advan. Chem. Phys.*, **21**, 113 (1971).
48. R. A. Marcus rewritten by H. Heydtmann, Unimolecular reaction rate theory; in *Chemische Elementarprozesse*, ed. H. Hartmann, Springer-Verlag, Berlin, Heidelberg and New York, p. 109 (1968).
49. B. Morris and R. D. Present, *J. Chem. Phys.*, **51**, 4862 (1969).
50. R. D. Present, *J. Chem. Phys.*, **31**, 747 (1959); E. Montroll and K. E. Shuler, *Adv. Chem. Phys.*, **1**, 361 (1958); K. E. Shuler, *J. Chem. Phys.*, **31**, 1375 (1959); H. Eyring, T. S. Ree, T. Ree and F. M. Wanlass, *Chem. Soc. Spec. Pub. No. 16*, p. 3 (1962); K. J. Laidler, *Theories of Chemical Reaction Rates*, McGraw-Hill, New York, Ch. 8 (1969).
51. R. D. Present, *J. Chem. Phys.*, **31**, 747 (1959), p. 116; D. G. Truhlar, W. L. Hase and J. T. Hynes, *J. Phys. Chem.*, **87**, 2664 (1983).
52. D. L. Bunker, *J. Chem. Phys.*, **40**, 1946 (1964).
53. E. Thiele, *J. Chem. Phys.*, **36**, 1466 (1962); **38**, 1959 (1963).
54. W. Forst and P. St. Laurent, *Can. J. Chem.*, **45**, 3169 (1967).
55. W. L. Hase in *Dynamics of Molecular Collisions*, Part B, ed. W. H. Miller, Plenum Press, New York (1976).

4 The Evaluation of Sums and Densities of Molecular Quantum States

Application of the RRKM theory depends critically on the ability to evaluate the quantities $\rho(E_{vr}^*)$ for a molecule A and $W(E_{vr}^+)$ for a postulated model of the corresponding transition state A^+. The significance of these quantities has already been discussed in some detail (section 3.11) and it will be realised that their evaluation depends in one way or another on an assessment of the distribution of vibrational–rotational quantum states at various energy levels. For the purpose of the present chapter the presence of the superscripts * and $^+$ is of little relevance, and we therefore discuss more generally the calculation of the density $\rho(E_{vr})$ of vibrational–rotational states at a non-fixed vibrational–rotational energy E_{vr}, and the sum of states up to this energy, $W(E_{vr})$, which is given by equation (4.1), where $P(\varepsilon)$ is defined in section 3.5.

$$W(E_{vr}) \equiv \sum_{\varepsilon=0}^{E_{vr}} P(\varepsilon) \qquad (4.1)$$

An important consequence of (4.1) is that if $P(E_{vr})$ is a continuous function, and for some of the classical approximations described below this is the case, then $P(E_{vr})$ can be equated with $\rho(E_{vr})\,dE_{vr}$ and equation (4.1) becomes

$$W(E_{vr}) = \int_0^{E_{vr}} \rho(x)\,dx \qquad (4.2)$$

Equation (4.2) can also be obtained from the integration of equation (3.33).

In section 4.1 the distributions of the vibrational and rotational states are separated for individual treatment. If there are no rotational degrees of freedom in the model under discussion $\rho(E_{vr})$ and $W(E_{vr})$ are simply replaced by $\rho(E_v)$ and $W(E_v)$ and one proceeds directly to section 4.2. Sections 4.2–4.4 discuss methods for calculating densities of vibrational states. So-called direct counting methods are described in section 4.2; with the advent of fast, powerful computers these methods are much more widely used and a direct count is often the method of choice, even for quite large molecules. Monte Carlo methods are considered briefly in section 4.3; this approach is not yet widely used, but has found an important application in calculation of sums and densities of states for non-separable degrees of freedom. Approximate, analytic methods are discussed in section 4.4, and the use of the Laplace transform in obtaining these approximations is described. Section 4.5 describes the treatment of rotational states on the basis of a classical approximation, a common simplification. It is noted in passing that this treatment converts the general equation (3.17) into the expression (4.81) derived by Marcus[1]. Section 4.6 deals with the treatment of vibrational–rotational systems.

4.1   SEPARATION OF VIBRATIONAL AND ROTATIONAL DEGREES OF FREEDOM

Consider first the sum $W(E_{vr})$. As shown in equation (4.1), this is the total number of vibrational-rotational quantum states with energies less than or equal to E_{vr}. This number can be obtained by summing separate contributions involving each of the possible vibrational levels, i.e. those with $E_v \leqslant E_{vr}$. The total number of vibrational–rotational states corresponding to a given vibrational level of energy E_v is the number $P(E_v)$ of vibrational states of this energy multiplied by the number of possible rotational states, which is the total number of rotational states with energy $E_r < E_{vr} - E_v$, denoted by

$$\sum_{E_r=0}^{E_{vr}-E_v} P(E_r) \equiv W_r(E_{vr} - E_v)$$

Thus the number of vibrational–rotational quantum states in which the vibrational energy is specifically E_v (and the total energy does not exceed E_{vr}) is $P(E_v)W_r(E_{vr} - E_v)$, and the total number of states counting all possible values of E_v is given by the convolution

$$W(E_{vr}) = \sum_{E_{vr}=0}^{E_{vr}} P(E_{vr}) = \sum_{E_v=0}^{E_{vr}} P(E_v) \sum_{E_v=0}^{E_{vr}-E_v} P(E_r)$$

$$= \sum_{E_v=0}^{E_{vr}} P(E_v)W_r(E_{vr} - E_v) \qquad (4.3)$$

It follows from equation (4.2) that the density of vibrational–rotational states, $\rho(E_{vr})$, can now be obtained as the differential of (4.3) with respect to E_{vr}. It is simplest to consider again the contribution from a particular vibrational state of energy E_v, and this is obtained by differentiation of the term $P(E_{vr})W_r(E_{vr} - E_v)$. Since E_v and E_{vr} are independent variables, the contribution to the density of states at E_{vr} is therefore

$$P(E_v)\frac{d}{dE_{vr}}W_r(E_{vr} - E_v) = P(E_v)\rho_r(E_{vr} - E_v)$$

The total density of vibrational–rotational states at E_{vr}, counting all possible vibrational levels, is therefore given by

$$\rho(E_{vr}) = \sum_{E_v=0}^{E_{vr}} P(E_v)\rho_r(E_{vr} - E_v) \qquad (4.4)$$

The above derivations of (4.3) and (4.4) can readily be visualised in terms of Figure 4.1, which shows the vibrational and rotational energies of each

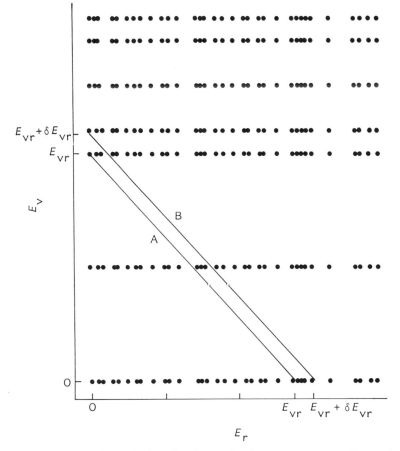

Figure 4.1 Distribution of vibrational–rotational quantum states illustrating the derivation of (4.2) and (4.3).

quantum state by plotting the states as points in the (E_v, E_r) plane. Since the vibrational levels are coarsely quantised, while the rotational levels are closely spaced, the diagram takes the form of a series of lines at constant E_v, each line comprising many closely spaced quantum states. If the vibrational levels are degenerate there are effectively $P(E_v)$ coincident lines at energy E_v, and each rotational point on the line at E_v occurs $P(E_v)$ times. The region of the plot corresponding to vibrational–rotational energies less than or equal to E_{vr} is the triangle to the left of the straight line A joining the intercepts $E_v = E_{vr}$ and $E_r = E_{vr}$ on the two axes; the equation of this line is $E_v + E_r = E_{vr}$. Thus the total number of quantum states with energy less than or equal to E_{vr} is the number of points in this triangle and is clearly given by (4.3). The density of states at energy E_{vr} is readily obtained by noting that the number of states with energy between E_{vr} and $E_{vr} + \delta E_{vr}$ is $\rho(E_{vr})\delta E_{vr}$, and is obtained from the diagram as the number of states between the lines A and B. On each horizontal line of states the number of rotational states in this band is

$$\rho_r(E_r)\delta E_{vr} = \rho_r(E_{vr} - E_v)\delta E_{vr}$$

and (4.4) follows by summation over all the relevant vibrational levels.

Finally, it is convenient to anticipate here a subsequent treatment of the vibrational states of an energised molecule A* in terms of a continuous function of density of states. Again it follows from equation (4.2) that $\rho_v(E_v) = (d/dE_v)W_v(E_v)$, in which case (4.4) can be replaced by an integration:

$$\rho(E_{vr}) = \int_0^{E_{vr}} \rho_v(E_v)\rho_r(E_{vr} - E_v)\, dE_v \qquad (4.5)$$

4.2 DIRECT COUNT OF VIBRATIONAL STATES

In this section, we examine methods for directly counting the states—a procedure which gives precise densities of states, given the initial assumptions regarding the state energies; for example, harmonic oscillators are generally assumed, especially for large molecules. We first describe the widely used Beyer–Swinehart[2] algorithm and then refer briefly to extensions which allow calculation of $W(E)$ and $\rho(E)$ for vibrational levels of anharmonic oscillators.

THE BEYER–SWINEHART (BS) ALGORITHM

We illustrate the method and the basic concepts that lie behind state distributions, by reference to a hypothetical bent triatomic molecule with three vibrational modes with reciprocal wavelengths $\tilde{\nu}_1 = 1500 \text{ cm}^{-1}$,

$\tilde{v}_2 = 1200$ cm^{-1} and $\tilde{v}_3 = 600$ cm^{-1}. The reasons for choosing these particular frequencies will become apparent when we consider the BS algorithm in more detail. We shall work in units of cm^{-1}, which are readily converted into kJ mol^{-1} by multiplying $N_A hc = 11.96 \times 10^{-3}$ kJ cm mol^{-1}. If the quantum numbers for the three vibrational modes are v_1, v_2 and v_3, then the vibrational energy is given by $(1500v_1 + 1200v_2 + 600v_3)$ cm^{-1} relative to the zero point energy.

Before examining the BS algorithm, it is useful to set out the energy states explicitly and hence evaluate $W(E)$ and $\rho(E)$ by longhand. We can then compare our results with the algorithm and hence understand its mechanism more clearly. Table 4.1 presents a systematic listing of all possible combinations of quantum numbers for which the total vibrational energy is ≤ 3000 cm^{-1}. In Table 4.2, these state energies have been combined to give the numbers of states occurring in 300 cm^{-1} energy ranges and, by summation, $W(E_v)$. The numbers of states in the 300 cm^{-1} ranges represent, in effect, densities of states with an energy unit of 300 cm^{-1}. Alternatively, by dividing these numbers by 300 cm^{-1} we obtain the average density of states $\langle \rho(E_v) \rangle$, (number of states per cm^{-1}) across the 300 cm^{-1} range. Note the

Table 4.1 Energies of vibrational quantum states for a hypothetical non-linear triatomic molecule with vibrational frequencies $\tilde{v}_1 = 1500$ cm^{-1}, $\tilde{v}_2 = 1200$ cm^{-1}, $\tilde{v}_3 = 600$ cm^{-1}

Quantum numbers			Total energy/ cm^{-1}	Label
v_1	v_2	v_3		
0	0	0	0	a
0	0	1	600	b
0	0	2	1200	c
0	0	3	1800	d
0	0	4	2400	e
0	0	5	3000	f
0	1	0	1200	g
0	1	1	1800	h
0	1	2	2400	i
0	1	3	3000	j
0	2	0	2400	k
0	2	1	3000	l
1	0	0	1500	m
1	0	1	2100	n
1	0	2	2700	o
1	1	0	2700	p
2	0	0	3000	q

Table 4.2 Numbers of vibrational quantum states in given energy ranges

Energy range/ cm^{-1}	Label from Table 4.1	States in range	Average density of states $\langle\rho(E_v)\rangle$/cm^{-1}	$W(E_v)$
0–299	a	1	3.33×10^{-3}	1
300–599	–	0	0	1
600–899	b	1	3.33×10^{-3}	2
900–1199	–	0	0	2
1200–1499	c, g	2	6.67×10^{-3}	4
1500–1799	m	1	3.33×10^{-3}	5
1800–2099	d, h	2	6.67×10^{-3}	7
2100–2399	n	1	3.33×10^{-3}	8
2400–2699	e, i, k	3	1.00×10^{-2}	11
2700–2999	o, p	2	6.67×10^{-3}	13
3000–3299	f, j, l, q	4	1.33×10^{-2}	17

increase in $\rho(E_v)$ with energy, even for a triatomic molecule at these modest energies, although this increase is somewhat structured.

We turn now to the BS algorithm proper and its implementation. We first state the procedure employed and then examine how it works, through our three-oscillator example, in Table 4.3.

1. We start by choosing a *grain size*, $\Delta\tilde{v}$; in our present example we choose $\Delta\tilde{v} = 300$ cm^{-1}.
2. We then reduce each frequency to an integer, R_i such that $R_i\Delta\tilde{v} \approx \tilde{v}_i$. In our present example, this procedure is exact, giving $R_1 = 5$, $R_2 = 4$, $R_3 = 2$. (This was the reason for our original choice of these particular frequencies.) For a real molecule, the choice of $\Delta\tilde{v}$ is determined by the final accuracy we require and the limitations of computer memory. For small to medium sized molecules (3–10 atoms, 3–24 oscillators) choices for $\Delta\tilde{v}$ of 1 cm^{-1} or 10 cm^{-1} are typical.
3. We now initialise an array $T(I)$, for $I = 1$ to M, such that $E_{max}/(N_A hc) = (M - 1)\Delta\tilde{v}$ and set all the elements except the first to zero; $T(1)$ is set to unity (Table 4.3). $E_{max}/N_A hc$ is 3000 cm^{-1} in the example, giving $M = 11$.
4. For each vibrational mode, we then perform the addition:

$$T(R_i + N) = T(R_i + N) + T(N)$$

for $N = 1$ to $M - R_i$, e.g. for $\tilde{v}_1$, $i = 1$ and $R_1 = 5$:

$$N = 1, T(6) = T(6) + T(1) = 0 + 1 = 1$$

We have updated the original element $T(6)$ by adding $T(1) = 1$ to give

Table 4.3 Implementation of the BS algorithm for the hypothetical triatomic molecule considered in Tables 4.1 and 4.2

$\Delta\tilde{v}(I-1) = (E_v/N_A hc)$ (cm^{-1})	0	300	600	900	1200	1500	1800	2100	2400	2700	3000
I	1	2	3	4	5	6	7	8	9	10	11
$T(I)$ (initial)	1	0	0	0	0	0	0	0	0	0	0
Add in $\tilde{v}_1$ states $T(5+N) = T(5+N) + T(N)$	1 a	0	0	0	0	1 m	0	0	0	0	1 q
Add in $\tilde{v}_2$ states $T(4+N) = T(4+N) + T(N)$	1 a	0	0	0	1 g	1 m	0	0	1 k	1 p	1 q
Add in $\tilde{v}_3$ states $T(2+N) = T(2+N) + T(N)$	1 a	0	1 b	0	2 c, g	1 m	2 d, h	1 n	3 e, i, k	2 o, p	4 f, j, l, q
Evaluate $W(E(I))$ $T(I) = T(I-1) + T(I)$	1	1	2	2	4	5	7	8	11	13	17

an entry $T(6) = 1$ in the array. The subsequent elements then remain unchanged, e.g.

$$N = 2, \ T(7) = T(7) + T(2) = 0 + 0 = 0$$

until we reach $N = 6$:

$$N = 6, \ T(11) = T(11) + T(6) = 0 + 1 = 1$$

where we now add in the current entry in element 6 and hence update element 11.

This procedure gives an array which has elements 1, 6 and 11 equal to unity. These elements correspond to energies of 0, 1500 and 3000 cm^{-1} and the unit elements to the states labelled a, m and q in Table 4.1

This procedure is repeated for $i = 2$ and 3, giving the arrays listed in Table 4.3. The elements are linked to the state energies in Table 4.1 via lower-case labels.

5. Each element $T(I)$ now contains the number of states $P(E(I))$ within the energy grain at energy $E(I)$, where $E(I) = N_A hc \Delta \tilde{v}(I - 1)$. The density of states is $P(E(I))/N_A hc \Delta \tilde{v}$ and the sum of states is obtained by performing the sum

$$T(I) = T(I - 1) + T(I)$$

for $I = 2$ to M; $W(E(I))$ is then contained in $T(I)$ (see Table 4.3).

THE STEIN–RABINOVITCH (SR) ALGORITHM

Stein and Rabinovitch[3] recognising both the power of the BS algorithm and its limitations — it can only be applied to harmonic oscillators — developed an extension which can be applied to degrees of freedom whose energy level distributions are not evenly spaced. They illustrated its use by reference to anharmonic oscillators, free rotors and hindered rotors[4]. We demonstrate its mode of operation by considering the general case in which the energy levels need not be equally spaced.

1. $\Delta \tilde{v}$, M and E_{max} are chosen as before.
2. Two arrays, T and AT, both of length M are initialised such that $T(1) = AT(1) = 1$, and all other members of both arrays are set to zero.
3. For the jth degree of freedom, the energy levels, $E_{j,k}$, are calculated up to the maximum energy E_{max}.
4. The $E_{j,k}$ are reduced to integers, $R_{j,k}$, such that $R_{j,k} \Delta \tilde{v} \approx E_{j,k}$.
5. The following sequence of additions is made for each k of the jth degree of freedom

$$AT(R_{j,k} + I) = AT(R_{j,k} + I) + T(I)$$

For $I = 1$ to $(M - R_{j,k})$

6. The contents of the array AT are then copied into the array T (overwriting the original contents of T) and steps (3), (4) and (5) are then repeated for the next degree of freedom, until all degrees of freedom have been considered.

7. The elements of T now contain the approximate densities of states and the corresponding sum of states can be obtained by the following summation:

$$T(I) = T(I) + T(I - 1) \quad \text{for } I = 2 \text{ to } I = M$$

The reader may verify this algorithm by applying it to the example given above with energy levels determined by solution of harmonic oscillator Schrödinger equation.

The errors intrinsic to the extended algorithm are generally less than in the BS technique itself, because the energy levels are rounded individually, thus eliminating cumulative errors. The SR algorithm is somewhat slower than BS and has greater memory requirements, since two arrays are required.

4.3 MONTE CARLO METHODS

The BS and SR algorithms are limited to *separable* vibrational modes. We have explicitly considered harmonic oscillators, in which the total vibrational energy for s oscillators $E(v_1, v_2 \ldots v_i, \ldots v_s)$ is given by

$$E + E_Z = \sum_{i=1}^{s} (v_i + \tfrac{1}{2}) hc\tilde{v}_i \tag{4.6}$$

where E_Z is the total zero point energy. More realistically, especially at the energies required for reaction, the levels should be treated as anharmonic, with total vibrational energy expressed by

$$E + E_Z = \sum_{i=1}^{s} (v_i + \tfrac{1}{2}) hc\tilde{v}_i + \sum_{i=1}^{s} \sum_{j \geqslant i}^{s} (v_i + \tfrac{1}{2})(v_j + \tfrac{1}{2}) hcx_{ij} \tag{4.7}$$

where the x_{ij} are the anharmonic constants. An approximation is often made that the cross-terms, x_{ij} $(i \neq j)$ may be neglected, giving

$$E + E_Z = \sum_{i=1}^{s} hc\{(v_i + \tfrac{1}{2})\tilde{v}_i + (v_i + \tfrac{1}{2})^2 x_{ii}\} \tag{4.8}$$

This results in vibrational level spacings, in a given mode, that are no longer equal so that the BS algorithm is inapplicable, but the SR algorithm can be used. Motion of this type is said to be separable—the vibrational modes are assumed to be independent. The anharmonic cross-terms, however, are crucial to the flow of energy between modes and make a significant contribution to the sum of states.

A new approach, based on Monte Carlo sampling[5,6] is required for non-separable motion. The idea is to sample an s-dimensional region of 'volume', V_s, defined by the vibrational quantum numbers, v_i; V_s contains a total number of states W_s. If we are interested in the sum of states, $W(E_v)$, then we randomly sample the values of v_i ($i = 1$ to s) and check to see if the energy E (equation 4.7) is less than or equal to E_v. If it is, we score a success. This procedure is repeated many times, and the sum of states is given by

$$W(E_v) = V_s(n/N) \qquad (4.9)$$

where N is the total number of trials and n the number of successes. The major difficulty is how to define V_s. We must ensure that the integration region is entirely contained within the sampling volume, but also keep it as small as possible, otherwise most of our trials are unsuccessful, n/N is small and the calculation is computationally inefficient. If we define the efficiency of the procedures as $\varepsilon = n/N$, then we wish to keep ε close to unity. There are several techniques available and an algorithm, proposed by Barker[7], is given in Appendix 5.

Doll[8], has described an alternative approach in which the density of states of the harmonic system is corrected for anharmonic effects using a classical correction factor which is evaluated using Monte Carlo techniques.

The techniques described by Barker[7] and by Doll[8] are particularly useful when the energy of the system can be described explicitly as a function of the vibrational quantum numbers. Such a description is often not possible, especially in the region of the transition state. In the dissociation of ethane, for example, the transition state is very loose and the modes which correspond to the torsion and the CH_3 rocking vibrations in the ground state molecule are transformed into interacting hindered rotors; a description based on normal mode vibrational frequencies and anharmonic constants is clearly inapplicable. Recently, considerable strides have been made in the calculation of *ab initio* potential energy surfaces and recourse may then be made to the calculation of the sum of classical states for the system on this surface. The sum is evaluated, once more, using a Monte Carlo method, and several techniques are available which have been developed for general application in the field of statistical mechanics[6]. Klippenstein and Marcus have discussed one such technique in the present context[9]; they have also demonstrated that a classical model reproduces the quantum mechanical sum of states to high accuracy[10]. We shall return to the use of this approach in Chapter 6, when discussing reactions occurring through 'loose' transition states.

4.4 CLASSICAL TREATMENT OF VIBRATIONAL STATES AND DERIVED SEMICLASSICAL APPROXIMATIONS

Although direct count methods are easily and efficiently implemented in full RRKM calculations, it is frequently useful to have available approximate analytic expressions for $W(E)$ and $\rho(E)$. We shall make use of such expressions in Chapter 8.

A natural approach to the derivation of simple expressions for the density and sum of vibrational states is to take the well-known expressions for classical harmonic oscillators and introduce modifications designed to overcome the failings of these equations when applied to quantised vibrations. Such modifications lead to what are generally called 'semiclassical' approximations, an example is the Marcus–Rice[11] approximation. In this section the more satisfactory Whitten–Rabinovitch[12] formulation is also discussed, and its extension to vibrational–rotational systems will be considered below.

TREATMENT OF CLASSICAL HARMONIC VIBRATIONS[13–15]

The simplest approximation to the sum and density of vibrational states is obtained by treating the molecule as a set of classical harmonic oscillators, the classical behaviour most easily being evaluated as a limiting case of the quantised behaviour.

The approach used here is based on the canonical partition function which is given by

$$Q(\beta) = \int_0^\infty \rho(E) \exp(-\beta E) \, dE \tag{4.10}$$

and is the equivalent of calculating the Laplace transform[13] (LT) of $\rho(E)$ with $\beta = 1/kT$ as the transform variable (β is used instead of $1/kT$ throughout this section as it greatly simplifies LT manipulations).

One important general result can be obtained by integrating (4.10) by parts to give

$$Q(\beta) = \beta \int_0^\infty W(E) \exp(-\beta E) \, dE \tag{4.11}$$

That is, the LT of $W(E)$ is obtained from that of $\rho(E)$ simply by dividing by β, a result that will be employed below.

Consider the case of a set of s uncoupled classical harmonic oscillators with frequencies ν_i. The classical partition function of a single oscillator is given by

$$Q_i = \frac{1}{\beta h \nu_i} \tag{4.12}$$

This formula may be obtained from the quantum mechanical result by allowing $\beta h\nu_i \to 0$ which corresponds to the case where states become very close and ultimately reach the classical limit. The overall partition function for the s oscillators is given by

$$Q = \prod_{i=1}^{s} Q_i = \frac{1}{\beta^s} \frac{1}{\displaystyle\prod_{i=1}^{s} h\nu_i} \tag{4.13}$$

The density of states $\rho(E)$ can be obtained from Q by an inverse LT (ILT). In general inversion can be a complicated procedure, but fortunately there exist many tables of ILTs and that required for (4.13) is well known (see equation 4.40 given later). Hence we can write down equation (4.14) straight away:

$$\rho(E) = \frac{E^{s-1}}{\Gamma(s)\displaystyle\prod_{i=1}^{s} h\nu_i} = \frac{E^{s-1}}{(s-1)!\displaystyle\prod_{i=1}^{s} h\nu_i} \tag{4.14}$$

where $\Gamma(s)$ is the gamma function described in Appendix 4. The sum of states can be obtained either by integrating (4.14) or using the relationship (4.11) to give

$$\mathscr{L}[W(E)] = \frac{Q}{\beta} = \frac{1}{\beta^{s+1}} \frac{1}{\displaystyle\prod_{i=1}^{s} h\nu_i} \tag{4.15}$$

where $\mathscr{L}$ refers to the LT. Equation (4.15) on inversion yields

$$W(E) = \frac{E^s}{\Gamma(s+1)\displaystyle\prod_{i=1}^{s} h\nu_i} = \frac{E^s}{s!\displaystyle\prod_{i=1}^{s} h\nu_i} \tag{4.16}$$

THE DIRICHLET INTEGRAL[16]

Before proceeding it is useful to consider an alternative approach to the above problem, elements of which are used in more advanced theories. The sum of states for quantum mechanical systems was previously defined in equation (4.1) as

$$W(E) = \sum_{\varepsilon=0}^{E} P(\varepsilon) \tag{4.1}$$

The energy of a classical oscillator can be written by analogy with the quantum mechanical result as

$$E_i = v_i h v_i \tag{4.17}$$

where v_i now represents a continuous quantum number and the zero point energy has been neglected. With this form of energy it is now possible to rewrite (4.1) as an integral over the set of v_i,

$$W(E) = \int \ldots \int dv_1 \, dv_2 \ldots dv_s \tag{4.18}$$

where the limits of the integral are subject to the restriction

$$E \geq \sum_{i=1}^{s} v_i h v_i \tag{4.19}$$

The integral (4.18) and the restriction (4.19) can be more compactly expressed in terms of the step function u as in

$$W(E) = \int_0^\infty \ldots \int_0^\infty u\left(E - \sum_{i=1}^{s} v_i h v_i\right) dv_1 \, dv_2 \ldots dv_s \tag{4.20}$$

where u is defined by

$$u = \begin{cases} 1 & \text{if } E \geq \sum_{i=1}^{s} v_i h v_i \\ 0 & \text{otherwise} \end{cases} \tag{4.21}$$

The step function u can be treated by LT. However, integrals of this form occur regularly and (4.20) is a specific example of the more general result due to Dirichlet[16]. The Dirichlet integral can be written as

$$I(y) = \int_0^\infty \ldots \int_0^\infty \prod_{i=1}^{s} x_i^{d_i-1} u\left(y - \sum_{i=1}^{s} (x_i/a_i)^{b_i}\right) dx_1 \, dx_2 \ldots dx_n \tag{4.22}$$

and it can be shown that the result of this integration is given by

$$I(y) = \frac{y^t}{\Gamma(t+1)} \prod_{i=1}^{s} \left[\frac{a_i^{d_i} \Gamma(d_i/b_i)}{b_i}\right] \tag{4.23}$$

where

$$t = \sum_{i=1}^{s} (d_i/b_i)$$

Clearly (4.20) is a special case of (4.22) with all $d_i = 1$, all $b_i = 1$ and $a_i = 1/h v_i$. Appropriate substitution yields (4.16).

The density of states can be tackled using an integral similar to (4.18) but with the restriction

$$E = \sum_{i=1}^{s} v_i \boldsymbol{h} v_i \tag{4.24}$$

This restriction can be expressed in terms of the Dirac δ function

$$\delta = \begin{cases} 1 & \text{if } E = \sum_{i=1}^{s} v_i \boldsymbol{h} v_i \\ 0 & \text{otherwise} \end{cases} \tag{4.25}$$

The integral corresponding to (4.22) for this case is then

$$L(y) = \int_0^\infty \cdots \int_0^\infty \prod_{i=1}^{s} x_i^{d_i-1} \delta\left(y - \sum_{i=1}^{s} (x_i/a_i)^{b_i} \right) dx_1 \, dx_2 \ldots dx_n \tag{4.26}$$

which can be shown to be

$$L(y) = \frac{y^{t-1}}{\Gamma(t)} \prod_{i=1}^{s} \left[\frac{a_i^{d_i} \Gamma(d_i/b_i)}{b_i} \right] \tag{4.27}$$

Again, if the above values of a_i, b_i and d_i are substituted then (4.14) is obtained. There are two points of interest that follow from equation (4.27):

1. It can be seen by inspection that $L(y)$ is the differential of $I(y)$ and so they satisfy the relationship (4.2).
2. If $a_i = b_i = d_i = 1$ then equation (4.26) becomes equivalent to (2.18), the value of which can be obtained from (4.14) by setting $\boldsymbol{h} v_i = 1$.

The validity of this classical approximation is indicated by some of the results for $W(E_V)$ in Table 4.4, which compares the classical results from (4.16) with the results of direct counts at various energies for $CHCl_3$. The other figures in the table will be referred to later. It is clear from these results that the classical approximation is poor at low energies, but better at

Table 4.4 $W(E_v)$ for $CHCl_3$ by various methods $[(v/cm^{-1})^{20}; 3033, 1205 (2), 760 (2), 667, 364, 260 (2), E_z = 50.92 \text{ kJ mol}^{-1}]$

E_v/E_z	Classical (4.31)	Marcus–Rice (4.33)	Whitten–Rabinovitch (4.36)	Exact count
1.00	0.030×10^3	15.5×10^3	8.18×10^3	8.20×10^3
2.00	0.155×10^5	5.97×10^5	4.58×10^5	4.57×10^5
3.00	0.597×10^6	7.94×10^6	6.87×10^6	6.85×10^6
5.00	0.592×10^8	3.05×10^8	2.87×10^8	2.86×10^8
8.00	0.407×10^{10}	1.17×10^{10}	1.14×10^{10}	1.14×10^{10}

the higher energies as would be expected from the Bohr correspondence principle, which implies that in the limit of high quantum numbers the classical and quantum theories become essentially equivalent[17]. The error at low energies is particularly clear in the case of a simple system containing s degenerate oscillators with the same frequency v. At energy levels of 0 and hv the classical expression predicts $W(E_v) = 0$ and $1/s!$ respectively, whereas the correct values are 1 and $s + 1$ respectively since there is one quantum state at $E_v = 0$ and there are s states at $E_v = hv$. It can be shown that this approximation must always underestimate the number of states at a given energy level[14] and, although the error decreases with increasing energy, it does not become small until impracticably high energies are reached. Better approximations are therefore required, and the development of these is discussed in the following sections.

It can easily be shown that insertion of the classical expressions (4.14) and (4.16) into the RRKM equations leads to Hinshelwood's expression for k_1/k_2 (equation 2.20) and to the widely used classical Kassel formula for $k_a(E)$ (equation 2.63). It follows that the accuracy of these results is limited by that of the implied classical treatment of the molecular vibrations.

SEMICLASSICAL (MARCUS–RICE) APPROXIMATION

Marcus and Rice[11] suggested that since a quantised oscillator with a variable energy E_v has a total vibrational energy above the classical ground state of $E_v + E_z$ (where E_z if the zero-point energy, $\frac{1}{2}hv$), it should be compared with a classical oscillator at energy $(E_v + E_z)$ rather than E_v. Thus the classical expressions (4.14) and (4.16) were modified to give the 'semi-classical' or Marcus–Rice expressions (4.28) and (4.29).

$$W(E_v) = \frac{(E_v + E_z)^s}{s!\Pi h v_i} \tag{4.28}$$

$$\rho(E_v) = \frac{(E_v + E_z)^{s-1}}{(s - 1)!\Pi h v_i} \tag{4.29}$$

where

$$E_z = \sum_{i=1}^{s} (\tfrac{1}{2} h v_i) \tag{4.30}$$

Table 4.4 includes some values of $W(E_v)$ calculated from (4.28) for $CHCl_3$, and the results confirm that this approach gives a more accurate approximation than the classical expression (4.16), although the number of states at any level is now always overestimated. This approach is an improvement, but is still far from accurate[18].

WHITTEN–RABINOVITCH APPROXIMATION[12]

Although the Marcus–Rice expression is not in itself highly accurate, it leads directly to the development of a greatly improved semiclassical type of approach to be described in the present section. Rabinovitch and Diesen suggested[19] that only an appropriate fraction of the zero-point energy should be added to E_v; thus an empirical factor a was introduced into (4.28) and (4.29), giving the expressions (4.31) and (4.32):

$$W(E_v) = \frac{(E_v + aE_z)^s}{s!\Pi h\nu_i} \tag{4.31}$$

$$\rho(E_v) = \frac{(E_v + aE_z)^{s-1}}{(s-1)!\Pi h\nu_i} \tag{4.32}$$

The factor a has a value between 0 and 1, and by comparison of the right-hand side of (4.31) with the result of direct counts it is possible to determine the precise value of a required to give the correct result for a given molecule at any energy level. These values have been tabulated quite extensively,[12,19–22] and the importance of this approach depends on the consequent ability to predict suitable values of a for other cases.

Thus Figure 4.2[12] shows the typical variation of a with energy for a

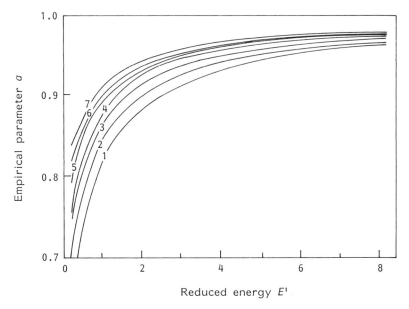

Figure 4.2 Values of a required for (4.31) to give correct values of $W(E_v)$ at various reduced energies for the molecules listed in Table 4.5. Reproduced by permission of John Wiley and Sons Ltd.

number of molecules.[a] The values always increase with increasing energy, approaching unity at high energies, and the values for different molecules lie in a fairly narrow band when plotted (as in Figure 4.2) against a reduced energy $E' = E_v/E_z$ rather than against E_v itself. Furthermore, the position in this band of the curve for a particular molecule is found empirically to depend only on the distribution of the molecular vibration frequencies; the values of a are lower for a molecule whose vibrations have widely different frequencies than for one containing many vibrations of similar frequency. This dispersion of the frequencies can be measured by their standard deviation, and Whitten and Rabinovitch[12] define a closely related *modified frequency dispersion parameter* β_R by

$$\beta_R = \frac{\sigma - 1}{\sigma} \frac{\langle v^2 \rangle}{\langle v \rangle^2} \qquad (4.33)$$

in which $\langle v \rangle$ and $\langle v^2 \rangle$ are the mean frequency and mean-square frequency of the molecule respectively. The value of the parameter β_R varies from about unity for molecules with largely similar frequencies, to about two for molecules with a wide spread of frequencies. For example, SF_6 with reciprocal wavelengths of 965 (3), 772, 642 (2), 617 (3), 525 (3) and 370 (3) cm^{-1} has $\beta_R = 1.03$, while $Ni(CO)_4$ with $\tilde{v}_i = 2057$ (3), 2040, 545 (3), 460 (2), 422 (3), 381, 300 (3) and 79 (5) cm^{-1} has $\beta_R = 2.04$. Table 4.5 shows the values of β_R for the molecules under consideration in Figure 4.2, and illustrates the very good correlation of the relative positions of the various curves with the values of β_R for the molecules concerned.

Table 4.5 Modified frequency dispersion parameters β_R for the molecules in Figure 4.4

Curve in Figure 4.4	Molecule	β_R
1	$Ni(CO)_4$	2.04
2	$CHBr_3$, C_4H_2	1.76, 1.72
3	$(CN)_2$, $C_2H_2Cl_2$	1.47, 1.45
4	C_5H_5N, $(CH_2)_3CO$	1.35, 1.34
5	$c\text{-}C_3D_6$, $(CH_2)_2O$	1.23, 1.22
6	CH_3F, B_2H_6	1.05, 1.15
7	SF_6	1.03

[a]The plots in Figure 4.2 are actually smoothed versions of a sawtooth type of variation. This behaviour arises because (4.31) predicts a smooth variation of $W(E_v)$ with energy, whereas the actual variation is stepwise (see Figure 3.7) and can therefore only be reproduced by having a stepwise variation of a with energy. Such effects are generally negligible for $E' > 1$, i.e. at energies where the density of states is high and $W(E_v)$ increases fairly smoothly with increasing energy (see section 3.11).

It remains to discover a simple analytical relationship between a, β_R and E', and Whitten and Rabinovitch achieved this by writing

$$a = 1 - \beta_R w(E') \tag{4.34}$$

where $w(E')$ is a unique function of E' shown in Figure 4.3 and described adequately by

$$\left.\begin{array}{ll}(0.1 < E' < 1.0) & w = (5.00E' + 2.73(E')^{0.5} + 3.51)^{-1} \\ (1.0 < E' < 8.0) & w = \exp(-2.4191(E')^{0.25})\end{array}\right\} \tag{4.35}$$

This completes what we shall call the Whitten–Rabinovitch approximation for $W(E_v)$, and the method of calculation can be summarised as follows. The molecular vibrational frequencies are used to calculate β_R (4.33) and E_z (4.30), then for any energy E_v of interest, the reduced energy $E' = E_v/E_z$ is used to obtain $w(E')$ (4.35), a (4.34), and hence $W(E_v)$ (4.31). The density $\rho(E_v)$ may be obtained directly from (4.32) or, more accurately[23], from the differential of (4.31), which is

$$\rho(E_v) = \frac{(E_v + aE_z)^{s-1}}{(s-1)!\Pi h\nu_i}\left[1 - \beta_R\left(\frac{dw}{dE'}\right)\right] \tag{4.36}$$

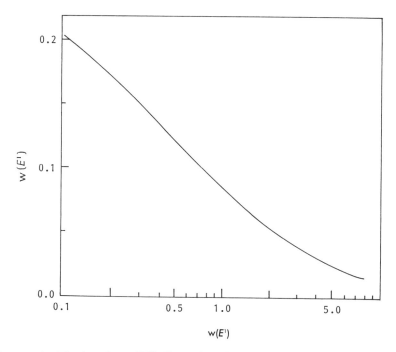

Figure 4.3 The function $w(E')$. Reproduced by permission of John Wiley and Sons Ltd.

where dw/dE' is obtained as (4.37) by differentiation of (4.35)

$$
\left.\begin{array}{ll}
(0.1 < E' < 1.0) & dw/dE' = -(5.00 + 1.365(E')^{-0.5})w^2 \\
(1.0 < E' < 8.0) & dw/dE' = -(0.60478(E')^{-0.75})w
\end{array}\right\} \quad (4.37)
$$

The Whitten–Rabinovitch approximation has been extensively tested by various authors, some typical results being shown in Table 4.4 and Figure 4.4. The results are always in very good agreement with the direct count for E_v greater than the zero-point energy E_z, and often for much lower energies. This is well illustrated by Figure 4.5 (from Whitten and Rabinovitch[12]), the important feature of which is the negligible error at energies

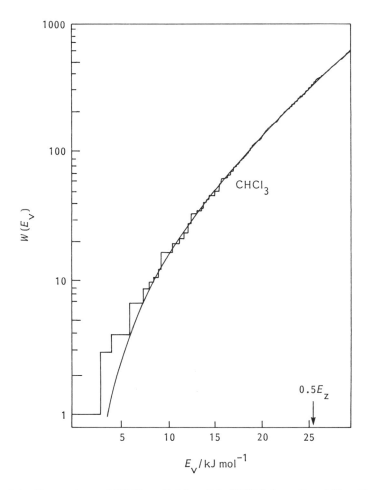

Figure 4.4 Comparison of Whitten–Rabinovitch $W(E_v)$ (equation 4.31 with direct count for chloroform. The smooth curve represents equation (4.31).

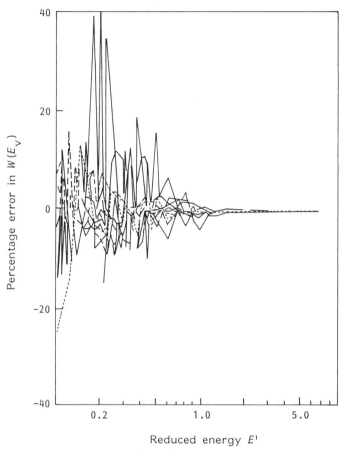

Figure 4.5 Error of the Whitten–Rabinovitch approximation at various reduced energies for the molecules listed in Table 4.5. Reproduced by permission of John Wiley and Sons Ltd.

above $E' = 1$. The errors at low energies can be avoided to some extent by using a 'truncated model'. This means that vibrations for which the quantum energy $h\nu_i$ is greater than the available energy E_v, and which therefore cannot be excited, are removed from the model. Thus SO_2, with $s = 3$ and frequencies of 1361, 1151 and 524 cm^{-1}, would be treated as such for $E_v > 1361$ cm^{-1}, but at energies between 1361 and 1151 cm^{-1} it would be treated as two vibrations with frequencies of 1151 and 524 cm^{-1}, and between 1151 and 524 cm^{-1} as a single vibration of frequency 524 cm^{-1}. Below 524 cm^{-1}, of course, $W(E_v) = 1$.

Other approaches of a semiclassical nature have also been suggested. In particular, Schlag and Sandsmark[24] obtained a useful approximation by manipulating the exact combinatorial expressions with suitable approximations; the result was a three-term expression in which the first term was the Marcus–Rice expression (4.28). Expressions derived by the inverse Laplace transformation technique (see below) can also often be approximated into a semiclassical form. Wahrhaftig and co-workers[14,25] developed a method in which an exact count is used for any oscillators in their ground state and a semiclassical type of approach is used for excited oscillators, but the results are still inferior to those of a number of other approaches[15].

The outstanding advantage of the Whitten–Rabinovitch approximation is its simplicity and ease of application. Its accuracy is good and compares favourably with that of other approaches, and the method has found wide application for harmonic oscillator models (including those with rotational degrees of freedom—see next section). Its deficiencies at low energies are seldom of significance, since direct count methods are so readily employed here.

DIRECT INVERSION OF AN APPROXIMATE PARTITION FUNCTION

The technique was first applied to the present problem by Haarhoff[26] and may be illustrated by his treatment of a collection of s quantised harmonic oscillators with frequencies v_i. The partition function for this system is given by the well-known expression

$$Q(\beta) = \prod_{i=1}^{s} \left\{ \frac{\exp(-\tfrac{1}{2}h v_i \beta)}{1 - \exp(-h v_i \beta)} \right\} \tag{4.38}$$

This expression (even for a single oscillator) is not readily subjected to inverse transformation in its exponential form. It may, however, be expressed as a power series in β; it emerges that only alternate powers appear, and the appropriate series for $Q(\beta)/\beta$ is

$$\frac{Q(\beta)}{\beta} = \left(\sum_{i=1}^{s} h v_i \right) \left(\frac{1}{\beta^{s+1}} - \frac{a_2}{\beta^{s-1}} + \frac{a_4}{\beta^{s-3}} - \frac{a_6}{\beta^{s-5}} + \ldots \right) \tag{4.39}$$

In this series a_2, a_4, etc. are constants involving the frequencies v_i together with numerical factors; for example

$$a_2 = (\langle v^2 \rangle / \langle v \rangle^2) E_z^2 / 6s$$

Haarhoff[26] gave detailed expressions for the first five terms in (4.39); their complexity increases rapidly, but the general term can be obtained from Thiele's formulation[27] in terms of the so-called 'Bernoulli numbers'.

The inverse transformation of $Q(\beta)/\beta$ is now considered term by term; it is a property of LTs that this is valid, i.e. that

$$\mathcal{L}^{-1}\{af(\beta) + bg(\beta)\} = a\mathcal{L}^{-1}\{f(\beta)\} + b\mathcal{L}^{-1}\{g(\beta)\}$$

where f and g are functions of β and a and b are constants. The infinite series in (4.39) consists of a limited number of terms having negative powers of β, after which all other terms have zero or positive powers of β. There are in fact $\frac{1}{2}s + 1$ or $\frac{1}{2}(s + 1)$ terms with negative powers of β according to whether s is even or odd, and the inverse transformation of there terms is readily accomplished (see equations 4.15 and 4.16 above), and is given by

$$\mathcal{L}^{-1}(\beta^{-m}) = E^{m-1}/(m - 1)! \quad (m = \text{positive integer}) \qquad (4.40)$$

Since (4.38) is based on energies measured from the classical energy zero, the result (4.41) emerges in terms of energies on the same scale, i.e. $E_v + E_z$ rather than E_v.

$$W(E_v) = \mathcal{L}^{-1}\{Q(\beta)/\beta\} = \left(\prod_i h\nu_i\right)^{-1}\left[\frac{(E_v + E_z)^s}{s!} - \frac{a_2(E_v + E_z)^{s-2}}{(s - 2)!} + \cdots\right]$$

$$= \frac{(E_v + E_z)^s}{s!\,\Pi h\nu_i}\left[1 - \frac{a_2'}{(E_v + E_z)^2} + \frac{a_4'}{(E_v + E_z)^4} - \cdots\right] \qquad (4.41)$$

$$[\tfrac{1}{2}s + 1 \text{ or } \tfrac{1}{2}(s + 1) \text{ terms}]$$

The transformation (4.40) is not valid for $m \leqslant 0$, but it is known that the inverse transformation of the remaining terms in (4.39) will produce a set of stepped functions of just the type required to reproduce the actual stepwise variation of $W(E_v)$ in a quantised system (see, for example, section 3.11). The terms in (4.41) represent a smooth-curve basis on which the stepped functions are superimposed, and Haarhoff's approximation assumes that this finite series provides a reasonable smooth-curve approximation to $W(E_v)$.

It is interesting to note that the first term in (4.41) is just the semiclassical expression (4.28), so that even this one term is a substantial improvement over the classical approximation. The second term in the brackets is given in detail by

$$-\frac{a_2'}{(E_v + E_z)^2} = -\frac{s}{6}\frac{\langle \nu^2 \rangle}{\langle \nu \rangle^2}\left(\frac{1}{1 + E_v/E_z}\right)^2 \qquad (4.42)$$

which provides an interesting connection [27] with the deduction of Whitten and Rabinovitch[12] that the negative correction needed to the Marcus–Rice expression was a more or less universal function of the frequency dispersion parameter (4.33) and the reduced energy $E' = E_v/E_z$.

Haarhoff's expression (4.41)[26] has been found very satisfactory in practice by a number of workers, particularly when the full series as developed by

Thiele[27] is used[26,28-31]. Haarhoff also derived the analytical equation (4.43) as an approximation to the series expression for the density of states[32], and extended the same treatment to vibrational–rotational systems, while Forst, Prášil and St Laurent[33] extended the series treatment to vibrational–rotational systems to give an excellent approximation.

$$\rho(E_v) = \left(\frac{2}{\pi s}\right)^{1/2} \frac{(1 - 1/12s)\lambda}{h\langle v\rangle(1 + \eta)}[(1 + \tfrac{1}{2}\eta)(1 + 2/\eta)^{1/2\eta}]^s[1 - 1/(1 + \eta)^2]^{\beta_0}$$

(4.43)

where

$$\eta = 1 + E_v/E_z, \qquad \lambda = \prod_{i=1}^{s}(\langle v\rangle/v_i)$$

and

$$\beta_0 = \frac{(s - 1)(s - 2)}{6s}\frac{\langle v^2\rangle}{\langle v\rangle^2} - \frac{s}{6}$$

4.5   CLASSICAL TREATMENT OF ROTATIONAL STATE DISTRIBUTIONS

The classical treatment of rotation may be approached as the limiting behaviour of quantised rotors at high energies, where certain approximations can be made to the quantum equations. On this basis we proceed to evaluate $W(E_r)$, the sum of the numbers of rotational quantum states at all energies up to E_r, and obtain $\rho(E_r)$ by differentiation. The treatment of hindered rotations is also discussed briefly.

QUANTISED ROTATIONS

It will be assumed that non-degenerate rotations (i.e. those with different moments of inertia) can be treated as being independent. Thus their energies are additive and all combinations of quantum states are possible; the validity of this assumption will be discussed below. The permissible energy levels of an independent, unhindered, quantised, rigid rotor are given by (4.44) for a one-dimensional or singly degenerate rotor[34,35], and by (4.45) for a two-dimensional or doubly degenerate rotor[36,37]

(Singly degenerate) $E_K = (h^2/8\pi^2 I)K^2$ $(K = 0, 1, 2, \ldots)$ (4.44)

(Doubly degenerate) $E_J = (h^2/8\pi^2 I)J(J + 1)$ $(J = 0, 1, 2, \ldots)$ (4.45)

Both terminologies are widely used; care must be taken not to confuse the degeneracy of a rotor with the degeneracies of its energy levels. In (4.44) and (4.45), h is Planck's constant, I the moment of inertia, and J and K are quantum numbers. For high values of the rotational quantum number the term $J(J + 1)$ in (4.45) can be approximated by J^2, and the energy in the ith rotation of a molecule with several rotational degrees of freedom is therefore given approximately by

$$E_i \approx (h^2/8\pi^2 I_i)J_i^2 \qquad (4.46)$$

which I_i and J_i are the appropriate moment of inertia and quantum number respectively.

For a one-dimensional rotor the number of states at any level except the ground state is two, corresponding to the two possible directions of rotation. For a two-dimensional rotor there are $(2J + 1)$ states at the Jth energy level[36,37]. Thus if the degeneracy of the ith rotor is d_i, the number of quantum states at energy E_i is given by

$$\begin{array}{ll} (d_i = 1, K \neq 0) & P(E_i) = 2 \\ (d_i = 2) & P(E_i) = (2J_i + 1) \end{array}\Bigg\} \qquad (4.47)$$

At high energies $2J_i + 1$ is approximately equal to $2J_i$ and the two cases can thus be combined in the single approximate expression

$$P(E_i) \approx 2J_i^{d_i-1} \qquad (4.48)$$

If there are altogether p non-degenerate rotors then, for a given set of quantum numbers J_i, the total rotational energy E_r is given by (4.49), and the number of rotational states at this energy by

$$E_r = \sum_{i=1}^{p} E_i \approx \sum_{i=1}^{p} \left(\frac{h^2}{8\pi^2} \frac{J_i^2}{I_i} \right) \qquad (4.49)$$

$$P(E_r) = \prod_{i=1}^{p} P(E_i) \approx \prod_{i=1}^{p} 2J_i^{d_i-1} \qquad (4.50)$$

$$(= 2J_1^{d_1-1}.2J_2^{d_2-1} \dots 2J_p^{d_p-1})$$

The total number of states at energies up to and including E_r is therefore (4.51), in which the summation is taken over all combinations of $J_1, \dots, J_p$ such that (4.52) is satisfied.

$$W(E_r) \equiv \sum_{E_r=0}^{E_r} P(E_r) = \sum_{J_p} \sum_{J_p-1} \dots \sum_{J_1} \left(\prod_{i=1}^{p} 2J_i^{d_i-1} \right) \qquad (4.51)$$

$$\sum_{i=1}^{p} (h^2/8\pi^2 I_i)J_i^2 \leqslant E_r \qquad (4.52)$$

CLASSICAL ROTATIONS

For rotors which are so highly excited as to permit a classical treatment, the summation (4.51) is replaced by a multiple integration

$$W(E_r) \approx \int_{J_p=0}^{J'_p} \int_{J_p-1=0}^{J'_p} \cdots \int_{J_1=0}^{J'_1} \left(\prod_{i=1}^{p} 2J_i^{d_i-1} \right) dJ_1 \, dJ_2 \ldots dJ_p \qquad (4.53)$$

The limits are again determined by (4.52), so that the lower limits are all zero and the upper limits J'_n are given by

$$(n = p) \qquad\qquad \frac{h^2 J_p'^2}{8\pi^2 I_p} = E_r$$

$$(n \neq p) \qquad\qquad \frac{h^2 J_n'^2}{8\pi^2 I_n} = E_r - \sum_{i=n+1}^{p} \frac{h^2 J_i^2}{8\pi^2 I_i}$$

It will be seen that these equations express the condition that the pth rotor cannot have energy E_p greater than the total available energy E_r, that the $(p-1)$th rotor cannot have more than the energy $(E_r - E_p)$ left after fixing E_p, and so on, the energy of the first rotor being limited to that which remains when all the other E_i are fixed.

The integral (4.53) can be expressed in the form of the Dirichlet integral (4.22) introduced in section 4.4 and consequently can be evaluated using equation (4.23). After appropriate substitution, equation (4.54) is obtained.

$$W(E_r) \equiv \sum_{E_r=0}^{E_r} P(E_r) = \frac{Q_r}{\Gamma(1 + \frac{1}{2}r)} \left(\frac{E_r}{kT} \right)^{\frac{1}{2}r} \qquad (4.54)$$

where

$$r = \sum_{i=1}^{p} d_i$$

The sum r of the degeneracies of the p rotors is simply the total number of rotational degrees of freedom involved, irrespective of whether they occur singly or in degenerate pairs. The quantity Q_r is the partition function for the rotational degrees of freedom under the same approximations ((4.49) and (4.50)) as were used in the derivation of (4.54) and is given by

$$Q_r = \left(\frac{8\pi^2 kT}{h^2} \right)^{\frac{1}{2}r} \prod_{i=1}^{p} \{ I_i^{\frac{1}{2}d_i} \Gamma(\frac{1}{2}d_i) \} \equiv \prod_{i=1}^{p} \left\{ \left(\frac{8\pi^2 I_i kT}{h^2} \right)^{\frac{1}{2}d_i} \Gamma(\frac{1}{2}d_i) \right\} \qquad (4.55)$$

It will be noted that the temperature dependence of Q_r is exactly cancelled by the term $(kT)^{-\frac{1}{2}r}$ in (4.55), so that $W(E_r)$ is temperature independent, as it obviously should be. Although the partition function is introduced into (4.54) mainly to simplify the expression, the result does suggest a useful approximate treatment of different rotational models (e.g. hindered rotations) by using (4.54) with the appropriate partition function expression inserted[1].

In addition to the result (4.54) for the sum of states, an expression is also required for the density of states, and this is obtained by simple differentiation of (4.54). Since $\Gamma(1 + \frac{1}{2}r) = \frac{1}{2}r\Gamma(\frac{1}{2}r)$ the result is

$$\rho(E_r) = \frac{\mathrm{d}}{\mathrm{d}E_r} W(E_r) = \frac{Q_r}{(kT)^{\frac{1}{2}r}\Gamma(\frac{1}{2}r)} E^{\frac{1}{2}r - 1} \tag{4.56}$$

VALIDITY OF THE CLASSICAL INDEPENDENT ROTOR TREATMENT

There are two main points which are worthy of consideration regarding the validity of the treatment outlined above. Firstly, some justification is needed for treating the various rotations independently of each other, and, secondly, the various mathematical approximations made in the derivation need to be examined.

The various points may conveniently by illustrated with reference to the overall rotations of a prolate symmetric top molecule, i.e. a molecule for which $I_A < I_B = I_C$. The stationary wave functions and energy levels of such a system are correctly obtained by solution of the Schrödinger equation for the molecule as a whole[38]. The energy levels are given by

$$\left.\begin{aligned} E_{J,K,M} &= BhJ(J + 1) + (A - B)hK^2 \\ J &= 0, 1, 2, \ldots, \qquad K = 0, \pm 1, \pm 2, \ldots, \pm J \\ (M &= 0, \pm 1,, \pm 2, \ldots, \pm J) \end{aligned}\right\} \tag{4.57}$$

where $A = h/8\pi^2 I_A$, etc. and J, K, M are integral quantum numbers in the ranges shown. It will be noted that there are three quantum numbers, only two of which (J, K) appear in the expression for the energy. For a given value of J there are $2J + 1$ acceptable values of M, so that each, J, K combination comprises not one but $2J + 1$ quantum states. The energy of a state $J, -K$ is the same as that of the state J, K, and the overall degeneracy of a given energy level E_r is therefore given by

$$\left.\begin{aligned} (K = 0) \qquad &P(E_r) = 2J + 1 \\ (K \neq 0) \qquad &P(E_r) = 2(2J + 1) \end{aligned}\right\} \tag{4.58}$$

The rotational energy levels for a given symmetric top can now be evaluated, together with their degeneracies, and plotted on a $W(E_r)$ versus energy diagram similar to Figures 3.7 and 3.8. Such a plot is shown in Figure 4.6 (curve a) for the methyl chloride molecule; the data were calculated by a direct-count technique (see section 4.2).

We now wish to examine the effects of regarding the rotational degrees of freedom as independent or separable. The symmetric top molecule would thus be regarded as equivalent to a two-dimensional rotor with moment of inertia I_B plus an independent one-dimensional rotor with moment of inertia I_A. Their energies will be additive and their degeneracies multiplicative, with no restriction on the permissible quantum numbers for the two rotations. For a one-dimensional rotor the energy levels are (4.44) and the degeneracies 1 ($K = 0$) or 2 ($K \neq 0$). For a two-dimensional rotor the energy levels are (4.45) and the degeneracy of a given level is ($2J + 1$); this degeneracy arises from the existence of a second quantum number J' which takes the values $0, \pm 1, \pm 2, \ldots, \pm J$, but does not enter the expression for

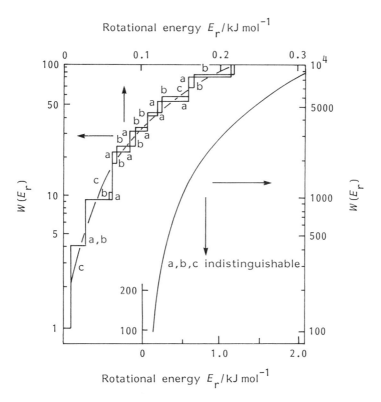

Figure 4.6 Distribution of rotational states for methyl chloride (A = 1.50 × 10¹¹ Hz, B = C = 1.33 × 10¹⁰ Hz) using rigorous treatment (a), independent rotor approximation (b) and classical approximation (equation 4.54) (smooth curve (c)).

the energy (cf. M for a symmetric top). Thus the energy levels of the symmetric top with its rotations separated are given by

$$E_{J,J',K} = BhJ(J + 1) + AhK^2$$
$$J = 0, 1, 2, \ldots, \qquad K = 0, 1, 2, \ldots$$
$$(J' = 0, \pm1,, \pm2, \ldots, \pm J)$$

(4.59)

and the degeneracy of a given level is again given by (4.58). The resulting $W(E_r)$ curve is plotted as curve b in Figure 4.6 for the methyl chloride molecule. It is clear that although the assumption of separability causes some slight redistribution of the energy levels, there is no significant systematic disturbance of the $W(E_r)$ curve. Current and Rabinovitch[39] reached similar conclusions for another symmetric top (having different moments of inertia) and also for the separation of an internal rotation from the overall rotation about the same axis. It thus appears that separability of rotations is a sufficiently accurate simplification for present purposes.

In order to obtain analytical results for the independent-rotor model, the approximations were introduced that $J(J + 1) \approx J^2$ and $2J + 1 \approx 2J$, and the summation (4.51) was replaced by the integration (4.53) (and similarly in derivation of the expression for Q_r). The accuracy of these approximations is illustrated for the present case by the smooth curve c in Figure 4.6

$$W(E_r) \equiv \sum_{E_r=0}^{E_r} P(E_r) = \left(\frac{8\pi^2 I_A}{h^2}\right)^{1/2} \left(\frac{8\pi^2 I_B}{h^2}\right) E_r^{3/2}$$

(4.60)

This curve, calculated from (4.60), is a good representation of the actual quantised behaviour even at very low rotational energies. Whitten and Rabinovitch[28,40] similarly checked the corresponding equation for the vibrational–rotational case against exact counts for some representative cases, and found only small discrepancies (generally less than 2% at energies above $1200 \, \mathrm{J \, mol^{-1}}$). Bearing in mind that the average rotational energy in three classical rotational degrees of freedom is $\frac{3}{2}RT = 3600 \, \mathrm{J \, mol^{-1}}$ at room temperature, it is clear that the classical approximation is generally adequate for rate calculations.

There is of course the wider question of whether the assumptions of rigid rotors and free rotations are adequate, and we shall return to this problem in Chapter 6.

SYMMETRY NUMBERS

Rotational symmetry reduces the sums and densities of rotational states and its effects must be recognised in direct count procedures. The effect arises because of the need for the molecule to conform to the Pauli principle when a rotational symmetry operation leads to an indistinguishable arrangement of the atoms in the molecule, e.g. the C_2 operation in $^{35}Cl^{35}Cl$. For homonuclear diatomic molecules, the total number of states is reduced by a factor of 2 and this is recognised in the partition function by the symmetry factor, $\sigma = 2$. For most homonuclear diatomic molecules, the effect on the sums and densities of states can similarly be accommodated by dividing the result by σ. For H_2 at low temperatures, however, where the rotational energy separations are comparable to kT, this simple procedure leads to errors. Instead, the nuclear spin weighting of each J state should be incorporated into the direct count procedure. This weighting is $\frac{1}{4}$ for even levels and $\frac{3}{4}$ for odd levels. A clear discussion of the origins of these weightings for H_2 may be found in ref. 41, while a discussion of polyatomic molecules is given in ref. 42. Table 4.6 lists the σ values for the common point groups; once again, the true sums and densities are readily calculated by dividing $W(E_r)$ and $\rho(E_r)$ by σ, a procedure that is only likely to lead to errors for hydrides at very low temperatures.

The division of the sums and densities of states by σ should always be employed in the master equation calculations described in Chapter 7. It should not be applied in RRKM calculations of the microcanonical rate constant if the path degeneracy, $L^{\ddagger}$, has been used, as suggested in section 3.9. In effect, the path degeneracy is equal to the inverse ratio of the symmetry numbers in the transition state, $A^{\ddagger}$ and in the molecule A:

$$L^{\ddagger} = (\sigma^{\ddagger}_A/\sigma_A)^{-1}$$

Thus, for the reaction

$$CH_4 \rightarrow (CH_3 - H)^{\ddagger} \rightarrow CH_3 + H$$

$\sigma_A = 12$, $\sigma^{\ddagger}_A = 3$ and $L^{\ddagger} = 4$. We can, therefore continue to correct for the effects of symmetry in our calculations of $\rho(E_r)$ and $W(E_r)$, exactly as described above, provided we omit $L^{\ddagger}$ in our calculation of $k_a(E_{vr})$. If, on

Table 4.6 Symmetry numbers for common point groups

Point group	σ
$C_{\infty v}$	1
$D_{\infty h}$	2
C_s	1
C_2, C_{2v}, C_{2h}	2
C_3, C_{3v}, C_{3h}	3
D_2, D_{2d}, D_{2h}	6
T, T_d	12
O_h	24

the other hand, we prefer to use $L^{\ddagger}$, then the symmetry corrections must not be made in $\rho(E_r)$ and $W(E_r)$ although they must be included in calculating numbers of states in the energy grains employed in the master equation described in Chapter 7.

INTERNAL ROTATIONS

Internal rotations present problems because of the difficulty of expressing the energies of the quantum states analytically. For a molecule such as ethane, the potential function may be expressed in the form

$$V(\theta) = \tfrac{1}{2}V_0(1 - \cos 3\theta) \tag{4.61}$$

where θ is the rotation angle and V_0 the barrier height. At high energies, $V_0 \ll E$ and the system approximates to a free internal rotor with sums and densities of states given by

$$W(E) = (8I_m E/\hbar^2)^{1/2} \tag{4.62}$$

and

$$\rho(E) = (2I_m/\hbar^2 E)^{1/2} \tag{4.63}$$

in the classical limit; I_m is the moment of inertia of the rotor relative to the molecular frame. For two symmetrical coaxial tops with moments of inertia, I_A and I_B, $I_m = I_A I_B/(I_A + I_B)$.

At low energies ($V_0 \gg E$), the potential reduces to a quadratic in θ and a torsional vibration treatment is appropriate, with

$$W(E) = \{W(E)\}_{\text{free}}(E/4V_0)^{1/2} \tag{4.64}$$

Troe[43] has argued that the transition from a hindered to a free rotor is relatively abrupt and that equation (4.62) can be used for $E > V_0$ and equation (4.64) for $E < V_0$. Troe then derived semiclassical expressions for the sums and densities of states for an internal rotor coupled to s harmonic oscillators:

$$W(E) \approx \left(\frac{2I_m}{\hbar^2 V_0}\right)^{1/2} [E + aE_z]^{s+1} \left\{\Gamma(s)\prod_{i=1}^{s}(h\nu_i)\right\}^{-1} \tag{4.65}$$

and

$$\rho(E) = \rho^{\text{free}}(E)\frac{\Gamma(s + \tfrac{1}{2})}{2\Gamma(\tfrac{3}{2})\Gamma(s + 1)}\left(\frac{E + aE_z}{V_0}\right)^{1/2} \tag{4.66}$$

for $E + aE_z < V_0$, where

$$\rho^{\text{free}}(E) = \left(\frac{8I_m}{\hbar^2}\right)^{1/2}[E + aE_z]^{s-1/2}\Gamma(\tfrac{3}{2})\left\{\Gamma(s + \tfrac{1}{2})\prod_{i=1}^{s}(h\nu_i)\right\}^{-1} \quad (4.67)$$

which corresponds to the total density of states of s harmonic oscillators coupled to one free internal rotor.

For $E + aE_z > V_0$

$$\rho(E) \approx \left(\frac{8I_m}{\hbar^2}\right)[E + aE_z]^{s+1}\left\{\Gamma(s + \tfrac{3}{2})\prod_{i=1}^{s}(h\nu_i)\right\}^{-1}$$

$$\times \left[1 - \sum_{v=0}^{s-1}\frac{\Gamma(s + \tfrac{3}{2})(v + \tfrac{5}{2})}{\Gamma(v + 1)\Gamma(s - v)\Gamma(\tfrac{3}{2})2(v + 2)(v + \tfrac{3}{2})}\right.$$

$$\times (-1)^v\left(\frac{V_0}{E + aE_z}\right)^{v+3/2}\Bigg] \quad (4.68)$$

and

$$\rho(E) \approx$$

$$\rho^{\text{free}}(E)\left[1 - \sum_{v=0}^{s-2}\frac{\Gamma(s + \tfrac{1}{2})(v + \tfrac{5}{2})(-1)^v}{\Gamma(v + 1)\Gamma(s - 1 - v)\Gamma(\tfrac{3}{2})2(v + 2)(v + \tfrac{3}{2})}\left(\frac{V_0}{E + aE_z}\right)^{v+3/2}\right]$$

$$(4.69)$$

These expressions are quite cumbersome and Troe proposed simplified expressions for high and low energies:

$$\rho(E) \approx \rho^{\text{free}}(E)[1 - \exp(-E/sV_0)], \quad E/V_0 > 3 \quad (4.70)$$

$$\rho(E) \approx \rho^{\text{free}}(E)[1 - \exp\{-[\Gamma(s + \tfrac{1}{2})/2\Gamma(\tfrac{3}{2})\Gamma(s + 1)](E/V_0)^{1/2}\}],$$

$$E/V_0 < 3 \quad (4.71)$$

This latter expression also, correctly, approaches $\rho^{\text{free}}(E)$ as E/V_0 becomes large; it is, however, less accurate than equation (4.70) at these high energies.

It is also appropriate here to give expressions for the partition functions for a torsion, a free internal rotor and for a hindered rotor:

$$Q_{\text{rotint}}^{\text{tors}} = [1 - \exp(-h\nu_{\text{tors}}/kT)]^{-1} \quad (4.72)$$

where

$$\nu_{\text{tors}} = n\sqrt{\frac{V_0}{8\pi^2 I_m}}$$

and n is the number of minima in the potential (e.g. $n = 3$ for ethane).

$$Q_{\text{rotint}}^{\text{free}} = \left(\frac{2\pi I_{\text{m}} kT}{\hbar^2} \right)^{1/2} \tag{4.73}$$

A reasonable approximation for the hindered rotor is

$Q_{\text{rotint}} \approx$

$$Q_{\text{rotint}}^{\text{free}} \left\{ \left[1 - \exp\left(-\frac{kT}{V_0} \right) \right]^{1/2} + \frac{\exp\left(-1.2kT/V_0 \right)}{Q_{\text{rotint}}^{\text{free}} \left\{ 1 - \exp\left[-\frac{n}{Q_{\text{rotint}}^{\text{free}}} \sqrt{\frac{\pi V_0}{kT}} \right] \right\}} \right\}$$

$$\tag{4.74}$$

4.6   COMBINED VIBRATIONAL ROTATIONAL SYSTEMS

In section 4.1 it was shown how vibrational and rotational degrees of freedom could be separated and thus treated independently of each other. In any system where active rotors are present then state densities must be combined. The key to this for $W(E_{\text{vr}})$ is equation (4.2), and for $\rho(E_{\text{vr}})$ is equation (4.4) and its continuous equivalent equation (4.5), all of which are convolutions of the appropriate rotational and vibrational functions.

The necessary convolution for the state density is achieved directly if a modification of the SR algorithm (see section 4.2) is employed. Effectively the SR algorithm convolutes together the state densities of individual degrees of freedom. In the case of oscillators there is only one state at a specified energy and so no complications arise from degeneracies. This is not the case with rotation which for a symmetric top (equations 4.57) does have degenerate levels.

This rotational degeneracy can be easily accommodated as follows: instead of initialising the arrays T and AT such that all elements are zero, except the first which is set to one, as in step (2) of the SR algorithm, the arrays are initialised with the state densities of the rotational degrees of freedom. For the case of a symmetric top, each energy level is calculated and reduced to an integer which is then used to identify the particular element of arrays T and AT to which it corresponds. The degeneracy of the level is then added to that array element. This process is repeated for all rotational energy levels which are less than or equal to the specified maximum in energy as prescribed in step (1) of the SR algorithm. The result is that the elements of arrays T and AT now contain the approximate rotational state densities. The algorithm then proceeds as before.

Figure 4.7 shows a plot of $\log \rho(E_{\text{vr}})$ versus E_{vr} for CH_3Cl, calculated using the modified SR algorithm. The input data are given in Table 4.7. The

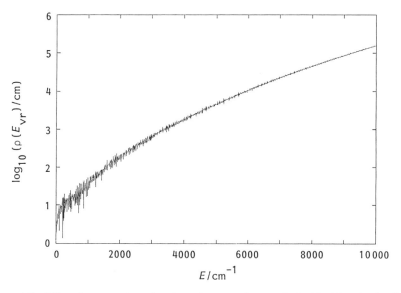

Figure 4.7 Vibration–rotation density of states for methyl chloride obtained by using the SR algorithm with a grain size of 10 cm^{-1}.

Table 4.7 Input data for the modified SR algorithm for CH$_3$Cl

---

Vibrational frequencies/cm^{-1}
    3044, 3044, 2968, 1488, 1488, 1355, 1018, 1018, 733
Rotational constants/cm^{-1}
    $A = 5.10$ cm^{-1}, $B = 0.443$ cm^{-1}

---

original SR[3,4] papers should be consulted for further examples. The algorithms are very efficient and can deal with large numbers of vibrational modes. Some care is needed in the rounding of frequencies or energies especially if large grains are used and the problem has been discussed by Stein and Rabinovitch[44].

WHITTEN–RABINOVITCH TREATMENT OF VIBRATIONAL–ROTATIONAL SYSTEMS

Whitten and Rabinovitch[40] extended their semi-classical treatment of vibrational state densities to incorporate rotation. They suggested,

$$W(E_{\mathrm{vr}}) = \frac{Q_{\mathrm{r}}}{(kT)^{\frac{1}{2}r}\Gamma(1 + s + \frac{1}{2}r)\Pi h\nu_i}(E_{\mathrm{vr}} + aE_z)^{s+\frac{1}{2}r} \qquad (4.75)$$

in which $a = 1 - \beta_R(E_{vr}/E_z)$, and β_R is extended from (4.33), now becoming

$$\beta_R = \left(\frac{s-1}{s}\right)\left(\frac{s + \frac{1}{2}r}{s}\right)\frac{\langle v^2\rangle}{\langle v\rangle^2} \tag{4.76}$$

The expression (4.75) reduces exactly to (4.31) if $r = 0$ and to (4.54) if $s = 0$ (and thus $E_z = 0$), but is otherwise an approximation of the true sum obtained from these two expressions. Whitten[28] obtained (4.75) by replacing the lower limit in the integral (4.77) by zero and replacing (4.33) by (4.76)

$$\int_{x=aE_z/(E_{vr}+aE_z)}^{1} x^{s-1}(1 - x)^{\frac{1}{2}r}\,dx \tag{4.77}$$

It is also obtained by treating the r rotational degrees of freedom with partition function Q_r in the same way as $\frac{1}{2}r$ vibrations with the same average zero-point energy as the genuine vibrations of the molecule, and with the product hv_i set equal to $(kT)^{1/2}/Q_r$. The change in the definition of β_R was found to give improved results in some test cases. The density of vibrational–rotational states is similarly obtained as (4.78) by differentiation of (4.75).

$$\rho(E_{vr}) = \frac{Q_r(E_{vr} + aE_z)^{s+\frac{1}{2}r-1}}{(kT)^{\frac{1}{2}r}\Gamma(s + \frac{1}{2}r)\Pi hv_i}\left[1 - \beta_R\left(\frac{dw}{dE'}\right)\right] \tag{4.78}$$

The resulting equations (4.75) and (4.78) have been tested in a number of cases[23,40], and appear to be good approximations for a wide range of models. Forst and Prášil[15] found a slightly modified version of (4.78) to be equally satisfactory.

RESULTS FOR $\rho(E_{vr}^*)$, $W(E_{vr}^+)$, AND k_{uni}

Finally, the results required for the density and sum of states used in RRKM theory, using the classical treatment of the active rotations, are now obtained by simple substitution of (4.56) and (4.54) into (4.4) and (4.2) respectively, and restoring the appropriate nomenclature; the results are given as equations (4.79) and (4.80).

$$\rho(E_{vr}^*) = \frac{Q_r^*}{(kT)^{\frac{1}{2}r}\Gamma(\frac{1}{2}r)}\int_{E_v^*=0}^{E^*}(E_{vr}^* - E_v^*)^{\frac{1}{2}r-1}\rho_v(E_v^*)\,dE_v^* \tag{4.79}$$

$$W(E_{vr}^+) = \frac{Q_r^*}{(kT)^{\frac{1}{2}r}\Gamma(1 + \frac{1}{2}r)}\sum_{E_v^+=0}^{E^+}[(E_{vr}^+ - E_v^+)^{\frac{1}{2}r}P(E_v^+)] \tag{4.80}$$

Finally, the expression for k_{uni} assuming classical active rotations is obtained by substituting (4.79) and (4.80) into the general expression (3.17) for k_{uni}.

The result is (4.81), in which $k_a(E_0 + E_{vr}^+)$ is given by (4.82) with (4.79) again substituted for $\rho(E_0 + E_{vr}^+)$.

$$k_{uni} = \frac{L^{\ddagger} Q_1^+ Q_r^+ \exp(-E_0/kT)}{h Q_1 Q_2 (kT)^{\frac{1}{2}r} \Gamma(1 + \frac{1}{2}r)}$$

$$\times \int_{E_{vr}^+=0}^{\infty} \frac{\left\{ \sum_{E^+=0}^{E_{vr}^+} (E_{vr}^+ - E_v^+)^{\frac{1}{2}r} P(E_v^+) \right\} \exp(-E_{vr}^+/kT)}{1 + k_a(E_0 + E_{vr}^+)/k_2[M]} dE_{vr}^+ \qquad (4.81)$$

$$k_a(E_0 + E_{vr}^+) = \frac{L^{\ddagger} Q_1^+}{Q_1} \frac{Q_r^+}{(kT)^{\frac{1}{2}r} \Gamma(1 + \frac{1}{2}r)} \frac{\sum_{E_v^+=0}^{E_{vr}^+} (E_{vr}^+ - E_v^+)^{\frac{1}{2}r} P(E_v^+)}{h\rho(E_0 + E_{vr}^+)}$$

$$(4.82)$$

The equations (4.81) and (4.82) are identical with those of Marcus[1], except for the treatment of reaction path degeneracy by the statistical factor $L^{\ddagger}$ and a minor point concerned with the possible existence of 'inactive' vibrations (see section 3.2).

When there are active rotations (4.79) can be integrated and (4.80) summed using values of the sum and density of the vibrational states derived by methods discussed in sections 4.2–4.4. The summation and integration must normally be carried out by numerical methods, although an analytical integration of (4.79) may occasionally be possible. This method of combining the distributions of vibrational and rotational states provides an exact result for the classical rotational model if the exact count of vibrational states is used (see section 4.2).

REFERENCES

1. R. A. Marcus, *J. Chem. Phys.*, **20**, 359 (1952).
2. T. Beyer and D. F. Swinehart, *Commun, Assoc. Comput. Machin.*, **16**, 372 (1973).
3. S. E. Stein and B. S. Rabinovitch, *J. Chem. Phys.*, **58**, 2438 (1973).
4. S. E. Stein and B. S. Rabinovitch, *J. Chem. Phys.*, **60**, 908 (1974).
5. J. M. Hammersley and D. C. Handscomb, *Monte Carlo Methods*, Chapman and Hall, London (1964).
6. J. P. Valleau and S. G. Whittington, *Modern Theoretical Chemistry*, Vol. 5, ed. B. J. Berne, Plenum Press, New York (1971).
7. J. R. Barker, *J. Phys. Chem.*, **91**, 3849 (1987).
8. J. D. Doll, *Chem. Phys. Letters*, **72**, 139 (1980).
9. S. J. Klippenstein and R. A. Marcus, *J. Chem. Phys.*, **87**, 3410 (1987).
10. S. J. Klippenstein and R. A. Marcus, *J. Phys. Chem.*, **92**, 5412 (1988).

11. R. A. Marcus and O. K. Rice, *J. Phys. and Colloid Chem.*, **55**, 894 (1951).
12. G. Z. Whitten and B. S. Rabinovitch, *J. Chem. Phys.*, **38**, 2466 (1963).
13. R. C. Tolman, *Foundations of Statistical Mechanics*, Oxford University Press, p. 492 (1938); J. C. Jaeger and G. H. Newstead, *An Introduction to The Laplace Transformation*, 3rd edn, Methuen, London (1969); G. A. Korn and T. M. Korn, *Mathematical Handbook for Scientists and Engineers*, McGraw-Hill, New York, (1961).
14. M. Vestal, A. L. Wahrhaftig and W. H. Johnston, *J. Chem. Phys.*, **37**, 1276 (1962); see comments in footnote 8, ref. 15.
15. W. Forst and Z. Prášil, *J. Chem. Phys.*, **51**, 3006 (1969).
16. H. Jeffreys and B. S. Jeffreys, *Method of Mathematical Physics*, 3rd edn, Cambridge University Press, p. 468 (1962).
17. J. C. Slater, *Quantum Theory of Matter*, McGraw-Hill, New York (1951).
18. C. S. Elliott and H. M. Frey, *Trans. Faraday Soc.*, **62**, 895 (1966); H. M. Frey and B. M. Pope, *Trans. Faraday Soc.*, **65**, 441 (1968).
19. B. S. Rabinovitch and R. W. Diesen, *J. Chem. Phys.*, **30**, 735 (1959).
20. B. S. Rabinovitch and J. H. Current, *J. Chem. Phys.*, **35**, 2250 (1961).
21. F. W. Schneider and B. S. Rabinovitch, *J. Amer. Chem. Soc.*, **84**, 4215 (1962).
22. R. W. Diesen, Ph.D. thesis, University of Washington (1958).
23. D. C. Tardy, B. S. Rabinovitch and G. Z. Whitten, *J. Chem. Phys.*, **48**, 1427 (1968).
24. E. W. Schlag and R. A. Sandsmark, *J. Chem. Phys.*, **38**, 2466 (1963).
25. J. C. Tou and A. L. Wahrhaftig, *J. Phys. Chem.*, **72**, 3034 (1968); J. C. Ton, L. P. Hills and A. L. Wahrhaftig, *J. Chem. Phys.*, **45**, 2129 (1966).
26. P. C. Haarhoff, *Mol. Phys.*, **6**, 337 (1963).
27. E. Thiele, *J. Chem. Phys.*, **39**, 3258 (1963).
28. G. Z. Whitten, Ph.D. thesis, University of Washington (1965).
29. M. J. Pearson, B. S. Rabinovitch and G. Z. Whitten, *J. Chem. Phys.*, **42**, 2470 (1965).
30. J. C. Tou, *J. Phys. Chem.*, **71**, 2721 (1967).
31. E. W. Schlag, R. A. Sandsmark and W. G. Valance, *J. Phys. Chem.*, **69**, 1431 (1965).
32. P. C. Haarhoff, *Mol. Phys.*, **7**, 101 (1963).
33. W. Forst, Z. Prášil and P. St Laurent, *J. Chem. Phys.*, **46**, 3736 (1967).
34. H. Eyring, D. Henderson, B. J. Stover and E. M. Eyring, *Statistical Mechanics and Dynamics*, Wiley, New York, 1964, p. 15.
35. T. M. Sugden and C. N. Kenney, *Microwave Spectroscopy of Gases*, Van Nostrand, London, p. 15 (1965).
36. *Ibid.*, p. 18.
37. J. E. Wollrab, *Rotational Spectra and Molecular Structure*, Academic Press, New York, 1967, p. 13.
38. *Ibid.*, p. 15; T. M. Sugden and C. N. Kenney, *Microwave Spectroscopy of Gases*, Van Nostrand, London, p. 48 (1965).
39. J. H. Current and B. S. Rabinovitch, *J. Chem. Phys.*, **38**, 783 (1963).
40. G. Z. Whitten and B. S. Rabinovitch, *J. Chem. Phys.*, **41**, 1883 (1964).
41. P. W. Atkins, *Molecular Quantum Mechanics*, 2nd edn, Oxford University Press, Oxford, p. 295 (1983).
42. G. Herzberg, *Molecular Spectra and Molecular Structure. II Infrared and Raman Spectra of Polyatomic Molecules*, Van Nostrand, New York, p. 406 (1945).
43. J. Troe, *J. Chem. Phys.*, **66**, 4758 (1977).
44. S. E. Stein and B. S. Rabinovitch, *Chem. Phys. Letters*, **49**, 183 (1977).

5 Numerical Application of the RRKM Theory

This chapter illustrates the steps involved in the numerical application of the 'Standard' RRKM theory to a specific example of a unimolecular reaction involving a tight transition state, namely the isomerisation of 1,1-dichloro-cyclopropane.

The general procedure in calculating the high-pressure activation parameters ($\Delta S^{\ddagger}$, A_∞, E_∞, E_0) from any postulated model of the reaction is first discussed in section 5.1 then the criteria for selection of a model follows in section 5.2 and discussion of the integration procedure in section 5.3. Section 5.4 contains the detailed treatment of the 1,1-dichlorocyclopropane isomerisation and section 5.5 illustrates the sensitivity of the calculated results to various parameters involved.

Discussion of the results of recent calculations using RRKM and other theories are compared with experiment in selected examples in Chapter 11.

5.1 CALCULATION OF ACTIVATION PARAMETERS FOR A POSTULATED MODEL OF THE REACTION

Before considering in detail the selection of a model for the reaction, it will be useful to consider how the postulated numerical properties of the reactant and transition state can be used to calculate the resulting activation parameters for the reaction. The high-pressure A-factor is particularly significant, and in fact the model is usually chosen so that the calculated A-factor agrees with the experimental value. The experimental high-pressure Arrhenius parameters are defined by (5.1) and (5.2). Since k_∞ is given in RRKM theory by the CTST expression (5.3) (cf. equation 3.21), it follows that the theoretical values of E_∞ and $\ln A_\infty$ are given by equations (5.4)–(5.7).

$$E_\infty = kT^2 \, \mathrm{d} \ln k_\infty / \mathrm{d}T \tag{5.1}$$

$$\ln A_\infty = \ln k_\infty + E_\infty/kT = \mathrm{d}(T \ln k_\infty)/\mathrm{d}T \tag{5.2}$$

$$k_\infty(\text{RRKM}) = L^\ddagger \frac{kTQ_1^+Q_2^+}{hQ_1Q_2} \exp\left(-E_0/kT\right) \tag{5.3}$$

where Q_1 and Q_1^+ are the partition functions of the adiabatic degrees of freedom of A and A^+ and Q_2 and Q_2^+ are the partition functions for the active degrees of freedom.

$$E_\infty(\text{RRKM}) = E_0 + kT + kT^2 \, \mathrm{d}(\ln Q_1^+ Q_2^+/Q_1 Q_2)/\mathrm{d}T \tag{5.4}$$

$$= E_0 + kT + \langle E^+ \rangle - \langle E \rangle \tag{5.5}$$

$$\ln A_\infty(\text{RRKM}) = \ln\left(L^\ddagger ekT/h\right) + \mathrm{d}(T \ln Q_1^+ Q_2^+/Q_1 Q_2)/\mathrm{d}T \tag{5.6}$$

$$= \ln\left(ekT/h\right) + \Delta S^\ddagger/R \tag{5.7}$$

where $\Delta S^\ddagger$ defined by this equation incorporates the value of $L^\ddagger$ and hence symmetry contributions[1]. Equation (5.5) is derived from (5.4) using the standard statistical-mechanical equation

$$\langle E \rangle = kT^2 \, \mathrm{d}\ln Q/\mathrm{d}T \tag{5.8}$$

The term kT in (5.5) may be identified with the average translational energy of the transition states in the reaction coordinate, and $\langle E^+ \rangle$ is the average internal energy in their other degrees of freedom, relative to the vibrational–rotational ground state of the transition state. The quantity here denoted $\langle E \rangle$ is the average internal energy of all A molecules relative to their ground state. It has sometimes been denoted $\langle E^* \rangle$ and incorrectly called the average energy of the energised molecules, although the correct expression and value for $\langle E \rangle$ has been used. Equation (5.5) is therefore in accord with Tolman's classic theorem[2] that the high-pressure Arrhenius activation energy of a unimolecular reaction is the average internal energy of the species which are reacting, minus the average internal energy of all the reactant molecules present. Equation (5.7) follows from (5.6) by virtue of the statistical-mechanical equation,

$$S = R \, \mathrm{d}(T \ln Q)/\mathrm{d}T = R \ln Q + RT \, \mathrm{d}\ln Q/\mathrm{d}T \tag{5.9}$$

where Q is the total partition function for vibration and rotation. In reverse, (5.7) is the usual equation for the calculation of the entropy of activation from the experimental value of A_∞; convenient numerical versions of these equations are as follows:

$$\log\left(A_\infty/\text{s}^{-1}\right) = 10.753 + \log\left(T/\text{K}\right) + \Delta S^\ddagger/19.15 \, \text{J K}^{-1}\,\text{mol}^{-1}$$

$$\Delta S^\ddagger/\text{J K}^{-1}\,\text{mol}^{-1} = 19.15 \log\left[1.766 \times 10^{-11}(A_\infty/\text{s}^{-1})(T/\text{K})\right]$$

Since the statistical factor is excluded from the present equations, the corresponding value of $\Delta S^\ddagger$ is the overall entropy of activation. In some applications it is useful to divide $\Delta S^{\ddagger}$ into vibrational and rotational contributions;

$$\Delta S_v^{\ddagger} = S_v(A^+) - S_v(A) \quad \text{and} \quad \Delta S_r^{\ddagger} = S_r(A^+) - S_r(A)$$

The expressions giving the entropy are usually taken to be those for quantum harmonic oscillators and classical rotations, (5.10)–(5.13), together with (5.9) (where $r = \sum_i d_i$ and d_i is the degeneracy of the ith rotor).

$$Q_v = \prod_{i=1}^{s} [1 - \exp(-hv_i/kT)]^{-1} \tag{5.10}$$

$$T \, d \ln Q_v/dT = \sum_{i=1}^{s} \{(hv_i/kT)[\exp(hv_i/kT) - 1]^{-1}\} = \langle E_v \rangle/kT \tag{5.11}$$

$$Q_r = (8\pi^2 kT/h^2)^{(1/2)r} \prod_{i=1}^{p} I_i^{(1/2)d_i} \Gamma(\tfrac{1}{2}d_i) \tag{5.12}$$

$$T \, d \ln Q_r/dT = \tfrac{1}{2}r = \langle E_r \rangle/kT \tag{5.13}$$

The equation (5.12) for independent classical rotations is not always strictly applicable and care is needed to be at least consistent; see section 4.5 and Appendix 2.

The appropriate critical energy E_0 for a given model may be postulated empirically, but is more often calculated from the observed value of E_∞ using (5.4) or (5.5), (5.11) and (5.13). If the model is chosen to reproduce the experimental A factor, it will then automatically predict correctly the experimental values of k_∞ and its temperature dependence at the temperature T used in the above calculations.

It is clear from the form of (5.4)–(5.7) that there are an infinite number of models which will reproduce a given set of Arrhenius parameters. Thus the Arrhenius parameters do not uniquely specify the model to be used, and it is usual to consider several different models for a given reaction. Factors influencing the choice of these models are discussed in the following sections.

5.2 SPECIFICATION OF A MODEL FOR THE REACTION

In order to carry out the calculation of k_{uni} as a function of pressure it is first necessary to set up models of the molecule A and the transition state A^+. The necessary parameters for A are the details of its internal motion (i.e. its vibrations and internal rotations), its overall moments of inertia and its collision diameter. These data can always be obtained in principle from experimental measurements, and can often be found in the literature or can be estimated with reasonable certainty by empirical methods. The same parameters (except for collision diameter) must also be specified for the transition state, but here there is, as yet, no possibility of experimental determination. The information might in principle be obtained by a complete solution of the Schrödinger equation for the system, which would in fact give all the information required, including E_0.

TREATMENT OF ROTATIONAL DEGREES OF FREEDOM; 'RIGID' AND 'LOOSE' TRANSITION STATES

The first step in the selection of a model is to decide how many rotational degrees of freedom will be included in the models of A and A^+, and which of these degrees of freedom will be taken to be active, i.e. to exchange their energy freely with the other degrees of freedom of the molecule. The overall rotations of reactant and transition state are always included in principle, even if they are assumed to have no effect by putting $Q_1^+/Q_1 = 1$. A change in the number of internal rotational degrees of freedom on formation of the transition state is commonly postulated to produce a model consistent with an unusually high or low A-factor[3]. This is possible since the density of rotational states is usually much higher than the density of vibrational states at the relevant energies, and the associated entropy is correspondingly higher (e.g. 40–50 compared with 0–8 $J K^{-1} mol^{-1}$ per degree of freedom). It should be realised, however, that changing a rotational degree of freedom from adiabatic to active (without changing the number of rotations) will have no effect on the high-pressure A-factor for the reaction[4]. The partition function for the rotation will be unchanged, and will merely enter (5.3) in Q_2 or Q_2^+ instead of in Q_1 or Q_1^+. Thus the magnitude of A_∞ will give guidance as to a possible change in the number of rotational degrees of freedom, but not as to the activity or otherwise of the rotations.

When choosing the number of rotations to be included it may be helpful to consider whether the transition state is likely to be rigid or loose. A loose transition state is one in which there is some degree of free internal rotation which was not present in the reactant molecule. An example is the transition state often assumed for the decomposition of ethane into two methyl radicals; in the transition state the incipient radicals have already a substantial degree of rotational freedom. Such transition states are usually postulated to explain high A-factors, and are generally associated with bond-fission processes which produce free radicals or atoms.

Isomerisation reactions, on the other hand, such as the isomerisation of cyclopropane to propene, decompositions of molecules to produce other molecules such as the decomposition of ethyl chloride to give ethene and hydrogen chloride, or radical decompositions producing a molecule plus a further radical, are all said to occur via rigid or tight transition states.

These processes often involve the partial transfer of an atom or group to a different point in the molecule through a cyclic configuration. It is difficult to see how a transition state thus formed can have any new internal rotations. The differences between rigid and loose transition states are illustrated by the different appearance of the potential energy profiles along the reaction coordinate. In Figure 5.1, the potential energy profile for a reaction involving a rigid transition state (Figure 5.1(a)) shows a definite energy barrier (Type I), whereas that involving a loose transition state (Type II) (in Figure 5.1(b)) shows that freely rotating product fragments can form the transition state without any increase in potential energy and it is likely that some of this energy will be retained as internal rotation in the transition state.

In general, therefore, loose transition states will be plausible for reactions involving bond fission to give separated fragments, and the transition states for these reactions will have one or more internal rotational degrees of freedom. The reactant molecule may or may not have internal rotations, but it will have fewer than the transition state. Rigid transition states will usually be assumed for reactions involving the simultaneous making and breaking of bonds. In some cases (e.g. six-centre rearrangements) the A-factor may be so low as to suggest that the reactant molecule has internal rotations which are lost in the transition state.

Finally, the rotations must be classified as active or adiabatic. There is little theoretical guidance here, but the situation is fairly well defined empirically. All internal rotations appear to be active; this seems to be a reasonable postulate and has produced results in good agreement with experiments for a number of systems. The overall rotations are often treated as adiabatic and their effect taken into account as discussed in section 3.10. However, there may be Coriolis coupling between certain internal modes and a particular overall rotation which would justify taking this particular rotation as active[5]. Such effects have been found to be important in interpreting the results of scattering trajectory calculations[6].

ASSIGNMENT OF NUMERICAL PROPERTIES TO THE TRANSITION STATE

After a chemical structure has been postulated for the transition state, and the treatment to be given to the rotational degrees of freedom has been settled, values must be assigned to the relevant vibration frequencies and moments of inertia of the transition state. This could be done fairly accurately if the geometry of the transition state were known (the determination of the moments of inertia of a molecule of known geometry is a standard procedure)[7] and if a complete set of force constants for the transition state could be constructed. Although these aims cannot be met

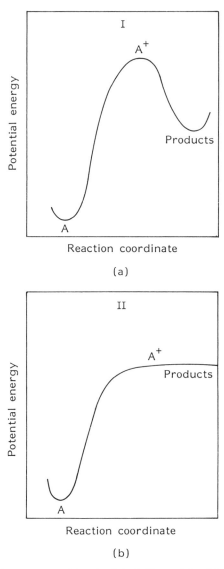

Figure 5.1 Potential curves resulting in (a) rigid or tight (Type I), and (b) loose (Type II) activated complexes. Reproduced by permission of John Wiley and Sons Ltd.

completely in practice, some progress can be made by consideration of analogous structures in normal molecules and free radicals, and by the use of semiempirical correlations of bond lengths and force constants with bond order. When all else fails, a purely empirical approach can be used (see next section) and this is frequently used as a final adjustment in conjunction with the more theoretically based methods.

The transfer of data from analogous molecules is useful provided that truly analogous structures can be found. It is well illustrated by the use of the known vibration frequencies of the CH_3 and CF_3 radicals for the essentially separated CX_3 fragments in the transition states for C_2H_6 and C_2F_6 dissociation. This leaves only the C–C distance to be specified, and this can be determined by a special procedure mentioned below. In other cases less complete assistance may be obtained in a similar way. For the isomerisation of methyl isocyanide, for example, the eight vibrations characteristic of the methyl group were assumed[8,9] to have the same frequencies in the transition state (I) as in the reactant molecule. Similarly the 12 CH_2 vibrations in the transition state (II) were assumed[10] to have the same frequencies as in the ethyl chloride molecule.

Vibrations of a transition state which are obviously different from those of the reactant molecule may nevertheless find approximate analogies in different molecules. For the transition state in reaction (5.14), for example, the B_1 ring deformation at 895 cm^{-1} was lowered to 650 cm^{-1} to correspond to a C–C=O bending vibration, and the ring-breathing vibration at 1266 cm^{-1} was raised to 1500 cm^{-1} to correspond to a C=O stretch[4].

$$(5.14)$$

When as much progress as possible has been made in the above manner there are usually some bond lengths and frequencies as yet undetermined. One way of estimating these, uses the empirical correlations of bond length and force constant with bond order which were developed by Badger[11], Pauling[12] and others, and have been summarised by Johnston[13] and Pauling[14]. For present purposes it is required to predict bond lengths and force constants from reasonably assigned bond orders, these usually being allocated on the assumption that the total bond order in the system (and sometimes at a given atom) is constant. This is not strictly true, of course, but the correlations can be adjusted to use the intuitive 'bond number' n rather than the MO theory bond order.[a] With this proviso the lengths R and stretching force constants F of bonds between a pair of atoms i and j are given by the following equations:

$$R = R_s - 0.60 \log n \tag{5.15}$$

$$F = F_s 10^{(R_s - R)/b_{ij}} = F_s n^{0.60/b_{ij}} \tag{5.16}$$

In these equations R_s and F_s are the length and force constant for the corresponding single bond, and b_{ij} is a constant determined by the rows of the periodic table in which atoms i and j occur; $b_{ij} = 0.60 \pm 0.05$ for most cases of interest. There are unfortunately no similarly well-established correlations for bending or torsional force constants, and the estimation of these has remained rather arbitrary[8,10,16,17].

To the extent to which the vibration of a given bond is independent of the rest of the molecule, its vibration frequency will be given by

$$v/v_s = \sqrt{(F/F_s)} \tag{5.17}$$

$$\text{or} \quad v = v_s n^{0.30/b_{ij}} \tag{5.18}$$

This simplification was used in the calculations on methyl isocyanide isomerisation[8,9], although the dependence of F on n was assumed in this case to follow Badger's original rule (5.19)[11],

$$F = a'_{ij}(R - b'_{ij})^{-3} \tag{5.19}$$

rather than (5.16); in fact these two equations give very similar values for F/F_s. Badger's rule and Pauling's equations are also said[17] to be the basis of the extensive tabulations by Benson and co-workers[17,18] of frequencies for the stretching and bending vibrations involving bonds of $\frac{1}{2}$ and $1\frac{1}{2}$ order. These values have been used with remarkable success in predicting the high-pressure A-factors of unimolecular reactions[17-19], and the transition state models thus derived have proved to be quite adequate for many fall-off calculations.

Although the simplification (5.17) or (5.18) can probably be used to derive vibration frequencies from the force constants without significant loss of accuracy, a more satisfying method involves the complete vibrational analysis of the transition state or of relevant parts of the transition state; in particular, the choice of reaction coordinate should emerge more naturally from this treatment. For example, in the calculations on ethyl chloride[10], detailed analyses were carried out for the ring structure (II) in which the

---

[a]For example, the bond number of the C–C bonds in benzene is 1.5, whereas[15] the bond order is 1.67, and the total bond order at each C atom is 4.33.

CH_2 groups were treated as the appropriate point masses. Various sets of bond numbers for the four bonds were assumed, e.g. 1.8, 0.8, 0.2, 0.2 for the C–C, C–Cl, Cl–H and H–C bonds respectively. The bond lengths were calculated from (5.15), apparently with 0.71 in place of the 0.60 although the latter is recommended[20], especially for bond numbers less than unity. The ring was assumed to be planar, with the HCC and CCCl angles equal, thus completely defining the geometry. Force constants were calculated from Badger's rule (5.19), and the five in-plane vibration frequencies of the ring were then calculated by the Wilson **FG** matrix method[21]. The normal mode in which the C–C and H–Cl distances decreased and the C–H and C–Cl distances simultaneously increased always had a very low frequency (e.g. 100 cm^{-1}) and was naturally chosen to represent the reaction coordinate. This left only the out-of-plane ring 'puckering' vibration frequency to be determined, and in the absence of any precedent this frequency was used as an adjustable parameter.

It is worth noting that Pauling's equation (5.15) and a similar correlation of bond energy with bond order for diatomic molecules form the basis of the bond energy–bond order (BEBO) method[22] for the empirical estimation of activation energies for some bimolecular transfer reactions.

One application of this method to a unimolecular reaction has involved the calculation of energy disposal in the unimolecular elimination of the species HX from an ion where X represents a hydrogen atom or a hetero-atom[23].

Special procedures exist for determining the location of the transition state and assigning its properties when loose transition states are involved and the potential energy profile is of Type II (Figure 5.1(b)). These matters are considered in some detail in Chapter 6.

EMPIRICAL APPROACH TO SPECIFICATION OF THE TRANSITION STATE

As has already been indicated, any attempt to deduce the properties of a transition state from first principles usually leaves some features undetermined, and such properties are often used as empirically adjustable parameters to make the model predict the correct value of A_∞. Many authors have made no attempt at all to deduce the properties of the transition state in the above ways, but have adopted a wholly empirical approach to the problem which is in fact useful, since it will be seen later that the calculated fall-off behaviour is surprisingly insensitive to the details of the model, providing it leads to the correct value of A_∞. In this approach, a plausible-looking transition state is constructed and the appropriate reaction-path degeneracy is selected. The partition function ratio Q_1^+/Q_1 for the adiabatic (overall) rotations is given some arbitrarily selected value, usually unity.

This is quite reasonable for a reaction proceeding through a rigid transition state because it will not be very different in size from the reactant molecule; since the moments of inertia enter as $(I_A^+ I_B^+ I_C^+ / I_A I_B I_C)^{1/2}$ quite substantial changes in molecular size would be required before a serious effect on Q_1^+/Q_1 resulted. Similarly, the simple treatment of the overall rotations by including the factor Q_1^+/Q_1 in k_{uni} is likely to be sufficiently accurate for such reactions (see section 3.10).

The entropy of activation is obtained from the experimental high-pressure A-factor using (5.7), and if necessary $R \ln(Q_1^+/Q_1)$ is subtracted to give the vibrational contribution $\Delta S_v^{\ddagger}$ to the entropy of activation. The vibration frequencies of the molecule are generally known, and the vibrational entropy $S_v(A)$ of the reactant at the selected reaction temperature can thus be calculated from the harmonic oscillator equations (5.9)–(5.11). It is now required to construct a transition state which has vibrational entropy $S_v(A^+) = S_v(A) + \Delta S_v^{\ddagger}$. One particular vibration of the molecule is chosen to represent the reaction coordinate and the corresponding frequency is removed, then the remaining frequencies are adjusted by trial and error in a largely empirical way to as to reproduce the required value of $S_v(A^+)$. At first sight this may appear involved in view of the complexity of (5.9)–(5.11). In fact a rapid convergence can be obtained by using in the first stage a coarse tabulation of harmonic oscillator entropies as a function of frequency and temperature and then a more precise tabulation as a function of the dimensionless quantity $h\nu/kT$. In the final stage one frequency or a number of frequencies can be adjusted to give a precise fit and this can be done by a simple computer program.

The detailed application of this approach to the isomerisation of 1,1-dichlorocyclopropane is described in section 5.4.

5.3   THE RRKM INTEGRATION

Figure 5.2 contains a general outline of the integration procedure; a detailed computer program is not given since there are no unusual difficulties. It is advantageous to have a number of possible treatments built into one program, and some of the alternatives are discussed in the present section.

The integration is carried out by a stepwise summation, and in view of the wide uncertainties involved in the model and its numerical parameters there is little to be gained by the use of a sophisticated integration procedure such as the Kutta–Merson method or even Simpson's rule. With machine integration it is easy in the present case to make the steps so small that a sufficiently accurate value for the integral is obtained from the simple summation (5.20) in which the summand for each step ΔE^+ has the value appropriate to the centre of the step. Equation (5.20) corresponds to (3.17)

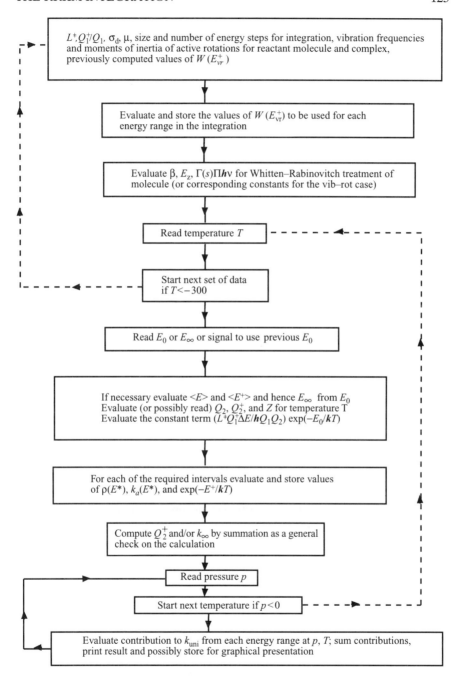

Figure 5.2   Flow chart for RRKM integration.

with $k_2[M]$ replaced by $\beta_c Z[M]$, where β_c is the collisional deactivation efficiency[24].

$$k_{uni} = \frac{L^{\ddagger}Q_1^+}{hQ_1Q_2} \exp(-E_0/RT)\Delta E^+ \sum_{i=1}^{i_{max}} \left[\frac{\{W(E_{vr}^+)\} \exp(-E^+/RT)}{1 + k_a(E^*)/\beta_c Z[M]} \right] \quad (5.20)$$

where

$$k_a(E^*) = L^{\ddagger}\frac{Q_1^+}{Q_1}\{W(E_{vr}^+)\}/h\rho(E^*)$$

$$E^* = E_0 + E^+, \qquad E^+ = (i + \tfrac{1}{2})\Delta E^+$$

and k_{uni} will be in s^{-1} if the energies E_0, E^+ and ΔE^+ are expressed in kJ mol^{-1} and the constants h and R are given the values

$$N_A h = 3.9903 \times 10^{-13} \text{ kJ mol}^{-1}\text{s}$$

$$R = 0.0083144 \text{ kJ mol}^{-1}\text{K}^{-1}$$

It is important in using equation (5.20) to take a self-consistent set of values for physical constants. The values used in the applications described in this chapter are taken from the IUPAC compilation[25], thus for example $N_A hc = 0.11963$ J m.mol^{-1}. Whenever the older units calories are used, these are 'thermochemical calories' and the conversion factor[26] is 1 cal $\equiv$ 4.184 J.

Adiabatic rotations are treated here by including the factor Q_1^+/Q_1; the extension to more refined treatments (see section 3.10) is obvious. In general a step length ΔE^+ of 0.4–2.0 kJ mol^{-1} seems to be small enough, and with machine computation it is simple enough to confirm this for a given case by checking the effect of varying the step length. The maximum energy to which the summation must be taken to obtain a sufficiently accurate result is also a matter for trial and error, and in any given case the question may be resolved by examining a print-out of the contributions to k_{uni} from the individual energy ranges. As a general guidance only, a maximum E^+ of $E_0 + skT$, where s is the number of active degrees of freedom involved, will probably give a result within 1% in typical cases.

Subroutines are conveniently used to evaluate and store values of $W(E_{vr}^+)$ and $\rho(E^*)$. Various methods of evaluating these important quantities have been fully discussed in Chapter 4. In one early RRKM computer program[27], $W(E_{vr}^+)$ is normally calculated by a direct count procedure and $\rho(E^*)$ by the Whitten–Rabinovitch approximation, although other options are available. Both $W(E_{vr}^+)$ and $\rho(E^*)$ are frequently calculated at the centre of the energy increment ΔE^+, thus in (5.20)

$$E^+ = (i + \tfrac{1}{2})\Delta E^+ \quad \text{where } i = 0, 1, 2, 3 \ldots$$

The ease of application of the SR algorithm (see Chapter 4) has led to its widespread use to calculate $\rho(E^*)$ in addition to $W(E_{vr}^+)$, and these calculations can likewise be delegated to a subroutine. Values of $k_a(E^*)$ are then computed from (5.20) and stored together with values of $\exp(-E^+/RT)$. Other quantities required in the evaluation of k_{uni} from (5.20) are the critical energy E_0, the partition functions Q_2 and Q_2^+ and the collision number Z.

The critical energy E_0 may be read in as such or calculated from E_∞ using (5.4), (5.11) and (5.13), with the given properties of the reactant and transition state. If a series of integrations at different temperatures is to be carried out it may be convenient to calculate E_0 from E_∞ at one temperature then use the same value of E_0 at the other temperatures. Of course E_0 is a true constant, and E_∞ varies with temperature according to (5.4). But the variation of E_∞ with temperature is experimentally inaccessible, and if the constant experimental value were used in (5.4) at different temperatures an apparent variation in E_0 would be created. It is thus useful to have these alternative sources of E_0 built into the program.

The partition functions Q_2 and Q_2^+ are obtained using (5.10) and (5.12). Previously calculated values could be read in, but since the frequencies and moments of inertia for at least the reactant are needed in the integration program (for $\rho(E^*)$) it is convenient to calculate Q_2 and Q_2^+ by subroutine in the same program. The collision number Z is given as a function of temperature by

$$Z = \sigma_d^2 N_A \left(\frac{8\pi RT}{\mu} \right)^{\frac{1}{2}} \qquad (5.21)$$

Here Z is given in molar units by this equation $(m^3\,mol^{-1}\,s^{-1})$ when the collision diameter σ_d is expressed in metres and the reduced mass in $kg\,mol^{-1}$. Note that the factor $\frac{1}{2}$ which normally appears before the right-hand side of this equation when it refers to like molecules is omitted here, since even when $M \equiv A$ we are concerned with collisions between two distinguishable entities A and A*. Since $M_A^* = M_A$, however, μ becomes $M_{A/2}$ in this case. Use of the SI units throughout equation (5.21), then provides a self-consistent set of units leading to the desired result.

Before actually carrying out the integration it is customary and worth while to perform a check on the setting up of the calculation by evaluating Q_2^+ and/or k_∞ by summation using the stored functions as above, and comparing the results with the accurately known values from the analytical expressions (5.10), (5.12) and (5.3). The approximate values are given by

$$Q_2^+ \text{ (by summation)} = \frac{\Delta E^+}{RT} \sum_{i=1}^{i_{max}} \{\Sigma P(E_{vr}^+)\}_i \exp(-E_i^+/RT) \qquad (5.22)$$

$$k_\infty \text{ (by summation)} = \frac{L^\ddagger Q_1^+}{h Q_1 Q_2} \exp\left(-E_0/\mathbf{R}T\right)\Delta E^+$$

$$\times \sum_{i=1}^{i_{max}} \{\Sigma P(E_{vr}^+)\}_i \exp\left(-E_i^+/\mathbf{R}T\right) \qquad (5.23)$$

These should closely reproduce the values from the exact expressions; it was shown in section 3.7 that Q_2^+ is given exactly by the integral to which the summation (5.22) is an approximation.

The integration can now proceed; for each pressure p the summand in (5.20) is evaluated for each energy range and the sum is formed. It is useful to have the option of printing out a selection of the contributions of the individual steps to the overall result, since this can give a further check that the integration is being taken to sufficiently high energy and can also give interesting information about the average energy of the molecules which are reacting at various pressures. In addition to printing out the results a graphical display is a useful adjunct and may perhaps include experimental data for direct comparison. (An output of this type is reproduced in Figure 5.3.)

5.4 APPLICATION OF RRKM THEORY TO THE ISOMERISATION OF 1,1-DICHLOROCYCLOPROPANE

As an example of the application of RRKM theory, a set of calculations based on those given in ref. 28 for the isomerisation of 1,1-dichlorocyclopropane to 2,3-dichloropropene will now be described in detail. The overall reaction

$$(5.24)$$

is homogeneous, first-order and unaffected by the presence of free-radical inhibitors, and has been concluded[29] to be a unimolecular reaction. The high-pressure Arrhenius equation (5.25) was established from studies at 20–120 Torr and 615–714 K, and

$$\log k_\infty/s^{-1} = 15.13 - 241.9 \text{ kJ mol}^{-1}/2.303\mathbf{R}\text{T} \qquad (5.25)$$

experimental data for the fall-off at pressures down to 0.2 Torr were obtained in a separate series of studies[28] at 697.6 and 632.4 K.

The reaction is assumed to proceed through a transition state of the type shown in (5.24), for which the statistical factor $L^\ddagger$ is 4 (see section 3.9). The

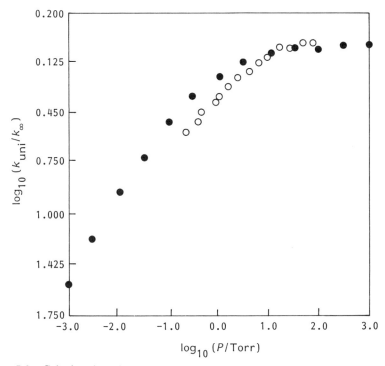

Figure 5.3 Calculated and experimental fall-off curves for 1,1-dichlorocyclopropane. Open circles, experimental data.

approach used for setting up models of the transition state was the empirical method of section 5.2, and Q_1^+/Q_1 was accordingly taken to be unity. It was assumed that there were no active rotations, and the vibrational contribution to the entropy of activation, calculated from (5.7), was $\Delta S_v^{\ddagger} = 17.85 \, \text{J K}^{-1} \text{mol}^{-1}$ per path at the selected temperature of 697.6 K. The vibrational entropy of the reactant molecules at 697.6 K, calculated from equations (5.9)–(5.11) with the frequency assignment shown in Table 5.1, was $S_v(A) = 95.66 \, \text{J K}^{-1} \text{mol}^{-1}$, and the required value of $S_v(A^+)$ was therefore $S_v(A) + \Delta S_v^{\ddagger} = 113.5 \, \text{J K}^{-1} \text{mol}^{-1}$. In one of the transition states considered[28] the reaction coordinate was taken to be the CCl_2 anti-symmetric stretching vibration at $717 \, \text{cm}^{-1}$. The remaining 20 frequencies were grouped as shown in Table 5.1, and a computer program was used to lower the frequencies of all but the C–H stretches in proportion to one another until $S_v(A^+)$ reached the required value with the frequencies shown in the third column. This model now reproduces the experimental value of A_∞, and with its derived value of E_0 (see section 5.1 and below) reproduces k_∞ and its temperature variation correctly. Other models for the same reaction will be considered in section 5.5.

Table 5.1 Vibration frequencies and vibrational entropies of 1,1-dichloro-cyclopropane and a postulated transition state

Vibrations of reactant molecule[a]		Corresponding vibrations of transition state[b]	
Frequency/ (cm^{-1})	S_v(697.6 K)/ J K^{-1} mol^{-1}	Frequency/cm^{-1} and (degeneracy)	S_v(697.6 K)/ J K^{-1} mol^{-1}
3106, 3096 3048, 3022	0.435	3068 (4)	0.435
1454, 1409 1292, 1238	8.322	1346 (4)	8.272
1164, 1130 1037, 952	13.06	1068 (4)	12.99
874, 825, 772 500, 443	32.32	445 (5)	46.57
717	5.782	Reaction coordinate	
404, 300, 272	35.74	216.8 (3)	45.22
Total vibrational entropy/ J K^{-1} mol^{-1}	[S_v(A)] 95.66		[S_v(A$^+$)] 113.5 (assuming $L^{\ddagger} = 1$, i.e. per path)
Overall vibrational entropy	95.66		125.0 ($L^{\ddagger} = 4$)

[a]The normal vibration frequencies are those quoted by Kay[30] and are not correct in detail,[31] although the errors will have little effect on the RRKM calculation and the data provide a perfectly valid illustration of the approach under discussion.
[b]Designated 'transition state B' in ref. 28 and section 5.5.

The RRKM theory can now be applied to predict the fall-off behaviour for this model. Considerable detail of the calculation will be given, to illustrate a number of points which need attention and to provide a comprehensive set of data against which the reader's own computer programs may be checked. The basic parameters are shown in Table 5.2. In this illustration there are 850 intervals for the integration (see comments in section 5.5), and the relevant functions for each energy range will be tabulated only for a few representative intervals. Values of $W(E_v^+)$ computed by direct count are listed at some corresponding energies in Table 5.3 together with the linearly interpolated values at the centre of each range. Selection of the required temperature (697.6 K) then permits calculation of the parameters in Table 5.4. In a similar way a selection of values of $\rho(E^*)$ and $k_a(E^*)$ at the centre

Table 5.2 Basic parameters

$L^{\ddagger} = 4$ $Q_1^+/Q_1 = 1$ $\sigma_d = 0.65$ nm $\mu = M_A/2 = 55.5 \times 10^{-3}$ kg mol^{-1}

Vibration frequencies of molecule and transition state from Table 5.1

Calculated values: $\beta_R = 1.4454$, $E_z = 161.8$ kJ mol^{-1}

Integration up to $E_{\max}^+ = 170$ kJ mol^{-1} in steps of $\Delta E^+ = 0.20$ kJ mol^{-1}

Table 5.3 Values of $W(E_v)$

$E^+/$kJ mol^{-1}	0.00		0.20		0.40		0.60		0.80
$W(E_v^+)$	1		1		1		1		1
$W(E_v^+)$ at centre of step		1		1		1		1	

$E^+/$kJ mol^{-1}	4.60		4.80		5.00		5.20	
$W(E_v^+)$	4		4		10		15	
$W(E_v^+)$ at centre of step		4		7		12.5		

$E^+/$kJ mol^{-1}	83.00		83.20		83.40
$W(E_v^+)$	2.4533×10^7		2.5049×10^7		2.5742×10^7
$W(E_v^+)$ at centre of step		2.4791×10^7		2.5397×10^7	

$E^+/$kJ mol^{-1}	169.40		169.60		169.80
$W(E_v^+)$	5.4651×10^{10}		5.5395×10^{10}		5.6144×10^{10}
$W(E_v^+)$ at centre of step		5.5023×10^{10}		5.5769×10^{10}	

Table 5.4 Further parameters

Temperature 697.6 K		$E_\infty = 241.88$ kJ mol^{-1}
$Q_2 - 115.84$	$Q_2^+ = 568.13$	$Z = 9.480 \times 10^6$ Torr^{-1}s^{-1}
		$= 4.121 \times 10^8$ m^3 mol^{-1}s^{-1}
$\langle E_v \rangle = 39.167$ kJ mol^{-1}	$\langle E_v^+ \rangle = 42.389$ kJ mol^{-1}	
whence $E_0 = 232.86$ kJ mol^{-1}		

of the corresponding step are shown in Table 5.5. Table 5.6 compares the values of Q_2^+ and k_∞ obtained by summation (5.22 and 5.23) with the accurate analytical values (5.10 and 5.3); the very satisfactory agreement confirms that the integration has been taken to a sufficiently high energy in small enough steps and is set up correctly. The integration now proceeds;

selected values of the contributions to k_{uni} from the different energy ranges at various pressures are given in Table 5.7, together with the final results for k_{uni} and k_{uni}/k_∞. The way in which the contributions to k_{uni} vary with energy and pressure is illustrated in Fig. 3.2 and discussed in section 3.8. The calculated variation of k_{uni} with pressure is compared with experimental results in Figure 5.3. The effects of variation of the model for the transition state and other parameters are discussed in section 5.5.

Table 5.5   Values of $\rho(E^*)$ and $k_a(E^*)$

$E^+/\text{kJ mol}^{-1}$	1.40	18.80	32.20
$10^{-10}\,\rho(E^*)/\text{kJ mol}^{-1}$	3.1802	8.1426	16.273
$k_a(E^*)/\text{s}^{-1}$	3.1523×10^2	1.1204×10^5	1.0165×10^6
$E^+/\text{kJ mol}^{-1}$	79.00	117.0	170.0
$10^{-10}\,\rho(E^*)/\text{kJ mol}^{-1}$	151.80	779.11	6164.8
$k_a(E^*)/\text{s}^{-1}$	1.024×10^8	1.003×10^9	9.0692×10^9

Table 5.6 Q_2^+ and k_∞ from summation and from analytical expressions

	By summation	Analytical	Error
Q_2^+	568.19	568.13	0.01%
$10^3\,k_\infty/\text{s}^{-1}$	1.0466	1.0465	

Table 5.7   Contributions to the integral and the final results

Energy range/ kJ mol^{-1}	10^6 (contribution to k_{uni}/s^{-1}) at pressures of			
	1000 Torr	10 Torr	0.1 Torr	0.001 Torr
0.00–0.20	0.0624	0.0624	0.0624	0.0603
8.00–8.20	0.5288	0.5287	0.5248	0.3017
16.00–16.20	1.6282	1.6273	1.5363	0.2331
24.00–24.20	3.1092	3.1000	2.3947	0.1008
32.00–32.20	4.0932	4.0517	2.0115	0.0392
40.00–40.20	4.4684	4.3393	1.1161	0.0148
48.00–48.20	4.1487	3.8627	0.4894	0.0055
64.00–64.20	2.5500	1.9099	0.0732	0.0007
80.00–80.20	1.1445	0.5345	0.0098	0.0000
96.00–96.20	0.4065	0.0977	0.0013	0.0000
120.00–120.20	0.0605	0.0052	0.0000	0.0000
169.80–170.00	0.0003	0.0000	0.0000	0.0000
10^3 (total k_{uni}/s^{-1})	1.0420	0.8884	0.3059	0.0269
k_{uni}/k_∞	0.9954	0.8491	0.2924	0.0258

5.5 SENSITIVITY OF THE RRKM CALCULATION TO COMPUTATIONAL DETAILS AND FEATURES OF THE MODEL

In this section comparisons will be made between the results of a number of RRKM calculations in order to define how (a) an accurate integration procedure can be selected for a given model, and (b) a model can be set up which fits the experimental results for a given reaction. The isomerisation of 1,1-dichlorocyclopropane again forms a suitable example for discussion.

DETAILS OF THE COMPUTATIONAL PROCEDURE

With the simple integration procedure described in sections 5.3 and 5.4 the variable features are the step length ΔE^+, the maximum energy E_{max}^+ to which the integration is taken, and the related question of whether or not the summation is artificially cut off at such an energy that the resulting k_∞ agrees with the accurate value from (5.3). The results of a series of calculations to illustrate these points are summarised in Table 5.8. All are based on the model described in section 5.4 (transition state B) and the first calculation, labelled 1 in Table 5.8 is identical with that of section 5.4.

It will be seen immediately that none of the variations tested in Table 5.8 introduces a marked change in either k_∞ or the k_{uni}/k_∞ values. In calculation 2 the integration was cut off at the first step at which the calculated k_∞ became greater than the accurate value. Since the error in k_∞ in calculation 1 was very small (0.01%) it is not surprising that the cut-off has no significant effect in this case. The upper limit of 170 kJ mol^{-1} for the integration is tested in calculations 3 and 4 by stopping at 125 kJ mol^{-1} and 85 kJ mol^{-1} respectively. Integration up to 125 kJ mol^{-1} gives results very close to those of calculation 1 and clearly this is an adequate upper limit.

Stopping at 85 kJ mol^{-1} does make a difference, although even this is not large. The value of k_∞ falls by about 5–6% at 697.6 K and 2–3% at 632.4 K, and the shape of the fall-off curve changes slightly. In Table 5.8 the values of k_{uni}/k_∞ are calculated using the values of k_∞ obtained by summation; this is self-consistent and tends to offset the errors introduced by a poor choice of integration parameters. The actual k_{uni} values have a maximum error, as above, at high pressures where the average energy of the reacting molecules is highest and where a low value of E_{max}^+ therefore has the most effect (cf. Table 5.7). At 10^{-3} Torr pressure, the error in k_{uni} is only 0.6–0.8%.

Integration in coarse steps of 4.0 kJ mol^{-1} (calculation 5) gives k_∞ values which are too high by 8–10%, but the k_{uni}/k_∞ values hardly change. When the integration is cut off (calculation 6) the error in k_∞ is correspondingly

Table 5.8 Effect of variation of details of the integration procedure

Calc.	E_{max}^+/ kJ mol⁻¹	ΔE^+/ kJ mol⁻¹	Cut-off	Temp/K	$10^5 k_\infty$/s⁻¹	k_{uni}/k_∞ at log p/Torr =						
						-3	-2	-1	0	1	2	3
1	170	0.20	No	632.4	[d]1.426	0.045	0.160	0.401	0.703	0.912	0.984	0.998
				697.6	104.7	0.026	0.102	0.292	0.589	0.849	0.968	0.995
2	170	0.20	Yes[a]	632.4	1.426	0.045	0.160	0.401	0.703	0.912	0.984	0.998
				697.6	104.6	0.026	0.102	0.292	0.589	0.849	0.968	0.995
3	125	0.20	b	632.4	1.426	0.045	0.160	0.401	0.703	0.912	0.984	0.998
				697.6	104.4	0.026	0.102	0.292	0.589	0.849	0.968	0.993
4	85	0.20	b	632.4	1.393	0.045	0.160	0.401	0.703	0.905	0.968	0.977
				697.6	98.53	0.026	0.102	0.292	0.586	0.836	0.927	0.940
5	170	4.00	No	632.5	1.579	0.056	0.194	0.469	0.783	0.940	0.968	0.971
				697.6	113.2	0.031	0.121	0.335	0.655	0.904	0.982	0.991
6	170	4.00	Yes[c]	632.4	1.449	0.056	0.194	0.469	0.783	0.940	0.968	0.971
				697.6	106.50	0.031	0.121	0.335	0.655	0.904	0.982	0.991

Note: All calculations are for complex B and associated parameters as in section 5.4.
[a]131.4 kJ mol⁻¹ at 632.4 K and 154.4 kJ mol⁻¹ at 697.6 K.
[b]Irrelevant since k_∞ (by summation without cut-off) < k_∞ (accurate from analytical expressions).
[c]64.0 kJ mol⁻¹ at 632.4 K and 80.0 kJ mol⁻¹ at 697.6 K.
[d]By summation. Accurate values from (5.3), 1.426 at 632.4 K and 104.7 at 697.6 K.

reduced. It is clear that a step size of 0.20 kJ mol^{-1} is perfectly adequate for these calculations.

In conclusion, the integration is surprisingly insensitive to variations in the computational detail and there is no practical difficulty in carrying out the integration in order to obtain an accurate result.

VARIATION OF THE MODEL

From Figure 5.3 it may be seen that the calculations based on transition state B did not give a particularly good representation of the experimental results. This is a typical state of affairs, and the question is then how the model can be modified to give better agreement. Some variations are first described which do not have much effect; the relevant results are summarised in Table 5.9 and details of the models are given in Table 5.10.

The obvious feature which can be varied is the assignment of vibration frequencies to the transition state, especially when the empirical approach of section 5.2. is used for determining these parameters. Calculations 7 and 8 are based on two transition states (A and C)[28] selected in a similar way to model B, i.e. to give the same value of A_∞, but differing in the details of their vibration frequencies (see Table 5.10). The fall-off data are virtually indistinguishable from those for calculation 1. A more extreme frequency pattern, with no particular physical significance, is tested in calculation 9. The transition state (model D) has all its frequencies except the C–H stretches in the narrow range $500-700 \text{ cm}^{-1}$. Again, however, the results are very similar to those of calculation 1.

The assignment of statistical factors is sometimes difficult, and the effect of a change in this parameter is investigated in calculation 10, which uses $L^\ddagger = 1$ and a slightly looser transition state (model E) to restore agreement with the experimental A_∞. Again the fall-off curve is scarcely altered. Since the equations used contain $L^\ddagger$ as the product $L^\ddagger Q_1^+/Q_1$, calculation 10 could also represent the effect of a marked change (in this case a decrease) in the molecular moments of inertia on formation of the transition state. Since Q_1^+/Q_1 is given by $(I_A^+ I_B^+ I_C^+/I_A I_B I_C)^{1/2}$ it appears that even a several-fold increase in all three moments of inertia would not affect the fall-off curves significantly, provided the A_∞ value is correctly predicted. The approximation used here exaggerates the effect on the fall-off curve (see section 3.10), and the treatment based on the assumption that $Q_1^+/Q_1 \approx 1$ is thus likely to be quite satisfactory for reactions having rigid transition states.

Calculation 12 introduces a major change in the type of model, by assuming that the transition state (model G) has an active internal rotation. The moment of inertia for this rotation is simply a 'typical value', since the calculation is presented as a general illustration of the effects of active rotations rather than a useful advance for the particular reaction under

Table 5.9 Variations of the model which have little effect on the results (integrations to 170 kJ mol⁻¹, step size 0.20 kJ mol⁻¹). For details of models see Table 5.10. All models give the same values of k_∞ at 632.4 K and 697.6 K within 0.7%

Calcn.	Transition state explored	Features	Temp./K	k_{uni}/k_∞ at $\log_{10} p/$Torr						
				−3	−2	−1	0	1	2	3
7	A		632.4	0.047	0.164	0.406	0.705	0.912	0.984	0.998
			697.6	0.027	0.105	0.295	0.589	0.847	0.968	0.995
1	B	Different frequency	632.4	0.045	0.160	0.401	0.703	0.912	0.984	0.997
			697.6	0.026	0.102	0.292	0.587	0.849	0.968	0.995
8	C	assignments	632.4	0.054	0.177	0.417	0.710	0.910	0.984	0.998
			697.6	0.030	0.111	0.301	0.589	0.843	0.966	0.995
9	D		632.4	0.061	0.187	0.427	0.713	0.910	0.984	0.998
			697.6	0.033	0.117	0.306	0.590	0.841	0.964	0.995
10	E	$L^{\ddagger} = 1$	632.4	0.046	0.163	0.404	0.705	0.912	0.984	0.998
			697.6	0.026	0.104	0.294	0.631	0.849	0.968	0.995
12	G[a]	Active rotation	632.4	0.038	0.152	0.402	0.719	0.929	0.991	0.999
			697.6	0.022	0.098	0.294	0.604	0.871	0.979	0.999

[a]Integration cut off at 91.2 kJ mol⁻¹ at 632.4 K and 111.8 kJ mol⁻¹ at 697.6 K.

Table 5.10 Details of models for Tables 5.9 and 5.11

Transition state	Vibration frequencies/cm^{-1}	$L^{\ddagger}Q_1^{\ddagger}/Q_1$	E_0/kJ mol^{-1}	$Q_2^{\ddagger}{}_{697.6\,K}$	$\Delta S^{\ddagger}_{697.6}$/J mol^{-1} K$^{-1\,a}$
A	3068 (4), 1383 (3), 1068 (4), 628.6 (2), 437.7 (4), 237.3 (3)	4	232.17	503.0	29.35
B	3068 (4), 1346 (4), 1068 (4), 445 (5), 216.8 (3)	4	232.85	568.1	29.37
C	3068 (4), 1008 (4), 800 (4), 636.1 (2), 442.9 (4), 248.3 (2)	4	230.29	365.7	29.36
D	3068 (4), 700 (4), 639.4 (4), 580 (4), 500 (4)	4	228.20	252.9	29.35
E	3068 (4), 1346 (4), 1068 (4), 385 (5), 171 (3)	1 4	230.83 238.49^b	1604.4 1604.4	29.35 40.87
G	3068 (4), 1346 (4), 1068 (4), 550 (5), 299.5 (2), + active internal rotation, $\sigma = 1$, $I = 10^{-45}$ kg m^2	1	237.48	5033.9	29.36

^{a}Overall entropy of activation, $\Delta S_{vr}^{\ddagger} + R \ln L^{\ddagger}$ ($R \ln 4 = 11.527$ J K^{-1} mol^{-1})
^{b}This value of E_0 corresponds to $E_\infty = 249.53$ kJ mol^{-1} at 697.6 K.

discussion. The vibration frequencies were selected in the usual way so as to produce the correct value of A_∞ when the entropy of the active rotation was included. The results of the calculation are virtually as before, the only difference being a very slight shift upwards of the fall-off curve at low temperature and low pressure. The insensitivity to this change in the model can be traced to the fact that the sum of states up to any given energy is not greatly affected (Figure 5.4). This follows from the construction of the two models to give the same k_∞; the remaining slight difference between the curves is compensated for when the different values of E_0 are used in (5.20). The net effect on k_{uni} is small, even at low concentrations, and the shift in the fall-off curve is considerably less than the likely experimental error. It is often loosely said that transition states with active rotations lead to fall-off at relatively high concentrations, but it is clear from the above calculations that this is not necessarily the case. The real cause lies in the unusually high $W(E_{vr}^+)$ values which follow when such a model is used to reproduce a high

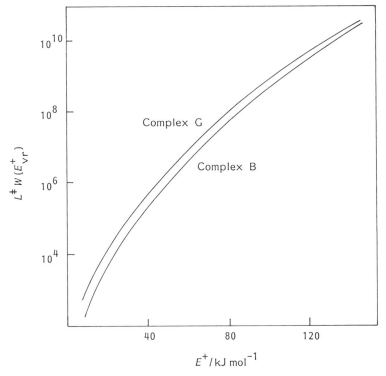

Figure 5.4 Comparison of $W(E_{vr}^+)$ as a function of E^+ for complex B (basic calculation) and complex G (which has an active rotation) (see Table 5.10). The $W(E_{vr}^+)$ values are multiplex by $L^{\ddagger}$ (4 and 1 respectively) to provide a valid comparison.

experimental A-factor, and the early fall-off is associated with the high A-factor rather than with the active rotations; this point will be further illustrated below.

The above calculations thus confirm the conclusion, which is now generally accepted, that the results of the RRKM calculation are strikingly insensitive to the details of the model used for the transition state. The same fall-off curve will be predicted by virtually all models considered *provided they give the same values of* $\Delta S^{\ddagger}$ *and hence* A_{∞}; changes in the vibration frequencies, statistical factor, moments of inertia of the adiabatic rotations or the number of active rotations will not lead to any significant shift of the fall-off curves.

One might wonder at this stage if there are any features at all to which the results are sensitive; there are, in fact, two features of the model and one additional factor. The first significant variable is the de-energisation rate constant $\beta_c Z$, where Z is given by (5.21). This term appears in the expression for k_{uni} only as the product $\beta_c Z[M]$, so that a change in $\beta_c Z$ simply causes an inversely proportional change in the concentration required to produce a given value of k_{uni} or k_{uni}/k_{∞}. Thus the plot of $\log k_{uni}/k_{\infty}$ versus $\log[M]$ is shifted horizontally by a distance equal to the change in $\log \beta_c Z$, without any change in shape. Uncertainties in $\beta_c Z$ can only arise from the values of β_c and the collision cross-section σ_d^2, and the latter can usually be guessed with reasonable certainty even if it is not known experimentally for the molecule in question. Table 5.11 (calculation 13) and Figure 5.5 show the effect of assuming $\beta_c = 0.4$, but otherwise the standard calculation, as in calculation 1, is used. It is seen that a significant shift of the curves is produced by this modest expedient, and the curve for 697.6 K is now a good fit to the experimental results. Unfortunately the curve for 632.4 K is now a worse fit, but this feature of the particular reaction should not mask the fact that $\beta_c Z$ is an important variable parameter, and in early work many authors introduced β_c values between 0.1 and 1 to obtain agreement with their results. Although a simple correction of this kind is adequate for many purposes, it is customary nowadays to allow for weak collisions by the more sophisticated master equation approach which is described in Chapter 7.

Secondly, and perhaps very significantly, the fall-off curves are quite sensitive to the values taken for the experimental A_{∞} and E_{∞} which the model is adjusted to fit. This is illustrated by calculation 14 (Table 5.11 and Figure 5.6), in which A_{∞} has been increased by a factor of four and E_{∞} has correspondingly been raised by 7.65 kJ mol^{-1}. These changes give the same value of k_{∞} at the experimental mean temperature, and values which are not drastically changed at the two temperatures used for the fall-off experiments. The effect of these changes is to produce a considerable displacement of the fall-off curve to higher concentrations. The change in the k_{∞} values

Table 5.11 Variations of model which have a significant effect on the results (integrations to 170 kJ mol⁻¹, step size 0.20 kJ mol⁻¹; parameters as in previous calculations except where otherwise stated)

Calcn.	Transition state	Features of calculation	Temp./K	10^5 k_∞/s⁻¹	k_{uni}/k_∞ at $\log_{10}$ p/Torr[a] =						
					−3	−2	−1	0	1	2	3
1	B	Standard	632.4	1.426	0.045	0.160	0.401	0.703	0.912	0.984	0.997
			697.6	104.7	0.026	0.102	0.292	0.587	0.849	0.968	0.995
13	B	$\beta_c = 0.4$	632.4	1.426	0.025	0.101	0.290	0.586	0.847	0.966	0.995
			697.6	104.7	0.014	0.061	0.200	0.465	0.760	0.938	0.991
14	E with $L^\ddagger = 4$	Different Arrhenius parameters[b]	632.4	1.331	0.025	0.099	0.284	0.574	0.838	0.964	0.995
			697.6	111.8	0.013	0.060	0.195	0.453	0.748	0.931	0.989

[a]Log p/Torr = log([M]/mol cm⁻³) + 1.638 at 697.6 K
Log p/Torr = log([M]/mol cm⁻³) + 1.596 at 632.4 K.
[b]Log A_∞/s⁻¹ = 15.73, E_∞ = 249.53 kJ mol⁻¹.

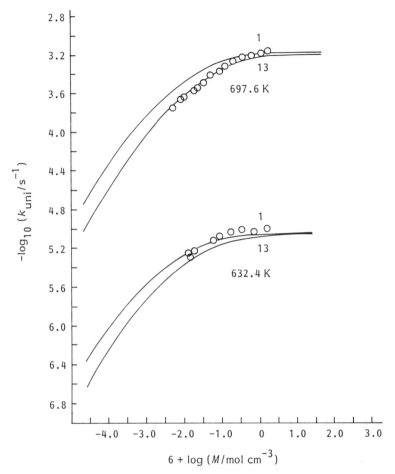

Figure 5.5 Effect of variation in β_c; calculations 1 and 13 use $\beta_c = 1.0$ and 0.4 respectively (see Tables 5.10 and 5.11).

produces simultaneous small vertical shifts, but these are less significant. The observed displacement along the $\log p$ axis is consistent with the value $\log p = 0.44$ predicted by (5.26), which is an approximate equation derived by Wieder and Marcus[32]:

$$\log p = (1 - \theta)\log A_\infty/A'_\infty$$

where

$$\theta = (s + \tfrac{1}{2}r - 1)kT/(E_0 + E_z + \langle E^+\rangle) \tag{5.26}$$

with the symbols defined as before.

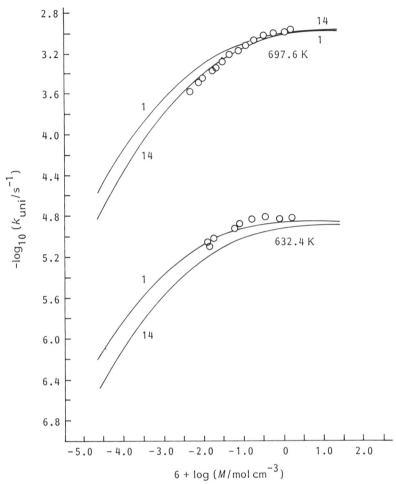

Figure 5.6 Effect of variations in A_∞ and E_∞; $\log_{10} A_\infty/s^{-1}$ and $E_\infty/kJ\,mol^{-1}$ are 15.13 and 241.9 kJ mol^{-1} respectively for calculation 1, and 15.73 and 249.5 kJ mol^{-1} respectively for calculation 14 (see Tables 5.10 and 5.11).

The present calculation, like 13, produces a good fit to the experimental results at 697.6 K, but is less satisfactory at 632.4 K. It is not suggested that the adjustments made to A_∞ and E_∞ are fully justified, but they are certainly not large in view of the extent of disagreement commonly found between different investigators' results for the same reaction. One particular source of error in A_∞ and E_∞ arises from the frequently unappreciated difficulty of obtaining the true high-pressure values[33]. The present calculations indicate that the isomerisation of 1,1-dichlorocyclopropane is significantly into the fall-off region in the pressure range 20–120 Torr used for the 'high-pressure' work, and the Arrhenius activation energy may be less than E_∞ by more than 4.2 kJ mol^{-1} in parts of this range. It seems that many other workers may have erred in the past in this way, and that incorrect values of A_∞ and E_∞ may be a common source of difficulty in fitting the RRKM theory to experimental observations. Frequently nowadays iterative methods are used to find the best A_∞ and E_∞ values to represent the experimental data.

A further factor which may lead to disagreement of the simple RRKM theory with experimental results for thermal isomerisation reactions is neglect of the back reaction converting energised product molecules into reactants. This factor which can be especially important for slightly endothermic or thermoneutral isomerisations has been discussed by Lin and Laidler[34] and applied to the experimental results of Rabinovitch and Michel[35] on the isomerisation of cis- to trans-but-2-ene.

In conclusion, the ease of carrying out the standard RRKM procedure described in this chapter and the ready availability of the necessary computer programs have led to its widespread application. When doing so, in addition to consideration of the important factors discussed above, it must always be remembered that although the standard calculation often fits experimental fall-off curves reasonably well, this should not be taken as endowing too much reality to the structure postulated for the transition state[36].

REFERENCES

1. S. W. Benson, *Thermochemical Kinetics*, Wiley, New York, p. 37 (1968).
2. R. C. Tolman, *Statistical Mechanics with Applications to Physics and Chemistry*, Chemical Catalog Co., New York, Ch. 21 (1927).
3. S. W. Benson, *Advan. Photochem.*, **2**, 1 (1964).
4. D. W. Setser, *J. Phys. Chem.*, **70**, 826 (1966).
5. See, for example, W. Forst and P. St Laurent, *Can. J. Chem.*, **45**, 3169 (1967); K. M. Maloney and B. S. Rabinovitch, *J. Phys. Chem.*, **73**, 1652 (1969).
6. C. G. Schlier, *Molec. Phys.*, **62**, 1009 (1987).
7. G. J. Janz, *Thermochemical Properties of Organic Compounds*, rev. edn, Academic Press, New York, p. 21 (1967).
8. F. W. Schneider and B. S. Rabinovitch, *J. Amer. Chem. Soc.*, **84**, 4215 (1962).
9. F. J. Fletcher, B. S. Rabinovitch, K. W. Watkins and D. J. Locker, *J. Phys. Chem.*, **70**, 2823 (1966).
10. J. C. Hassler and D. W. Setser, *J. Chem. Phys.*, **45**, 3246 (1966).
11. R. M. Badger, *J. Chem. Phys.*, **2**, 128 (1934); **3**, 710 (1935); *Phys. Rev.*, **48**, 284 (1935); parameters updated by D. R. Herschbach and V. W. Laurie, *J. Chem. Phys.*, **35**, 458 (1961).
12. L. Pauling, *J. Amer. Chem. Soc.*, **69**, 542 (1947).
13. H. S. Johnston, (a) *Gas-phase Reaction Rate Theory*, Ronald Press, New York, Ch. 4 (1966); (b) *Advan. Chem. Phys.*, **3**, 131 (1961).

14. L. Pauling, *The Nature of the Chemical Bond*, 3rd edn, Cornell University Press, Ithaca, Ch. 7 (1960).
15. A. Streitwieser, *Molecular Orbital Theory for Organic Chemists*, Wiley, New York (1962); L. Salem, *The Molecular Orbital Theory of Conjugated Systems*, Benjamin, New York (1966).
16. H. S. Johnston, *Gas-phase Reaction Rate Theory*, Ronald Press, New York, p. 82 (1960).
17. H. E. O'Neal and S. W. Benson, *J. Phys. Chem.*, **71**, 2903 (1967).
18. S. W. Benson, *Thermochemical Kinetics*, 2nd edn, Wiley, New York (1976).
19. H. E. O'Neal and S. W. Benson, *J. Phys. Chem.*, **72**, 1866 (1968); S. W. Benson and H. E. O'Neal, *Kinetic Data on Gas-phase Unimolecular Reactions*, Nat. Bur. Standards, NSRDS-NBS 21 (1970).
20. L. Pauling, *The Nature of the Chemical Bond*, 3rd edn, Cornell University Press, Ithaca Ch. 7 (1960).
21. E. B. Wilson, J. C. Decius and P. C. Cross, *Molecular Vibrations*, McGraw-Hill, New York (1955).
22. H. S. Johnston, *Gas-phase Reaction Rate Theory*, Ronald Press, New York, pp. 177–83 (1966).
23. J. R. Christie, P. J. Derrick and G. J. Rickard, *J. Chem. Soc., Faraday II*, **74**, 304 (1978).
24. J. Troe, *J. Chem. Phys.*, **66**, 4745 (1977).
25. I. Mills, T. Cvitas, K. Homann, N. Kallay, K. Kuchitsu, *Quantities, Units and Symbols in Physical Chemistry*, Blackwell Scientific Publications, Oxford (1988).
26. *Ibid.*, p. 23.
27. W. L. Hase and D. L. Bunker, *Quantum Chemistry Program Exchange* Cat. No. QCPE-234.
28. K. A. Holbrook, J. S. Palmer, K. A. W. Parry and P. J. Robinson, *Trans. Faraday Soc.*, **66**, 869 (1970).
29. K. A. W. Parry and P. J. Robinson, *J. Chem. Soc. (B)*, 49 (1969).
30. M. I. Kay, Ph.D. thesis, Rensselaer Polytechnic Institute (1962).
31. J. M. Freeman and P. J. Robinson, *Can. J. Chem.*, **49**, 2533 (1971).
32. G. M. Wieder and R. A. Marcus, *J. Chem. Phys.*, **37**, 1835 (1962).
33. I. Oref and B. S. Rabinovitch, *J. Phys. Chem.*, **72**, 4488 (1968).
34. M. C. Lin and K. J. Laidler, *Trans. Faraday Soc.*, **64**, 94 (1968); see also R. A. Marcus, *J. Chem. Phys.*, **43**, 2658 (1965).
35. B. S. Rabinovitch and K. W. Michel, *J. Amer. Chem. Soc.*, **81**, 5065 (1959).
36. R. E. Weston, *Int. J. Chem. Kinet.*, **18**, 1259 (1986).

6 Reactions with Loose Transition States

The RRKM theory presented in Chapter 3 assumes that the geometry and frequencies of the transition state are independent of energy. Such an approach is acceptable provided the transition state occurs at a well-defined energy maximum along the reaction coordinate. For many reactions, and especially those involving dissociation of a stable molecule to form two radicals (e.g. $C_2H_6 \rightarrow CH_3 + CH_3$), there is no well-defined maximum and a variational approach has to be employed. In this chapter we examine both microcanonical and canonical forms of such theory. In the former $k_a(E^*)$ is determined by locating the position of the minimum sum of states along the reaction coordinate for energy E^*, with the transition state getting tighter, i.e. moving to smaller internuclear separation, as E^* increases. In the canonical analysis the transition state is located at the minimum value of $\Delta G^{\ddagger}$ and the position depends on temperature. In the present chapter we shall also examine alternative techniques for calculating microcanonical rate constants for potential energy surfaces without a well-defined maximum.

6.1 CORRELATION OF VIBRATIONAL AND ROTATIONAL MOTION

Figure 5.1 shows two potential energy curves. Figure 5.1(a) shows a Type I potential with a constrained transition state—such transition states are often termed *tight* because they occur at comparatively small internuclear separations, with coordinates not too dissimilar from those of the stable molecule. The Type II surface (Figure 5.1(b)) shows no potential maximum and leads, as we shall see, to loose, energy-dependent transition states.

CENTRIFUGAL BARRIERS

A preliminary understanding of the factors affecting the location of the transition state can be obtained from a consideration of the dissociation of a diatomic molecule. For $J = 0$, a Morse function provides a simple representation of the potential energy as a function of bond length, r:

$$V(r) = D_e[1 - \exp(-\beta_M(r - r_e))]^2 \qquad (6.1)$$

where β_M is the Morse parameter, D_e the dissociation energy, relative to the bottom of the well and r_e is the equilibrium bond length. For $J > 0$, the effective potential energy, $V_{eff}(r)$, is obtained by adding the rotational energy, since the angular momentum is conserved:

$$E(J) = hc BJ(J + 1) \qquad (6.2)$$

where $B = h/(8\pi^2 \mu r^2 c)$. The resulting energy curves are shown in Figure 6.1. Because $E(J)$ varies less strongly with r than $V(r)$, $V_{eff}(r)$ shows maxima which move to smaller r as J increases. If we consider the combination of two atoms with $J > 0$, then at larger r the atoms encounter a repulsive potential, because $E(J)$ dominates. At smaller r, attractive bonding dominates. The location of the centrifugal barrier to recombination may be determined by setting $dV_{eff}(r)/dr$ to zero. The transition state is located at the barrier maximum and its location clearly depends on J.

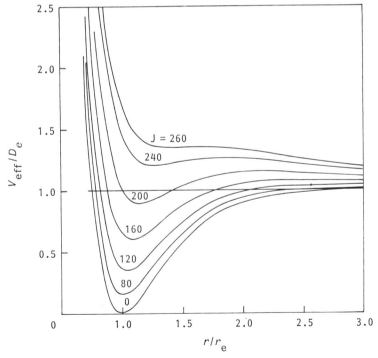

Figure 6.1 Effective potential energy curves for a diatomic molecule. From W. Forst, *Theory of Unimolecular Reactions* (1973), Fig. 3–4. Reproduced by permission of Academic Press.

The picture is more complex for polyatomic molecules and may be illustrated by reference to ethane dissociating into two methyl radicals. To a good approximation, we may correlate the vibrational and rotational modes of ethane with the vibrational and rotational modes of the methyl radicals. This correlation is shown in Figure 6.2[1].

One group, the *conserved* modes, retain their character and have frequencies which differ comparatively little in the molecule and the radical fragments. The six C–H stretching vibrations are good examples of conserved modes. The frequencies of two conserved modes change quite significantly on dissociation: the components of the near-degenerate umbrella vibrations

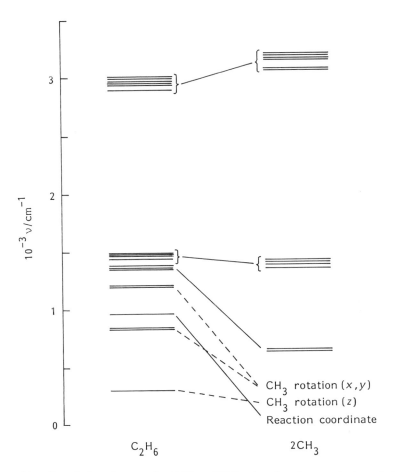

Figure 6.2   Correlation diagram for $C_2H_6 \rightarrow 2CH_3$, showing the correlation of vibrational frequencies between the molecule and the radical products. The correlations of the transitional modes are shown as dashed lines.

in ethane (1388 and 1379 cm^{-1}) correlate with the out-of-plane bending vibrations (606 cm^{-1}) in the two methyl radicals. The vibrational force field is quite different in the two species because of the change in hybridisation.

The *transitional* modes, in contrast, are qualitatively different in the molecular reactant and in the radical products. The $H_3C \cdots CH_3$ torsion correlates with rotation of the CH_3 radicals around the molecular (z) axis, while the four rocking vibrations also correlate with methyl radical rotations, about the x and y axes in the molecular frame.

It is appropriate to complete the correlations. The last contribution to the rotation of the methyl radicals derives from the rotation of C_2H_6 about the z axis, while the other rotations correlate with orbital translational motion of the fragments. The final relative translational motion is provided by the C–C stretch which corresponds to the reaction coordinate.

The variation in the vibrational frequencies of the ethane molecule with C–C bond length have to be taken into account when determining the position of the transition state. The problem is more complex than we found for the dissociation of a diatomic molecule because the vibrational quantum numbers, unlike J, are not conserved. Indeed the basis of the RRKM model is that the intramolecular motion is ergodic, although the angular momentum of the two methyl radicals needs to be taken into account in the angular momentum conservation. In the next section we consider variational techniques which have been developed to locate the transition states.

6.2 THE MINIMUM SUM OF STATES CRITERION

Later in this chapter we shall examine models, including the Gorin[2] model, which was the first canonical variational approach to describe loose transition states. In this section we shall concentrate on a microcanonical approach.

Keck[3] was first to recognise that if the location of the transition state was varied to minimise the rate constant, then, for classical systems, that estimate provides a rigorous upper limit to the true rate constant. His classical approach derived from earlier work by Wigner[4] and Horiuti[5]. The next significant advance was made by Bunker and Pattengill[6] in a comparison between RRKM calculations and Monte Carlo simulations. They argued that the transition state should be located at a 'bottleneck' between the reactants and the products, which they chose to identify as the position of minimum state density. Wong and Marcus[7] employed the same criterion for bimolecular reactions.

The basis of the minimum-state density approach was criticised by Quack and Troe[8] and also by Garrett and Truhlar[9] who provided a clear comparison with the more exact sum of states criterion. The basis of the latter may be appreciated from an examination of the microcanonical rate constant for a given rotational state, J, namely $k_a(E^*, J)$, which is given by equation (6.3),

$$k_a(E^*, J) = \cdot \frac{W(E^\ddagger, J)}{h\rho(E^*, J)} \qquad (6.3)$$

where E^* is the total rovibrational energy and the symmetry numbers have been included in W and ρ, so that $L^\ddagger$ (section 3.9) is omitted.

The state density, $\rho(E^*, J)$, depends on the vibrational frequencies of the molecule and is invariant for a given E^*. The sum of states, $W(E^\ddagger, J)$, on the other hand, varies as the molecule dissociates because the energy available for redistribution decreases as $V_{eff}(r)$ increases and the vibrational frequencies of the transition state (primarily of the transitional modes) vary with r. The microcanonical version of the classical criterion proposed by Keck[3] can be satisfied by minimising $k_a(E^*, J)$ and hence by minimising $W(E^\ddagger, J)$: the variational transition state is located at the minimum in the sum of states and $W(E^\ddagger, J)$, determined at that geometry, is then used to calculate $k_a(E^*, J)$ via equation (6.3).

In order to effect this calculation, it is necessary to describe the dependence of the vibrational frequencies on r. In simple implementations this is usually achieved using an exponential correlation. For the conserved modes, the r-dependent vibrational frequency is calculated from equation

$$v(r) = v(r_e)\exp(-\alpha(r - r_e)) + v_\infty[1 - \exp(-\alpha(r - r_e))] \qquad (6.4)$$

where $v(r_e)$ is the vibrational frequency in the stable molecule and v_∞ that of the corresponding mode in the separated fragments. The constant α is of order $\approx 1 \text{ Å}^{-1}$. For the transitional modes, all of which correlate with rotations, v_∞ is set to zero.

Armed with this explicit dependence, it is relatively straightforward to calculate the sum of states as a function of r, locate the minimum and hence calculate $k_a(E^*, J)$. This minimisation is achieved numerically, calculating $W(E^\ddagger, J)$ on a grid of spacing say 0.1 Å and locating the minimum by interpolation or by fitting a cubic spline. Alternatively, having approximately located the minimum, a finer grid can be used to define it more precisely. The minimum value of W is then identified with $W(E^\ddagger, J)$ allowing k_∞ to be calculated quite straightforwardly:

$$k_\infty = \frac{kT}{hQ_{vr}} \int_0^\infty \sum_{j=0}^\infty (2J + 1)W(E^\ddagger, J)\exp(-E^*/kT)\,dE^* \qquad (6.5)$$

where the symmetry factors have been incorporated in the rotational sums of states and the partition function, and hence $L^\ddagger$ does not appear.

Some comments on this simple model are in order before examing more realistic models.

1. The model has not clarified the distinction between the total angular momentum of the molecule, J, and the orbital angular momentum, l. In a diatomic molecule, $J = l$, but in a polyatomic molecule, the angular momenta of the fragments, j_1 and j_2 contribute to J:

$$|J - k| \leq l \leq J + k \tag{6.6}$$

where

$$|j_1 - j_2| \leq k \leq j_1 + j_2 \tag{6.7}$$

Thus far, we have implicitly assumed that $k = 0$.

2. In setting ν_∞ to zero for the transitional modes, we have essentially assumed that the methyl radical rotational energies are continuous rather than quantised and we have, in consequence, overestimated the densities of states at large r.

3. The correlation of the transitional modes is facilitated through a single parameter, α. In most implementations, α is used as an adjustable parameter.

APPLICATION TO THE CH$_3$ + H REACTION

It is instructive, at this stage, to examine in greater detail the use of a related approach in which some of these issues are addressed. Brouard, Macpherson and Pilling[10] examined the association reaction $CH_3 + H \rightarrow CH_4$. Clearly, this reaction is the reverse of the unimolecular dissociation of CH_4 and the two are related via the equilibrium constant, $K_c(T)$, for the reaction. It is thus possible to compare experimental measurements of the association reaction rate with theoretical predictions of the unimolecular dissociation by the application of $K_c(T)$. This approach is frequently used in analysis of gas kinetic rate data (see Chapter 11), and was adopted by Brouard, Macpherson and Pilling in their study of $CH_3 + H \rightarrow CH_4$, where $k_a(E^*)$ for CH_4 dissociation was calculated. Rotational effects were included in their calculations but $k_a(E^*, J)$ was not explicitly evaluated; the calculations of $k_a(E^*)$ were required for use in a master equation fit of experimental association data (Chapter 7).

1. A Morse curve (equation 6.1) was employed for the reaction coordinate potential energy, with $\beta_M = 1.88 \text{ Å}^{-1}$

2. As CH_3–H dissociates, the tetrahedral symmetry is lost and it was assumed that the molecule retains C_{3v} symmetry. The rotational constants A and B were calculated as a function of r by diagonalizing the inertial tensor, assuming a constant C–H bond length in the CH$_3$ moiety

and an r-dependent H–C$\cdots$H bond angle θ, calculated previously by Hase et al.[11]:

$$\theta/\text{deg} = \begin{cases} 109.47 - 9.53(r - r_e)/\text{Å} & r_e \leqslant r \leqslant 3.137 \text{ Å} \\ 90 & r > 3.137 \text{ Å} \end{cases} \tag{6.8}$$

3. The molecular constants for CH_4 and CH_3 are shown in Table 6.1; the division into conserved and transitional modes and the appropriate correlations are indicated.

Two models were employed. In the first the simple exponential correlation described above was used. The second was designed to overcome the overestimate of the rotational densities of states at large r and used a correlation based on that employed in the statistical adiabatic channel model (SACM) developed by Quack and Troe[8,12] (section 6.4). The correlation was used to construct more realistic expressions for r-dependent rovibrational energies to use in the sum of states minimisation. The correlations are shown in Figure 6.3 for $J = 3$. Two energy ladders, for reactants and products (E_R, E_P) were constructed for the transitional modes via the equations:

$$E_R = n_R \nu_2 + (A_R + B_R)K^2 \tag{6.9a}$$

$$E_P = B_P j(j + 1) + (A_P + B_P)k^2 \tag{6.9b}$$

Table 6.1 Molecular constants for CH_4 and CH_3

CH_4	$CH_3 + H$
Conserved modes ν/cm^{-1}	
3020 (3)	3044 (1)
	3162 (2)
1306 (2)	1396 (2)
Transitional modes ν/cm^{-1}	
1306 (1)	606 (1)[a]
1534 (2)	Product rotation x, y
Rotational constants A, B/cm^{-1}	
5.29 (3)	9.58 (2)
	4.74 (1)
Reaction coordinate	
2917 cm^{-1}	Translation, z
$\Delta H_0^0 = 36072$ cm^{-1}[b], Morse parameter[c],	
$\beta_M = 1.88$ Å^{-1} and $r_e = 1.094$ Å[c].	

References. Except where otherwise indicated: Herzberg[38].
[a]Yamoda, Hirota and Kawaguchi[39].
[b]Heneghan, Knot and Benson[40].
[c]Gray and Robiette[41].

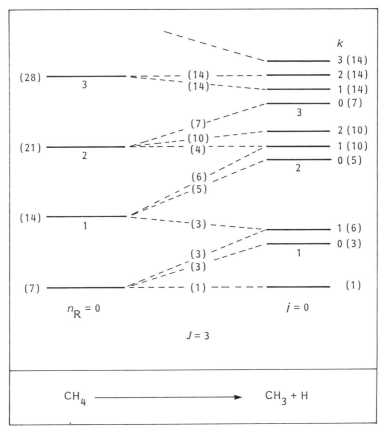

Figure 6.3 Rovibrational energy correlations for $J = 3$ for the reaction $CH_4 \to CH_3 + H$. The degeneracy of each level is given in brackets. Here n_R is the quantum number for the doubly degenerate rocking vibration in the (distorted) methane molecule that correlates with rotation. The diagram refers to CH_4 subject to only a minor distortion, so that the $K = 0 \to 3$ states are essentially seven-fold degenerate. The additional degeneracy of the CH_4 states derives from the $(n_R + 1)$ degeneracy of the vibrational states. The correlations are made purely on the basis of energy and the number of states contained in a specific correlation are shown bracketed on the dashed correlation lines. Reprinted with permission from Pilling *et al.*, *J. Phys. Chem.*, **93**, 4047 (1989). Copyright 1989 American Chemical Society.

where $n_R \nu_2$ is the energy of the reactant bending modes relative to the zero point energy and J, K, j and k are the rotational quantum numbers. CH_4 is a spherical top, but becomes a symmetric top as it is distorted along the reaction coordinate. Thus two rotational quantum numbers are required for the (distorted) reactant, R, (J, K) and CH_3 product, P, (j, k). The restrictions $0 \leqslant K \leqslant J$ and $0 \leqslant k \leqslant j$ apply. Levels with $K > 0$ are doubly degenerate. Brouard, Macpherson and Pilling[10] correlated each reactant state, for a specific J, with each successive product state, without regard for symmetry restrictions, and applied the interpolation parameter, described by equation (6.4) to give

$$E(r) = E_R \exp(-\alpha(r - r_e)) + E_P[1 - \exp(-\alpha(r - r_e))] + B_c P(P + 1)$$

(6.10)

where P is an effective orbital angular momentum quantum number and is an interpolation between J and l (equation 6.6). The effective rotational constant, B_c, was calculated explicitly as a function of r and θ, θ being determined from equation (6.8).

For high J, the number of states increases rapidly. Brouard, Macpherson and Pilling[10] approximated the correlation by assuming that l takes an average value J, thus avoiding the detailed sum over l. With this approximation, P in equation (6.10) is replaced by J.

Schemes similar to that shown in Figure 6.3 were constructed for $J = 0$ to J_{max} and used in the sum of states minimisation.

4. Angular momentum conservation was accommodated, in both the correlated and uncorrelated models, by ensuring that the angular momentum quantum number, J, only takes values which are energetically accessible in both the reactant, R, and transition state, i.e.

$$B_R J(J + 1) + E_v^* \leqslant E_{vr}^* \qquad B_c J(J + 1) + E_v^\ddagger \leqslant E_{vr}^\ddagger \qquad (6.11)$$

where $*$ and $\ddagger$ refer to the energised molecule and the transition state respectively, and E_v and E_{vr} refer to vibrational and rovibrational energies respectively (see equation (6.14) below for the relationship of E_v and E_{vr}). Brouard, Macpherson and Pilling[10] assumed that angular momentum about the main symmetry axis is active, so that K is not conserved.

5. An effective potential energy curve, $V_{eff}(r)$, was generated by adding the r-dependent zero point potential energies of all the vibrational modes except the reaction coordinate to the Morse potential. An exponential correlation was employed for both conserved and transitional modes. The form of $V_{eff}(r)$, which does not incorporate rotation (contrast with Figure 6.1), is shown schematically in Figure 6.4.

6. The r-dependent sums of rovibrational states were then evaluated using the direct count procedure described in section 4.2 for both the correlated and uncorrelated models.

7. Checks were made on the effect of tunnelling through the energy barriers generated by centrifugal effects and from the r-dependence of the

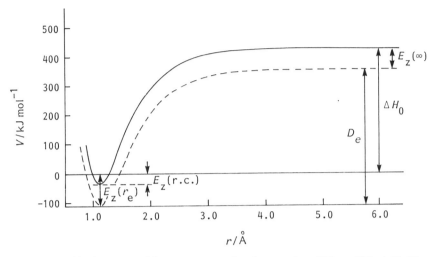

Figure 6.4 Effective potential energy curve for the reaction $CH_4 \rightarrow CH_3 + H$. Here E_z is the total zero point energy and $E_z(rc)$ that in the C–H stretching vibration (reaction coordinate).

transitional frequencies. The curvature, $(d^2V/dr^2)_{r^\ddagger}$, was calculated at the channel maximum in the correlated version of the calculation, for representative channels. An approximate transmission coefficient was then determined for a local parabolic approximation to the potential surface around $r^\ddagger$:

$$P(x) = \left[1 + \exp\left(\frac{-2\pi x}{\hbar\omega^*}\right)\right]^{-1} \qquad (6.12)$$

where $\omega^* = [(d^2V/dr^2)_{r^\ddagger}/\mu]^{1/2}$ and x is the translational energy along the reaction coordinate. The ratio of the quantum to the classical rate constant, ϕ, is given by

$$\phi = \int_{-\infty}^{\infty} P(x) \exp\left(-x/kT\right) dx/kT \qquad (6.13)$$

Representative values for ω^* and ϕ are shown in Table 6.2.

It is clear that quantum mechanical effects are quite small, even for high quantum numbers, where the curvature is most pronounced. Brouard, Macpherson and Pilling[10] concluded that tunnelling effects could be neglected in their calculations.

Microcanonical and high-pressure canonical association rate constants (for

Table 6.2 Ratio of quantum to classical dissociation rate coefficients, ϕ, for specific rovibrational states and $\alpha = 0.801 \text{ Å}^{-1}$

J	v_4	v_2	j, k	$\omega^*/10^{13}\,s^{-1}$	degeneracy	ϕ
45	6	5	11, 8	7.8	9	1.16
30	0	5	3, 2	4.0	2	1.04
15	0	0	2, 0	2.0	4	1.01

comparison with experiment) can now be calculated from the minimised sums of states:

$$k_a(E^*_{vr}) = \frac{W(E^\ddagger_{vr})}{h\rho(E^*_{vr})} \tag{6.14}$$

where $E^\ddagger_{vr} = E^*_{vr} - V_{eff}(r^\ddagger)$ and

$$k_{r,\infty} = \frac{K_c(T)}{Q_{vr}(T)} \int_0^\infty k_a(E^*_{vr})\rho(E^*_{vr}) \exp(-E^*_{vr}/kT)\,dE^*_{vr} \tag{6.15}$$

where $k_{r,\infty}$ is the association rate constant and $K_c(T)$ is the equilibrium constant.

The only variable parameter in the calculations is α. For large values of α, the transitional modes vary strongly with r in the region of the transition state and the zero value of the $r \to \infty$ asymptote for the v_2 mode renders the uncorrelated model unrealistic. A detailed comparison of correlated and uncorrelated calculations was made and it was found that the two estimates of $k_{r,\infty}$ agree to better than 20% for $\alpha \leqslant 0.9 \text{ Å}^{-1}$. At higher values of α, $k_{r,\infty}$ increases much more sharply with α for the uncorrelated model which must be considered seriously flawed under these circumstances. In a detailed comparison with experiment, α values within the valid range were obtained, so that the simpler uncorrelated model could be used. Brouard, Macpherson and Pilling[10] also compare the correlated calculations with the values obtained from the SACM model; the two agree to within 10% for $0.5 \leqslant \alpha/\text{Å}^{-1} \leqslant 1.0$, with SACM giving the lower high-pressure rate constant.

Brouard, Macpherson and Pilling[10] used the uncorrelated model to analyse experimental data for $CH_3 + H$, using the energy-grained master equation (Chapter 7). Figure 6.5 shows a fit and the experimental data at 400 K, with $\alpha = 0.89 \text{ Å}^{-1}$ and the energy transfer parameter $\langle \Delta E \rangle_d = 285 \text{ cm}^{-1}$ (see Chapter 7). The fit gives $k_{r,\infty} = 4.5 \times 10^{-10} \text{ cm}^3 \text{ mol}^{-1} \text{s}^{-1}$. Experimental data were obtained over a temperature range 300–600 K and $k_{r,\infty}$ was found to be essentially constant and in the range $(4.4–4.7) \times 10^{-10} \text{ cm}^3 \text{ mol}^{-1} \text{s}^{-1}$. More detailed calculations by Aubanel and Wardlaw[13], based on flexible transition state theory and a realistic potential give lower

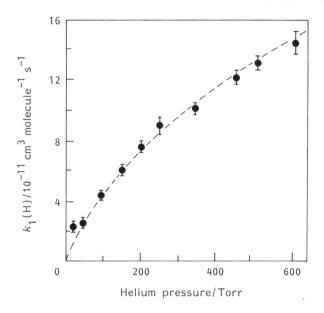

Figure 6.5 Fit of CH$_3$ + H data[10] at 400 K. Reprinted with permission from Pilling *et al.*, *J. Phys. Chem.*, **93**, 4047 (1989). Copyright 1989 American Chemical Society.

rate constants which increase slightly with temperature (section 6.3). A comparison of the reactions CH$_3$ + D and CH$_3$ + H is presented in Chapter 10.

6.3 FLEXIBLE TRANSITION STATE THEORY (FTST)

The major difficulty with the analysis described in section 6.2 is that α is simply employed as an adjustable parameter and the description of the transitional modes, and their dependence on r, is a heuristic one based on computational simplicity rather than detailed analysis. This problem has been addressed by Wardlaw and Marcus[14] using FTST.

The advantage of the FTST approach is that it incorporates a realistic representation of the angular potential and, through this, is able to address the interaction of the transitional modes which have large amplitudes in the region of the transition state. Some applications of FTST have employed *ab initio* potentials, but it is instructive to examine first the treatment of the C$_2$H$_6$ → 2CH$_3$ reaction for which a more intuitive potential was employed.

The potential was represented as a function of the bond distance r and the angles θ_1 and θ_2 between the C$_3$ axes of the two CH$_3$ fragments and the bond axis. The bonded interaction potential V_B was represented by a Morse potential modified to account for orientations:

$$V_B = V_{eff,M} \cos^2 \theta_1 \cos^2 \theta_2 \qquad (6.16)$$

In effect the cosine squared terms allow for the variation with orientation in the overlap of the singly occupied p orbitals on the methyl radicals; $V_{eff,M}$ is an effective Morse potential.

A non-bonded potential, V_{NB}, was then added to account for interaction between all non-bonded HH and CH pairs, using a Lennard-Jones potential, V_{LJ}:

$$V_{NB} = \sum_{i,j=1}^{4} V_{LJ}(r_{ij}) \qquad (6.17)$$

where r_{ij} are distances between non-bonded atoms, and i and j label the atoms on different CH_3 fragments. An interpolated structure for CH_3 using exponential correlations for the HCH angle the $r(C-H)$ bond distance was used in calculating V_{NB}.

Wardlaw and Marcus[14] wrote the sum of states $W(E^{\ddagger}, J)$ as a convolution:

$$W(E^{\ddagger}, J) = \int_0^{E'} W_v(E' - E)\rho(E, J)\,dE \qquad (6.18)$$

where $\rho(E, J)$ is the density of states for the transitional modes for angular momentum J, $W_v(E' - E)$ is the sum of states for the conserved modes for energy less than or equal to $E' - E$ and E' is the available energy at a given r ($E' = E^{\ddagger} - V_{rc}$, where V_{rc} is the reaction coordinate potential). The two types of modes were assumed uncoupled and W_v determined by direct count, with the r dependence of the conserved mode frequencies defined via the usual exponential interpolation with α as a variable parameter. The density $\rho(E, J)$ was determined classically, using a Monte Carlo method, as a $2n + 5$ dimensional phase space integral, where n is the number of transitional degrees of freedom.

The approach is computationally intensive, but has the advantage of incorporating a realistic treatment of the interacting transitional modes. Comparison with a quantum mechanical determination of $\rho(E, J)$ by Klippentstein and Marcus[15] shows that the classical evaluation is realistic. Figure 6.6 shows $k_a(E^*, J)$ vs E^* and J.

Based on this analysis, Wagner and Wardlaw[16] fitted the experimental data of Macpherson, Pilling and Smith[17] and Slagle et al.[18] for the association reaction $CH_3 + CH_3$. They used a modified strong collision approximation to describe the pressure dependence and treated α, the interpolation

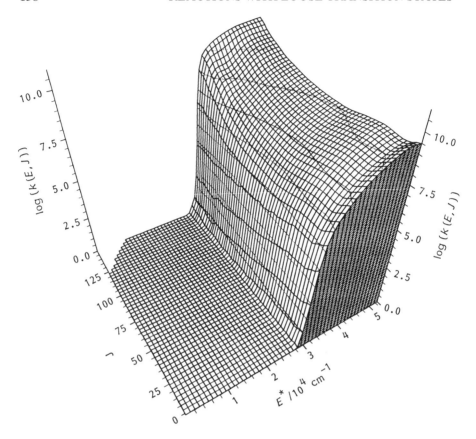

Figure 6.6 The energy and angular momentum dependence of $k(E, J)$, for $C_2H_6 \rightarrow 2CH_3$.

parameter for the conserved modes and β_c the Troe energy transfer parameter (see section 8.1), as variable parameters. The high-pressure recombination rate constant is given by

$$k_{r,\infty} = \frac{g_e kT}{h Q_{trans} Q_{vr}} \int_0^\infty \sum_{j=0}^{J_{max}} W(E^\ddagger, J) \exp(-E/kT) \, dE \qquad (6.19)$$

(cf. equation 6.15) where g_e is the ratio of the electronic partition function at the transition state to that for the separated reactants ($g_e = \frac{1}{4}$). The vibration rotation partition function for the reactants is Q_{vr} and $Q_{trans} = (2\pi \mu kT)^{3/2}/h^3$. Once again, the symmetry number is incorporated implicitly in the partition function and sum of states. Figure 6.7 shows the resulting temperature dependence for $k_{r,\infty}$. The negative temperature dependence obtained by Wagner and Wardlaw[16] is not universally accepted[19].

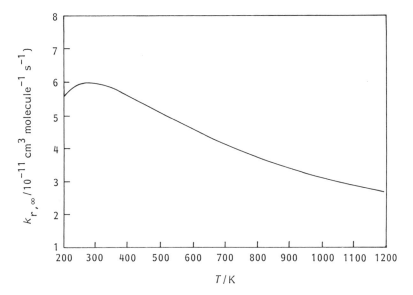

Figure 6.7 Temperature dependence of $k_{r,\infty}$ for the reaction $CH_3 + CH_3 \rightarrow C_2H_6$ using FTST[16].

Aubanel and Wardlaw[13] applied detailed FTST analysis of $k_{r,\infty}$ to the association reaction, $CH_3 + H$. They employed an *ab initio* calculation of the potential energy surface at 100 geometries covering a wide range of $CH_3 \cdots H$ orientations and bond distances. The surface was fitted by Hase *et al.*[10] to generate an analytic function for use in the calculation of the sums and densities of states. A minimum energy potential analysis, proposed by Miller, Handy and Adams[20], was also employed to determine the r-depend-ence of the frequencies of the conserved modes directly from the shape of the surface, rather than via the correlation parameter α. There were thus no variable parameters in the model.

Table 6.3 lists the calculated values for $k_{r,\infty}$, based on equation (6.19). The rate constant $k_{r,\infty}$ increases with temperature before eventually reach-ing a limit of $\approx 3.6 \times 10^{-10}$ cm^3 mol^{-1} s^{-1}. This behaviour contrasts markedly with that for $CH_3 + CH_3$. Aubanel and Wardlaw[13] also examined the effects of anharmonicity (Table 6.3) and found a significant rate en-hancement which increases with temperature. This effect derives from the abnormal behaviour of the umbrella mode, which has a negative anharmoni-city, i.e. the level spacing increases with vibrational quantum number. The anharmonicity for this mode decreases as r decreases so that there is an anharmonicity enhancement of $W(E^{\ddagger}, J)$. The enhancement of the canon-ical rate coefficient is greater at higher temperatures because configurations

Table 6.3   Rate constants for the reaction
$CH_3 + H \rightarrow CH_4$

	$k_{r,\infty}/10^{-10}$ cm^3 molecule^{-1} s^{-1}			
T/K	k_{μ}	k_{μ}^{anh}	$k_{\mu,D}$	$k_{\mu,D}^{anh}$
300	2.36	2.67	1.72	1.90
400	2.65	3.05	1.90	2.18
500	2.88	3.33	2.07	2.38
600	3.03	3.60	2.12	2.57
800	3.05		2.30	
1000	3.43	4.23	2.38	3.02
1250	3.53		2.47	
1500	3.58		2.48	
1750	3.60		2.48	
2000	3.58		2.48	
2500	3.67			

Notes: A superscript anh indicates anharmonicity corrections included. A subscript D indicates a rate constant for $CH_3 + D \rightarrow CH_3D$

with smaller separations dominate the integral over the sums of states (equation 6.19). As mentioned earlier Brouard, Macpherson and Pilling[10] obtained a temperature-independent experimental value for $CH_3 + H$ of 4.7×10^{-10} cm^3 mol^{-1} s^{-1} which is significantly larger than the calculated values of $k_{r,\infty}$ shown in Table 6.3. The origin of this discrepancy is not clear. Aubanel and Wardlaw[13] also calculated rate constants for $CH_3 + D$. In this case they obtained a much better agreement with the experimental value of 1.75×10^{-10} cm^3 mol^{-1} s^{-1} over the temperature range 300–400 K. These discrepancies are discussed further in Chapter 10.

The overriding difficulty of FTST is that, as formulated by Marcus and Wardlaw[21], it is a computationally intensive technique and as such does not lend itself readily to the analysis and interpretation of experimental data. At the centre of the Wardlaw–Marcus formulation there is a multidimensional phase space integral which can only be tackled using Monte Carlo methods. This difficulty has been alleviated to some extent by the recent work of Smith[22] who reformulated FTST in such a way that the dimensionality of the phase space integration was halved. The key to this reduction was the observation that integration over the momentum variables can be performed analytically. Earlier studies at the canonical level by Kassel[23] and others had made extensive use of this property, but Smith[22] was the first to apply it to microcanonical calculations of J-dependent sums of states. The advantage of this reformulation is that there is a substantial decrease in the computational effort to calculate a rate coefficient to the same accuracy. Indeed in some cases it is possible to avoid the use of Monte Carlo methods altogether and use more accurate methods of integration, and even in those cases where this is not possible and Monte Carlo has to be used a greater accuracy can be achieved with a less drastic increase in computational requirement.

Both the Marcus–Wardlaw and Smith formulations are restricted to systems where the reaction coordinate is the centre of mass distance of the dissociating fragments. For some reactions, such as the dissociation of ethane to methyl radicals, this restriction does not appear to be of great significance. However, recent work by Klippenstein[24] suggests that this may not be the most appropriate choice of reaction coordinate for all systems and that a reaction coordinate based on the distance between the two atoms that form the bond that breaks in the dissociation process is more realistic. Klippenstein has developed a form of FTST in which this bond is taken to be the reaction coordinate and has found that for some systems, notably the dissociation of NCNO, there is a significant lowering of the microcanonical rate coefficient and thus a concomitant effect on the canonical rate coefficient.

6.4   STATISTICAL ADIABATIC CHANNEL MODEL (SACM)

Quack and Troe[8,12] developed a model in which the individual channel eigenvalues are determined explicitly as a function of r, a channel being defined as an effective potential energy curve that connects a specific reactant quantum state with a specific product quantum state(s). The microcanonical rate constant, $k_a(E^*, J)$ is given by the usual form:

$$k_a(E^*, J) = \frac{W(E^{\ddagger}, J)}{h\rho(E^*, J)} \qquad (6.3)$$

but now $W(E^{\ddagger}, J)$ is the number of 'open' channels, i.e. those with energy maxima, V_{max}, below energy E^*. The model assumes adiabaticity and excludes reaction, at energy E^*, on curves such as the one labelled (c) in Figure 6.8. The problem of calculating $k_a(E^*, J)$ thus reduces to one of constructing the adiabatic potential curves, $V_a(r)$, and counting the number of open channels. The $V_a(r)$ are constructed by adding the channel eigenvalues, $E_a(r)$, to the r-dependent reaction coordinate potential energy function $V_{rc}(r)$. In early applications, the Morse potential was employed for $V_{rc}(r)$. The calculation of the channel eigenvalues requires a detailed correlation between reactant and product states of the type illustrated in Figure 6.3 for $CH_4 \rightarrow CH_3 + H$. The following rationale is applied:

1. The vibrational quantum numbers of the conserved modes, $v_{R,i}$ and $v_{P,i}$ have a one-to-one correlation.

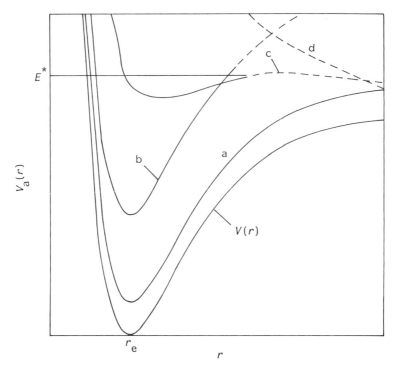

Figure 6.8 Adiabatic potential energy curves illustrating open (curve a) and closed (curves b, c) channels (Reproduced with permission).

2. The overall angular momentum quantum number, J, and its projection, M, are conserved.
3. The remaining quantum numbers, i.e. for the transitional vibrational modes, for the fragment rotational modes and for orbital motion, are not uniquely correlated because they are not precisely defined in the intermediate range. There are, however, symmetry and angular momentum conservation restrictions.

Quack and Troe[8,12] illustrated the application of the SACM by reference to the dissociation of a bent triatomic molecule. One of the reactant stretching vibrations is the reaction coordinate and the other $\nu_{R,s}$ is a conserved vibration, correlating with the vibration of the diatomic product, $\nu_{P,s}$. Thus the conserved quantum numbers are J, M and $\nu_{R,s}(= \nu_{P,s})$. The transitional modes are the bending vibration, with quantum number $\nu_{R,b}$ and the rotation associated with quantum number K (Quack and Troe[8,12] approximated the triatomic molecule to a symmetric top) which correlate with the rotational quantum number, j, of the diatomic molecule and the orbital quantum number, l. Thus the correlation can be written as

$$(J, M, \nu_{R,s}, [\nu_{R,b}, K]) \rightarrow (J, M, \nu_{P,s}(= \nu_{R,s}), [j, l]) \qquad (6.20)$$

where the square brackets indicate the transitional modes. Angular momentum conservation requires (cf. equation 6.6)

$$|J - j| \leq l \leq J + j$$

Using the exponential correlations discussed previously, the channel eigenvalues are given by

$$E_a(r) = [(\nu_s + \tfrac{1}{2})h\nu_{R,s} + (\nu_b + \tfrac{1}{2})h\nu_{R,b} + (A_R - B_R)K^2]\exp(-\alpha(r - r_e))$$
$$+ [(\nu_s + \tfrac{1}{2})h\nu_{P,s} + A_P j(j + 1)][1 - \exp(-\alpha(r - r_e))]$$
$$+ B_s(r)P(P + 1) \qquad (6.21)$$

where $P = J\exp(-\alpha(r - r_e)) + l[1 - \exp(-\alpha(r - r_e))]$ and A_R and $B_R(= (B_R + C_R)/2)$ are the rotational constants of the triatomic, assumed to behave as a symmetric top. Here A_P is the rotational constant of the diatomic product and $B_s(r) = (B(r) + C(r))/2$ as a function of the reaction coordinate, r, and is evaluated from the geometry of the dissociating triamotic molecule, for a fixed bond angle.

It is instructive to examine each of the terms in the channel eigenvalue equation (6.21). The first term represents the stretching and bending energies in the reactant plus the K^2 component of the rotational energy. The contribution of this term to the channel eigenvalue is assumed to decay exponentially with r. The second term corresponds to the rovibrational energy in the diatomic product, which is assumed to approach the $r \rightarrow \infty$ asymptote exponentially. Finally, the term in $B_s(r)$ is an exponential correlation between the J-dependent term in the rotational energy of the reactant the orbital energy of the products. Figure 6.9 shows several channel eigenvalues for NO_2 with $J = 10$, $\nu_s = 0$ for a few ν_b, K combinations.

In effect, SACM is a state-resolved version of transition state theory which examines a reaction in considerable detail. The approach described above, which is based on the original version described by Quack and Troe[8,12], is capable of extension to accommodate a more complete description of the potential, such as the *ab initio* potential described above for CH_4. The major problem presented by SACM is the complexity of the correlations that are required to describe the dissociation of larger molecules[25].

6.5 CANONICAL RATE CONSTANTS

It is frequently useful to calculate the canonical high-pressure limiting rate constants, k_∞, to compare with experimental rate constants or to use in

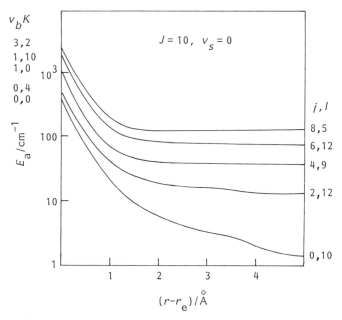

Figure 6.9 Adiabatic curves for $J = 10$ and $v_3 = 0$ for the reaction $NO_2 \rightarrow NO + O$ (Reproduced with permission).

appropriate Troe fall-off calculations (see Chapter 8). The rate constant k_∞ can be calculated from equation (6.5), but with considerable effort, and a general canonical approach is desirable.

From the perspective of a thermodynamic formulation, the transition state occurs at the maximum in the Gibbs free energy change:

$$k_\infty = \frac{kT}{h} \exp\left(-\Delta G^{\ddagger}/RT\right) \qquad (6.22)$$

emphasising the interplay between the increasing entropy as the molecular vibrational frequencies decrease along the reaction coordinate, and the transition state loosens, and the increasing potential energy. More practically, and equivalently, a partition function expression can be formulated (see equations 1.7 and 3.7):

$$k_\infty = \frac{kT}{h} \frac{Q^{\ddagger}}{Q} \qquad (6.23)$$

where $Q^{\ddagger}$ is the total partition function of the transition state and Q is that of the stable molecule. The partition $Q^{\ddagger}$, which contains the reaction coordinate potential energy term, is determined as the minimum in $Q'(r)$, where $Q'(r)$ is the partition function for the molecule as a function of r; the mode corresponding to the reaction coordinate has, of course, been removed from $Q'(r)$. In effect, $Q^{\ddagger}$ is a Boltzmann average over the microcanonical variational transition states at a specific temperature. The variation of $Q^{\ddagger}$ with T depends on the tightening of the canonical transition state with temperature, reflecting the increasing contributions from higher energy states at higher temperatures.

Quack and Troe[26] developed a canonical version of SACM by defining $Q^{\ddagger}$ as a sum over the channel maxima:

$$Q^{\ddagger} = \sum_i \exp\left(-V_{\text{max},i}/kT\right) \tag{6.24}$$

The interpolation formula employed by Quack and Troe[26] typifies the variational approach followed by others and is the one employed here; $Q'(r)$ may be factorised by assuming separability of the various forms of motion:

$$Q'(r) = Q'_e(r)Q'_{pd}(r)Q'_{vr}(r)Q'_z(r) \tag{6.25}$$

where $Q'_e(r)$ is the electronic partition function, $Q'_{pd}(r)$ is the partition function of the pseudo-diatomic molecule formed from the dissociating reactant, $Q'_{vr}(r)$ is the rovibrational partition function excluding the centrifugal term and $Q'_z(r)$ is introduced to account for the variation in the zero-point vibrational energy with r. $Q'_e(r)$ is given by

$$Q'_e(r) = \exp\left(-V_{rc}/kT\right) \tag{6.26}$$

where V_{rc} is the reaction coordinate potential energy and is usually expressed as a Morse function.

Here $Q'_{pd}(r)$ is the centrifugal partition function and describes the influence of rotation of the dissociating molecule, which is approximated to a diatomic molecule, i.e. $Q'_{pd}(r)$ is a two-dimensional rotational partition function:

$$Q'_{pd}(r) = \frac{kT}{hc\,B_s(r)} \tag{6.27}$$

where $B_s(r)$ is the average of the two small rotational constants of the distorted molecule, i.e. $B_s(r) = (B(r) + C(r))/2$.

Quack and Troe[26] used an exponential interpolation of the logarithm of the rovibrational partition function between reactants and products, reflecting, although not exactly, the exponential interpolation of vibrational frequencies in the microcanonical model:

$$\ln Q'_{vr}(r) = [\ln Q'_{vr}(r_e) - \ln Q'_{vr}(\infty)]\exp\left(-\gamma(r - r_e)\right) + \ln Q'_{vr}(\infty) \tag{6.28}$$

where $Q'_{vr}(r_e)$ is the total partition function of the reactant, $Q_{vr}(r_e)$, with the contributions of the reaction coordinate and the centrifugal term removed:

$$Q'_{vr}(r_e) = \frac{S_e Q_{vr}(r_e)}{Q'_{pd}(r_e)}[1 - \exp(-h\nu_{rc}/kT)]^{-1} \qquad (6.29)$$

where ν_{vc} is the vibrational frequency of the mode corresponding to the reaction coordinate and S_e is the symmetry correction, which will be discussed below. Note that $Q_{vr}(r_e)$ is the total partition function and so includes the full rotational contribution, including the symmetry number (see section 4.3 and contrast with the use of $L^{\ddagger}$ in Chapter 3) and the vibrational partition function corresponding to the reaction coordinate.

Similarly,

$$Q'_{vr}(\infty) = S_{\infty} Q_{vr}(A)Q_{vr}(B) \qquad (6.30)$$

where $Q_{vr}(A)$ and $Q_{vr}(B)$ are the total partition functions for the fragments, A and B, and S_{∞} is the symmetry correction for the fragments.

The symmetry number correction, $S_e(r)$, takes account of the decrease in symmetry as the reactant molecule moves along the reaction coordinate:

$$S_e = \sigma_e/\sigma(r) \qquad (6.31)$$

where σ_e is the symmetry number in the equilibrium geometry and $\sigma(r)$ that in the distorted geometry, e.g. for CH_4, $\sigma_e = 12$, while $\sigma(r) = 3$, assuming that C_{3v} symmetry is maintained; $S_{\infty} = 2$ if $A \equiv B$, otherwise $S_{\infty} = 1$.

The term $Q'_z(r)$ accounts for the change in zero-point vibrational energy:

$$kT \ln Q'_z(r) = (E_z(r_e) - E_z(\infty))\exp(-\gamma(r - r_e)) + E_z(\infty) \qquad (6.32)$$

where $E_z(r_e)$ is the total zero-point vibrational energy in the reactant, less that for the reaction coordinate, i.e.

$$E_z(r_e) = \sum_i h\nu_i/2 \qquad (6.33)$$

and $E_z(\infty)$ is the total zero-point vibrational energy for the products, so that (cf. Figure 6.4)

$$\Delta H_0^0 = D_e + E_z(\infty) - E_z(r_e) - h\nu_{rc}/2 \qquad (6.34)$$

Finally, the exponential correlation parameter γ is clearly related to α in equation (6.4) but is not equal to it. The partition function, $Q'_{vr}(r)$ contains contributions both from vibrations and from internal rotations which do not vary in the same way with r. Indeed the overall transposition from exponentially varying eigenvalues to exponentially varying logarithmic partition functions is a heuristic one. Quack and Troe[26] investigated the problem numerically for selected examples and proposed the correlation shown in equation (6.28) with $\gamma = 0.75\alpha$.

The canonical approach cannot incorporate the level of detail that is feasible using a microcanonical model. In addition, only k_∞ can be calculated, so that fall-off curves cannot be calculated except by approximate methods. However, one advantage of the approach is that it does not employ the problematic asymptotics of the simple minimum sum of states calculation, because the total partition function is used in the correlation.

Quack and Troe[26] applied the technique to a range of association reactions, calculating $k_{r,\infty}$ from k_∞ via the equilibrium constant. They used $\gamma = 0.75$ Å^{-1}, corresponding to $\alpha = 1.0$ Å^{-1}, their 'universal' parameter. Table 6.4 shows results for $CH_3 + H$ and $CH_3 + CH_3$, and includes a comparison with SACM. Troe[27] has also developed a simple, semi-empirical expression which depends parametrically on α/β_M (cf. section 8.2). Typically $\beta_M \approx 2$ Å^{-1} where β_M is the Morse parameter (equation 6.1). The expression can either be used to obtain a rapid estimate of k_∞, using the 'universal' estimate of $\alpha/\beta_M = 0.5$, or to fit to experiment by varying α/β_M.

The partition function correlation technique has been used, in essence, by others with some changes in the details of the correlations. Hase and co-workers[28], in particular, have exploited the method using more detailed information from *ab initio* potential energy surface calculations; usually separating vibrational, internal rotational and external rotational terms. For $CH_3 + H$, they used the *ab initio* surface calculated by Hirst[11], and fitted the partition function for hindered internal rotation by quadrature. For $CH_3 + H$ and for $CH_3 + CH_3$, Leblanc and Pacey[29] modelled the r-dependent hindering potential for internal rotation by a function which allows Q'_{vr} to be calculated analytically.

Table 6.4 A comparison of SACM and maximum free energy (A) treatments with experiment for a number of reactions

| Reaction | T/K | $k_{r,\infty}/10^{-11}$ cm^3 molecule^{-1} s^{-1} | |
		SACM	Max. A $\gamma = 0.75$ Å^{-1}
$H + CH_3$	300	0.012	5.97
	2200	1.4	17.25
$CH_3 + CH_3$	300	5.1	4.65
	1300	4.8	6.23

6.6 CANONICAL FLEXIBLE TRANSITION STATE THEORY (CFTST)

The maximum free energy model is easy to implement and physically transparent. It is also feasible to extend the model to include more complex information on the angular potential. The approach has been used by Robertson, Wagner and Wardlaw[30] in developing a canonical version of FTST. They applied the model specifically to association reactions $A + B \rightarrow AB$, and we shall follow their analysis, although we shall also quote the final results for dissociation as well. The model provides insight into the determinants of the canonical rate constants.

The association rate constant is given by

$$k_{r,\infty} = g_e \frac{kT}{h} \frac{\sigma}{\sigma^{\ddagger}} \frac{Q'(r^{\ddagger}, T)}{Q_R(T)} \exp\left(-\Delta V_{rc}(r^{\ddagger})/kT\right) \qquad (6.35)$$

where g_e is the ratio of the electronic degeneracies in the separated reactants and the transition state, $\sigma/\sigma^{\ddagger} = L^{\ddagger}$, $Q_R(T)$ is the total partition function for the reactants and $Q'(r^{\ddagger}, T)$ that for the transition state, with the mode corresponding to the reaction coordinate removed. $\Delta V_{rc}(r) = V_{rc}(r) - V_{rc}(\infty)$, i.e. the difference between the reaction coordinate potential energy (see section 6.4) and the corresponding quantity for reactants. As $\Delta V_{rc}(r)$ and $Q'(r^{\ddagger}, T)$ both depend on r, an optimal value, $r^{\ddagger}$, has to be found variationally such that $k_{r,\infty}$ is minimised. The value of $r^{\ddagger}$ will depend on T so the minimisation has to be performed for each temperature of interest. Note that the symmetry numbers have been written explicitly, and have been removed from Q' and Q_R.

Here $Q'(r)$ may be expressed in the form

$$Q'(r) = Q_{AB,trans} Q'_c(r) Q'_{tr}(r) \qquad (6.36)$$

where $Q_{AB,trans}$ is the translational partition function of the AB system, $Q'_c(r)$ is the partition function for the conserved modes and $Q'_{tr}(r)$ is the partition function for the transitional modes plus external rotation. Robertson, Wagner and Wardlaw[30] showed that $Q'_{tr}(r)$ can be rigorously expressed in the form:

$$Q'_{tr}(r) = Q'_{pd}(r) Q'_{fr,A}(r) Q'_{fr,B}(r) \Theta(r, T) \qquad (6.37)$$

where $Q'_{pd}(r)$ is the partition function for the pseudo-diatomic molecule A–B (i.e. $Q'_{pd}(r) = 8\pi^2 \mu r^2 kT/h^2$, where μ is the reduced mass, $m_A m_B/(m_A + m_B)$). The $Q'_{fr}(r)$ terms are free rotor partition functions for the A and B species and the explicit dependence on r allows for a change in geometry e.g. for CH_3, in $CH_3 + H$ reaction, from planar to tetrahedral. The term $\Theta(r, T)$ is a hindering function consisting of integrals over the relative orientation of the fragments, weighted by a Boltzmann term in the r and angular coordinate potential energy.

Minimising $k_{r,\infty}(r, T)$ gives

$$k_{r,\infty} = g_e \frac{kT}{h} \frac{\sigma}{\sigma^\ddagger} \frac{\exp(-\Delta V_{rc}(r^\ddagger)/kT)}{Q_{trans}} \left\{ \frac{Q_{vib}^\ddagger}{Q_{vib,A} Q_{vib,B}} \right\}$$

$$\times \left\{ \frac{Q_{pd}^\ddagger Q_{fr,A}^\ddagger Q_{fr,B}^\ddagger}{Q_{fr,A}(\infty) Q_{fr,B}(\infty)} \right\} \Theta^\ddagger \qquad (6.38)$$

where the $\ddagger$ signifies that a function has been determined at the canonical transition state ($r = r^\ddagger$). The term Q_{trans} in equation (6.38) is

$$(Q_{trans,AB}/Q_{trans,A} Q_{trans,B})^{-1} = (2\pi\mu kT)^{3/2}/h^3$$

It is instructive to examine successive approximations of equation (6.38). It is often found that the conserved mode frequencies vary little along the reaction coordinate, so that $Q_{vib}^\ddagger = Q_{vib,A} Q_{vib,B}$ and

$$k_{r,\infty} = g_e \frac{kT}{h} \frac{\sigma}{\sigma^\ddagger} \frac{\exp(-\Delta V_{rc}(r^\ddagger)/kT}{Q_{trans}} \left\{ \frac{Q_{pd}^\ddagger Q_{fr,A}^\ddagger Q_{fr,B}^\ddagger}{Q_{fr,A}(\infty) Q_{fr,B}(\infty)} \right\} \Theta^\ddagger \quad (6.39)$$

If the shapes of the fragments do not change (or if, for example, we neglect the change in the moments of inertia of CH_3 with r), then $Q_{fr}^\ddagger = Q_{fr}(\infty)$ and

$$k_{r,\infty} = g_e \frac{kT}{h} \frac{\sigma}{\sigma^\ddagger} \frac{\exp(-\Delta V_{rc}(r^\ddagger)/kT)}{Q_{trans}} Q_{pd}^\ddagger \Theta^\ddagger \qquad (6.40)$$

Finally, if the transition state is located where the internal rotors are free, then $\Theta^\ddagger$ is unity. Expanding $Q_{pd}^\ddagger$, we obtain

$$k_{r,\infty} = g_e \frac{\sigma}{\sigma^\ddagger} \pi(r^\ddagger)^2 \left(\frac{8kT}{\pi\mu} \right)^{1/2} \exp(-\Delta V_{rc}(r^\ddagger)/kT) \qquad (6.41)$$

which is simply the collision theory expression for association with the necessary inclusion of the electronic degeneracy and rotational symmetry factors. The collision diameter, $r^\ddagger$, corresponds in this case to the canonic-ally averaged maximum in the centrifugal potential, since the only r depend-ent term in $Q'(r)$ is the partition function for the pseudo-diatomic molecule, A–B. This description corresponds to that of Gorin[2] for recombination of two structureless particles. The distance $r^\ddagger$ can be determined explicitly if a simple model potential, such as the Lennard-Jones function, is employed.

The inclusion of $\Theta^\ddagger$ allows for hindrance of the transition state internal rotors; $\Theta^\ddagger$ is bounded above by unity and below by the harmonic oscillator approximation. It depends on the transitional mode potential, V_{tr} and can be calculated from either a full *ab initio* potential or from a simple model. In an earlier form of this model, Aubanel, Robertson and Wardlaw[31] used the potential of Leblanc and Pacey[29] for $CH_3 + H$, in which the hindering potential varies as $\sin^2 \psi$, where ψ is the angle between the C_3 axis of the CH_3 and the $H_3C\cdots H$ axis. Smith and Golden[32] have, in principle, used a similar approach in their hindered Gorin model.

The inclusion of the partition functions for internal rotation and vibration in equation (6.38) allows a refinement of the model to incorporate changes in geometry and vibrational frequencies between the fragments and the transition state.

The corresponding limiting high-pressure dissociation rate constant can be obtained by multiplying (6.38) by the equilibrium constant

$$K_c(T) = \frac{Q_R}{Q_{AB}} \exp\left(-\Delta E_0^0/kT\right) \tag{6.42}$$

where Q_{AB} is the total partition function for AB and $\Delta E_0^0 = V_{rc}(\infty) + \Delta E_z$ is the (molecular) energy change at zero Kelvin, ΔE_z being the change in total zero-point energy on dissociation. Thus

$$k_\infty = \frac{kT}{h}\frac{\sigma_{AB}}{\sigma^\ddagger}\exp\left(-(V_{rc}(r^\ddagger) + \Delta E_z)/kT\right)\left\{\frac{Q_{vib}^\ddagger}{Q_{vib,AB}}\right\}\left\{\frac{Q_{pd}^\ddagger Q_{fr,A}^\ddagger Q_{fr,B}^\ddagger}{Q_{rot,AB}}\right\}\Theta^\ddagger$$

$$\tag{6.43}$$

where σ_{AB} is the symmetry number for AB and $Q_{vib,AB}$ and $Q_{rot,AB}$ are the total vibrational and rotational partition functions for AB. The physical interpretation of equation (6.43) follows from the discussion of the association rate constant, but this expression is even more physically transparent because the correlation of the modes of motion is more direct.

The CFTST model represents a potentially powerful and physically appealing approach to calculating rate constants on Type II surfaces. It is flexible, providing a continuous link between free and highly hindered rotor limits and can be used with very approximate or detailed description of the potential surface. In the canonical form it can only generate $k_\infty(T)$ and cannot be used directly to describe pressure-dependent rate constants. Inverse Laplace transformation (ILT) (see section 6.7) of $k_\infty(T)$ allows access to the microcanonical rate constants required for such a description via master equation models (Chapter 7). A related discussion of hindered rotational models has been given by Pitt, Gilbert and Ryan.[33]

Finally, in this section, we describe the application by Robertson, Wagner and Wardlaw[30] of the CFTST model to the $CH_3 + H$ reaction. Figure 6.10 shows a contour plot of the potential energy function, V_{tr}, for $r = 2.9$ Å, corresponding to a high temperature ($1000 \leqslant T/K \leqslant 2000$) value for $r^\ddagger$. The angle of rotation of the CH_3 radical about its axis is ϕ and ψ is the orientation of the C_3 axis to the reaction coordinate. The latter is taken as the line joining the centres of mass of the fragments. Because the CH_3 radical is slightly non-planar at the transition state, its centre of mass, which

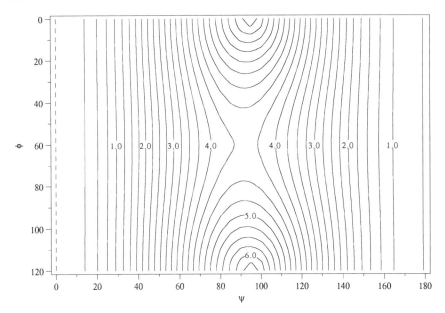

Figure 6.10 Potential energy surface for the transitional modes of the reaction $CH_3 + H \rightarrow CH_4$. From Robertson *et al.*, *J. Chem. Phys.*, **103**, 2917 (1995). Reproduced with permission from the American Institute of Physics.

is also used as the origin for ψ and ϕ, does not quite coincide with the carbon atom. There are several interesting features of the surface:

1. V_{tr} peaks at $\psi \approx 90°$, corresponding to the H atom reactant approaching towards the H atoms in CH_3 (i.e. where the CH_3 'plane' contains the reaction coordinate).
2. For $\psi = 90°$, the minimum occurs for $\phi = 60°$, where H approaches between two C–H bonds.
3. For $\psi = 0, 180°$, V_{tr} is approximately independent of ϕ and the potential is a minimum, except that it is slightly lower for $\psi = 0$, where the CH_3 radical is splayed away from the approaching H. This asymmetry disappears if CH_3 retains its planar geometry throughout the reaction.

Figure 6.11 shows the results of the CFTST calculations and compares them with the canonical calculations of Aubanel and Wardlaw[13] using the full FTST.

It is of interest to examine the magnitude of the terms in equation (6.38). The dominant contribution to the vibrational term derives from the umbrella mode of CH_3 which, as previously noted, changes markedly along the reaction coordinate. Despite the change in geometry of CH_3, the free rotor partition function ratio, $Q_{fr}^{\ddagger}/Q_{fr}(\infty)$ is within 1% of unity and the only significant term in the second bracketed function in equation (6.38) is $Q_{pd}^{\ddagger}$.

The most interesting component of the rate constant expression is the hindering function, $\Theta^{\ddagger}$, which is dependent on V_{tr}.

Figure 6.11 shows that the agreement between the CFTST calculations based on equation (6.38) (CFTST–HR) and the full CFTST calculations[13] (CFTST0) is excellent. At least as far as $k_{r,\infty}(T)$ is concerned, this comparatively easily implemented version of the theory gives results which are just as good as the full FTST model, which requires considerably more computer time.

Figure 6.11 also shows the calculated rate constants in the free rotor ($\Theta^{\ddagger} = 1$) and harmonic oscillator limits. These rate constants represent the

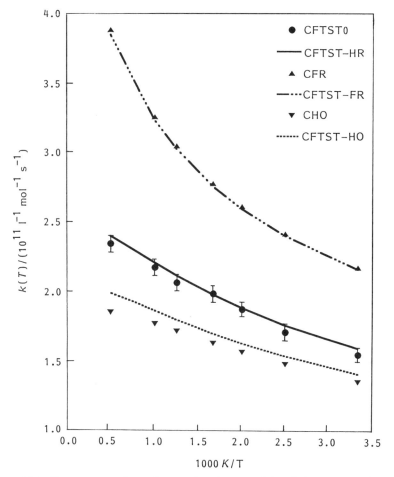

Figure 6.11 Temperature dependence of various models of the rate coefficient of the reaction $CH_3 + H \rightarrow CH_4$. From Robertson *et al.*, *J. Chem. Phys.*, **103**, 2917 (1995). Reproduced with permission from the American Institute of Physics.

upper and lower bounds on $k_{r,\infty}(T)$. The exact location of $k_{r,\infty}(T)$ within these bounds is determined by V_{tr}. The free rotor model results (CFTST–FR) are compared with the usual method employed in variational TST calculations (CFR) in which the external and internal rotational modes are assumed to be separable. The agreement is within 1%. The agreement between the harmonic limit (CFTST–HR) and the conventional harmonic model (CHO) is less satisfactory.

6.7 INVERSE LAPLACE TRANSFORMATION

Reference has been made in the last section to ILT of $k_\infty(T)$ as a route to microcanonical rate constants. The relationship was first noted by Slater[34] and later used extensively by Forst[35] and Pritchard[36]. The dissociation rate constant, $k_\infty(T)$ is given by

$$k_\infty = \frac{1}{Q_{\mathrm{vr,AB}}(T)} \int_0^\infty k_a(E^*)\rho_{\mathrm{AB}}(E^*)\exp(-E^*/kT)\,\mathrm{d}E^* \qquad (6.44)$$

where $Q_{\mathrm{vr,AB}}$ is the rovibrational partition function of the dissociating reactant, AB. Thus, $k_a(E^*)$ is linked to $k_\infty(T)$ by ILT:

$$k_a(E^*)\rho_{\mathrm{AB}}(E^*) = \mathcal{L}^{-1}[Q_{\mathrm{vr,AB}}(T)k_\infty(T)] \qquad (6.45)$$

A simple, physically appealing treatment, which demonstrates the significance of these equations, was provided by Pritchard[36]. Setting $k_\infty(T) = A_\infty \exp(-E_\infty/kT)$, equation (6.44) can be written as

$$A_\infty Q_{\mathrm{vr,AB}}(T) = \exp(E_\infty/kT)\int_0^\infty k_a(E^*)\rho_{\mathrm{AB}}(E^*)\exp(-E^*/kT)\,\mathrm{d}E^* \qquad (6.46)$$

From which it follows that

$$A_\infty Q_{\mathrm{vr,AB}}(T) = \int_{-E_\infty}^\infty k_a(E^*)\rho_{\mathrm{AB}}(E^*)\exp(-(E^*-E_\infty)/kT)\,\mathrm{d}(E^*-E_\infty) \qquad (6.47)$$

Noting that $k_a(E^*)=0$ for $E^* < E_\infty$, the limits of equation (6.47) may be changed to give

$$A_\infty Q_{\mathrm{vr,AB}}(T) = \int_0^\infty k_a(E^*)\rho_{\mathrm{AB}}(E^*)\exp(-(E^*-E_\infty)/kT)\,\mathrm{d}(E^*-E_\infty) \qquad (6.48)$$

in which $A_\infty Q_{\mathrm{vr,AB}}(T)$ is the Laplace transform of $k_a(E^*)\rho_{\mathrm{AB}}(E^*)$, but with the integration variable shifted by E_∞. Taking the inverse

$$k_a(E^*) = \frac{A_\infty \rho_{\mathrm{AB}}(E^*-E_\infty)}{\rho_{\mathrm{AB}}(E^*)} \qquad (6.49)$$

a very simple relationship, originally derived by Forst[35], in which $k_a(E^*)$ can be calculated without reference to the transition state, via experimental values of A_∞ and E_∞ and calculations of the densities of states. The expression has a physical interpretation too, since the energy dependence of $k_a(E^*)$ is reflected in the ratio of the shifted and actual densities of states at E^*, while the overall magnitude is determined primarily by A_∞ which implicitly contains transition state information.

The treatment is only valid provided $k_\infty(T)$ is strictly described by the simple Arrhenius expression over the range $0 < T/K < \infty$. One of the drawbacks in its application is the extreme sensitivity of $k_\infty(T)$ to temperature, so that it is difficult to measure over a wide range of temperatures and hence determine E_∞ precisely.

Davies, Green and Pilling[37] proposed an alternative approach utilising measurements of recombination and association rate constants, $k_{r,\infty}(T)$, instead. These reactions are much less sensitive to T and can, in principle, be measured over a very wide range of temperatures, with more precise determinations of the temperature dependence than is possible for dissociation. Association reactions primarily sample energies above the dissociation limit and the population density at energies significantly below the dissociation limit is very small, except very near the high-pressure limit. They are, therefore, exactly the range of energies contributing to the reaction.

The new Laplace transform relation is developed via the equilibrium constant (equation 6.42),

$$k_\infty(T) = k_{r,\infty}(T) \frac{Q_{vr,R}}{Q_{vr,AB}} \exp\left(-\Delta E_0^0/kT\right) \tag{6.50}$$

Defining E_R as the rovibrational energy of the fragments relative to their zero point, i.e. $E_R = E^* - \Delta E_0^0$ and writing $k_{r,\infty}(T) = A_{r,\infty} \exp\left(-E_{r,\infty}/kT\right)$ the inverse Laplace transform is

$$k_a(E^*) =$$

$$\frac{A_{r,\infty} C'}{\rho_{AB}(E^*)} \int_0^\infty \rho_R(E_R) \mathscr{L}^{-1}\left[(kT)^{3/2} \exp\left(-(E_R + E_{r,\infty} + \Delta E_0^0)/kT\right)\right] dE_R \tag{6.51}$$

where $C' = (2\pi\mu)^{3/2}/h^2$ and derives, together with the term in $(kT)^{3/2}$ from the translational partition functions. The total (i.e. convoluted) density of states of the fragments A and B at the combined energy E_R is $\rho_R(E_R)$. Evaluating the inverse gives

$$k_a(E^*) = \frac{2A_{r,\infty} C'}{\pi^{1/2} \rho_{AB}(E^*)} \int_0^{E_R - E_{r,\infty} - \Delta E_0^0} \rho_R(E_R)[E^* - (E_R + E_{r,\infty} + \Delta E_0^0)]^{1/2} dE_R$$

$$\text{for } E^* \geqslant E_{r,\infty} + \Delta E_0^0 \tag{6.52}$$

Davies, Green and Pilling[37] tested this approach for $CH_3 + CH_3$ and used it to illustrate the inadequacies of simple variational transition state models through a direct comparison of experimentally derived and model-based microcanonical rate constants. The inversion technique has also been employed to fit experimental data for $CH_3 + NO$, $CH_3 + CH_3$ and $i\text{-}C_3H_7$ decomposition, using a master equation approach (see Chapter 7). The functional dependence of $k_{r,\infty}$ on T was extended to the modified Arrhenius form $A_{r,\infty} T^n \exp(-E_{r,\infty}/kT)$, which can be inverted for $n_\infty \geqslant -1.5$. As noted in section 6.6, the technique also provides a mechanism for obtaining microcanonical rate constants from canonical models. Note, however, the rotational averaging implicit in $k_{r,\infty}$ which cannot be reversed in the inversion: it is not possible to construct the J dependence of the microcanonical rate constant using this approach.

REFERENCES

1. J. W. Davies, D.Phil thesis, Oxford (1989).
2. E. Gorin, *Acta Physiochim. URSS*, **9**, 691 (1938).
3. J. C. Keck, *Adv. Chem. Phys.*, **13**, 85 (1967).
4. E. Wigner, *J. Chem. Phys.*, **5**, 720 (1937).
5. J. Horiuti, *Bull. Chem. Soc. Jpn.*, **13**, 210 (1938).
6. D. L. Bunker and M. Pattengill, *J. Chem. Phys.*, **48**, 772 (1968).
7. W. H. Wong and R. A. Marcus, *J. Chem. Phys.*, **55**, 5625 (1971).
8. M. Quack and J. Troe, *Ber. Bunsenges. Phys. Chem.*, **78**, 240 (1974).
9. B. C. Garrett and D. G. Truhlar, *J. Chem. Phys.*, **70**, 1593 (1979).
10. M. Brouard, M. T. Macpherson and M. J. Pilling, *J. Phys. Chem.*, **93**, 4047 (1989).
11. W. L. Hase, S. L. Mondro, R. J. Duchovic and D. M. Hirst, *J. Am. Chem. Soc.*, **109**, 2916 (1987).
12. M. Quack and J. Troe, *Ber. Bunsenges. Phys. Chem.*, **79**, 170 (1975).
13. E. E. Aubanel and D. M. Wardlaw, *J. Phys. Chem.*, **93**, 3117 (1989).
14. D. M. Wardlaw and R. A. Marcus, *J. Phys. Chem.*, **90** 5383 (1986).
15. S. J. Klippenstein and R. A. Marcus, *J. Chem. Phys.*, **87**, 3410 (1987).
16. A. F. Wagner and D. M. Wardlaw, *J. Phys. Chem.*, **92**, 2462 (1988).
17. M. T. Macpherson, M. J. Pilling and M. J. C. Smith, *Chem. Phys. Lett.*, **94**, 430 (1983); *J. Phys. Chem.*, **89**, 2268 (1985).
18. I. R. Slagle, D. Gutman, J. W. Davies and M. J. Pilling, *J. Phys. Chem.*, **93**, 2455 (1988).
19. D. L. Baulch, C. J. Cobos, R. A. Cox, C. Esser, P. Frank, T. Just, J. A. Kerr, M. J. Pilling, J. Troe, R. W. Walker and J. Warnatz, Evaluated kinetic data for combustion modeling, *J. Phys. Chem. Ref. Data* **21** 411 (1992).
20. W. H. Miller, N. C. Handy and J. E. Adams, *J. Chem. Phys.*, **72**, 99 (1980).
21. D. M. Wardlaw and R. A. Marcus, *J. Chem. Phys.*, **85**, 3462 (1985).
22. S. C. Smith, *J. Chem. Phys.*, **97**, 2406 (1992) and *J. Phys. Chem.*, **97**, 7034 (1993).
23. L. S. Kassel, *J. Chem. Phys.*, **4**, 276 (1936).

24. S. J. Klippenstein, *Chem. Phys. Lett.*, **170**, 71 (1990); *J. Chem. Phys.*, **94**, 6469 (1991).
25. A. I. Maergoiz, E. E. Nikitin and J. Troe, *J. Chem. Phys.*, **95**, 5117 (1991).
26. M. Quack and J. Troe, *Ber. Bunsenges. Phys. Chem.*, **81**, 329 (1977).
27. J. Troe, *J. Chem. Phys.*, **75**, 226 (1981).
28. S. L. Mondro, S. R. Vande Linde and W. L. Hase, *J. Chem. Phys.*, **84**, 3783 (1989); **86**, 1348 (1987); W. L. Hase and X. Hu, *Chem. Phys. Lett.*, **156**, 115 (1989).
29. J. F. Leblanc and P. D. Pacey, *J. Chem. Phys.*, **83**, 4511 (1985).
30. S. H. Robertson, A. F. Wagner and D. M. Wardlaw, *J. Chem. Phys.*, **103**, 2917 (1995).
31. E. E. Aubanel, S. H. Robertson and D. M. Wardlaw, *J. Chem. Soc. Farad. Trans.*, **87** 2291 (1991).
32. G. P. Smith and D. M. Golden, *Int. J. Chem. Kinet.*, **10**, 489 (1978).
33. I. G. Pitt, R. G. Gilbert and K. R. Ryan, *J. Phys. Chem.*, **99** 239 (1995).
34. N. B. Slater, *Proc. Leeds Phil. Lit. Soc.*, **6**, 259 (1955).
35. W. Forst, *J. Phys. Chem.*, **76**, 342 (1972).
36. H. O. Pritchard, *The Quantum Theory of Unimolecular Reactions*, Cambridge University Press (1984).
37. J. W. Davis, N. J. B. Green and M. J. Pilling, *Chem. Phys. Lett.*, **129**, 373 (1986).
38. G. Herzberg, *Molecular Spectra and Molecular Structure; Electronic Spectroscopy of Polyatomic Molecules*, Van Nostrand, New York (1966).
39. C. Yamoda, E. Hirota and K. Kawaguchi, *J. Chem. Phys.*, **75**, 5256 (1981).
40. S. Heneghan, P. A. Knoot and S. W. Benson, *Int. J. Chem. Kinet.*, **13**, 677 (1981).
41. D. L. Gray and A. G. Robiette *Mol. Phys.*, **37**, 1901 (1979).

7 Master Equation Analysis of Collisional Energy Transfer

The RRKM analysis developed in Chapter 3 and employed in Chapter 5 was based on the strong collision assumption, that collisional energisation and de-energisation are essentially single-step processes. Some recognition of the inadequacy of this approach was made by introducing a collisional efficiency parameter, β_c into equation (3.17). Comparatively small amounts of energy ($\sim 2\ \mathrm{kJ\,mol^{-1}}$) are transferred on collision, particularly with inert gas atoms, and a realistic analysis of thermal reactions and, especially, of chemical and photo-activation processes must accommodate *stepwise* collisional energy transfer. In this chapter we shall develop a practical means of analysis based on the so-called *energy-grained master equation*.

Section 7.1 develops the master equation for a molecular system with an effectively continuous distribution of energy states. Such an approach is well suited to analytic solution in idealised situations, but not for numerical solution of practical systems. Accordingly, in section 7.2, the energy-grained master equation is formulated in which the energy states are bundled into discrete 'grains' and a matrix technique is employed. In section 7.3, the basis of the application of the master equation to chemical activation systems is presented and is extended in sections 7.4 and 7.5 to include equilibration and isomerisation reactions. Methods of solution are discussed in section 7.6, using specific examples of dissociation, and association reactions. Finally, in section 7.7 a diffusion approach to describing energy transfer is presented.

It is impossible in this short chapter to give a full account of all the issues concerned, and the reader is referred to the excellent review by Oref and Tardy[1] for a more detailed treatment.

7.1 SYSTEMS WITH CONTINUOUS STATE DISTRIBUTIONS

We begin by considering the unimolecular reaction of molecule A:

$$A \xrightarrow{k_{uni}} products$$

$$\frac{-d[A]}{dt} = k_{uni}[A]$$

We shall be primarily concerned with high energies of A in the region of and above the dissociation limit, where, even for triatomic molecules, we may assume that the energy levels are so closely spaced that they may be treated as a continuum, described by a density of states, $\rho(E)$. We shall assume that collisions are simple, binary, independent events and that both energy transfer and reaction depend only on the total energy, E, of the molecule. In particular, we shall, for the most part, neglect the dependence of the microcanonical rate coefficient for dissociation on angular momentum.

We denote the energy of the molecule before and after collision as x and y respectively and the energy-dependent population density of A molecules at time t as $n(y, t)$. The time-dependence of $n(y, t)$, in the absence of reaction, is given by the equation

$$\frac{dn(y, t)}{dt} = \begin{matrix} \text{rate of collisional} \\ \text{energy transfer into} \\ \text{states in range } y \to y + dy \end{matrix} - \begin{matrix} \text{rate of collisional} \\ \text{energy transfer out of} \\ \text{states in range } y \to y + dy \end{matrix}$$

$$\frac{dn(y, t)}{dt} = \omega \int_0^\infty P(y|x)n(x, t)\,dx - \omega n(y, t) \tag{7.1}$$

where $P(y|x)$ is the conditional probability of transfer from x to y and ω is the collision frequency $(= Z[M])$.

In order to solve equation (7.1), $P(y|x)$ must be specified. There is comparatively little information available on its detailed form; experimental investigations are described in section 9.1. Theoretical investigations can be broadly divided into two types—those based on classical mechanics and those based on quantum mechanics.

In the classical mechanical approach collisions are modelled by following the classical trajectory of the colliding species. The potential energy surface on which the collision takes place has to be specified and if the interaction of the internal modes with external forces is present then inelastic collisions can be modelled. The transition density $P(y|x)$ can be estimated by running a very large number of trajectories for different initial reactant internal energies and collision parameters. Information on the final energy of the reactant as a function of the initial energy is then collected and used to construct $P(y|x)$. Gilbert and Lim[2] have investigated azulene/Ar collisions using this approach and Lendvay and Schatz[3] have investigated CS_2/CO collisions similarly.

The quantum mechanical investigation has been mostly confined to small molecule systems[4], the main reason being that the Schrödinger equation has to be solved, which in general is a much more difficult problem. Clary and Kroes[5] have tackled large molecule/atom collision systems by using the family of sudden approximations, the basis of which is the idea that energy transfer occurs on a time scale which is short compared to the periods of internal motion (vibrational or rotational). These approximations allow collision cross-sections of energy transfer processes to be calculated, and, in principle, these cross-sections can then be used to calculate $P(y|x)$ by averaging them over the appropriate Maxwell–Boltzman distribution.

For routine analysis neither of these approaches is practical and heuristic models are used, usually having a number of parameters, often with a physical interpretation. There are a number of such models, eg. stepladder[6], Gaussian[6] and biased random walk[7]. For the present we shall adopt the so-called *exponential down* model[8], in which it is assumed that energy removed from the molecule is exponentially distributed and depends only on the energy difference $(x - y)$:

$$P(y|x) = C(x)\exp\{-\alpha(x - y)\} \qquad x \geqslant y \qquad (7.2)$$

where $C(x)$ is a normalisation factor and, provided $x \gg 0$, $\alpha = \langle \Delta E \rangle_d^{-1}$, the average energy transferred in a downward direction. This model (sometimes referred to at the exponential gap model) reflects the common-sense notion that large energy jumps are less likely than small ones.

The probability distribution for 'upward' transitions, in which the energy of A is increased, is related to that for downward transitions via detailed balance:

$$P(y|x)\rho(x)\exp(-x/kT) = P(x|y)\rho(y)\exp(-y/kT) \qquad (7.3)$$

where $\rho(x)\exp(-x/kT)$ is the Boltzmann factor for energy level x, so that, for $x < y$,

$$P(x|y) = C(y)\exp\{-\alpha(y - x)\} \text{ (cf. 7.2)}$$

and, from equation (7.3),

$$P(y|x) = \frac{C(y)\rho(y)}{\rho(x)}\exp\{-(\alpha + \beta)(y - x)\} \qquad x < y \qquad (7.4)$$

where $\beta = 1/kT$. Detailed balance ensures that the long-time solution of equation (7.1) is the Boltzmann distribution, $f(y)$.

As well as detailed balance $P(y|x)$ constrained by the normalisation condition,

$$\int_0^\infty P(y|x)\,dy = 1 \tag{7.5}$$

which expresses the physical notion that collision does not destroy the unimolecular reactant. The normalisation condition equation (7.5) is used to determine $C(x)$ (see section 7.6).

Finally, reaction may be incorporated into the master equation (7.1) via an energy-dependent first-order loss term:

$$\frac{dn(y, t)}{dt} = \omega\int_0^\infty P(y|x)n(x, t)\,dx - \omega n(y, t) - k(y)n(y, t) \tag{7.6}$$

where $k(y)$ is the microcanonical rate coefficient for reaction at energy y.

7.2 ENERGY-GRAINED MASTER EQUATION (EGME)

Equation (7.6) is appropriate for analytic solution, but such an approach is only generally feasible with simple analytic forms for $\rho(x)$, $k(x)$ and in the limit $\omega \to 0$, i.e. at the low-pressure limit. For application to realistic systems, numerical solution is generally required, necessitating a reformulation of the master equation. The usual approach is to partition the energy levels into a contiguous set of 'grains'. The width of the grains, δ, must be less than $\langle \Delta E \rangle_d$ for a convergent solution and must also be sufficiently small that $P(x|y)$, $C(y)$, $k(y)$ and $\rho(y)$ do not vary significantly across a grain. In these circumstances, each grain may be thought of as an energy-grained state, i, of degeneracy N_i, where N_i is the total number of states within the grain:

$$N_i = \int_{y''}^{y'} \rho(y)\,dy \tag{7.7}$$

associated with an appropriately weighted energy, E_i and rate constant, k_i:

$$E_i = \int_{y''}^{y'} y\rho(y)\,dy \bigg/ \int_{y''}^{y'} \rho(y)\,dy \tag{7.8}$$

$$k_i = \int_{y''}^{y'} k(y)\rho(y)\,dy \bigg/ \int_{y''}^{y'} \rho(y)\,dy \tag{7.9}$$

where y' and y'' are the upper and lower limits of the grain.

The discretised energy transfer probability distribution takes the form

$$P_{ij} = C_j \exp\{-\alpha(E_j - E_i)\} \qquad j \geqslant i \tag{7.10a}$$

$$P_{ij} = C_i \frac{N_i}{N_j} \exp\{-(\alpha + \beta)(E_i - E_j)\} \qquad j < i \tag{7.10b}$$

for transfer from state j to state i. The normalisation condition is

$$\sum_{i=1}^{m} P_{ij} = 1 \tag{7.11}$$

where the upper limit, $i = m$, corresponds, to a finite energy, $m\delta$, where δ is the grain width. $m\delta$ replaces $y = \infty$ for practical computing purposes; m is determined by trial and error. The determination of the normalisation constants is discussed in section 7.6.

The densities of states, $\rho(y)$, may be calculated using the Whitten–Rabinovitch approximation[9] (section 4.4), in which case a continuous function is obtained and the analysis given above applies. Alternatively, the Stein–Rabinovitch (SR)[10] algorithm (section 4.2) may be used, which results in discrete sums and densities, calculated using a preset energy width. We shall refer to the grains employed in the SR[10] analysis as cells, to distinguish them from the grains in the master equation. Generally speaking, the cell widths $(1–10 \text{ cm}^{-1})$ are smaller than the grain width, δ ($\sim 100 \text{ cm}^{-1}$). Defining the number of states in a cell contained within grain i as $N'_{i,l}$ then

$$N_i = \sum_l N'_{i,l} \tag{7.12}$$

while equations (7.11) and (7.12) become:

$$E_i = \sum_l E'_{i,l} N'_{i,l} / \sum_l N'_{i,l} \tag{7.13}$$

$$k_i = \sum_l k'_{i,l} N'_{i,l} / \sum_l N'_{i,l} \tag{7.14}$$

where E'_{il} and k'_{il} are the energy and rate constant of the lth cell in grain i.

We are now in a position to formulate the energy-grained master equation (EGME), based on (7.6):

$$\frac{dn_i}{dt}(t) = \omega \sum_{j=1}^{m} P_{ij} n_j(t) - \omega n_i(t) - k_i n_i(t) \tag{7.15}$$

Equation (7.15) describes the rate of change of the population density of grain i. It is one of a set of coupled differential equations and this set can be more compactly written in matrix form:

$$\frac{d\mathbf{n}}{dt} = [\omega(\mathbf{P} - \mathbf{I}) - \mathbf{k}]\mathbf{n} = \mathbf{M}\mathbf{n} \tag{7.16}$$

where $\mathbf{M} = \omega(\mathbf{P} - \mathbf{I}) - \mathbf{k}$, $\mathbf{P}$ is the matrix of transition probabilities with elements P_{ij}, $\mathbf{I}$ is the identity matrix, $\mathbf{k}$ is a diagonal matrix of microcanonical rate coefficients, k_i and $\mathbf{n}$ is the population vector, with elements n_i. Figure 7.1 demonstrates the form of the matrix $\mathbf{M}$ in (7.16). Here $\mathbf{M}$ may be

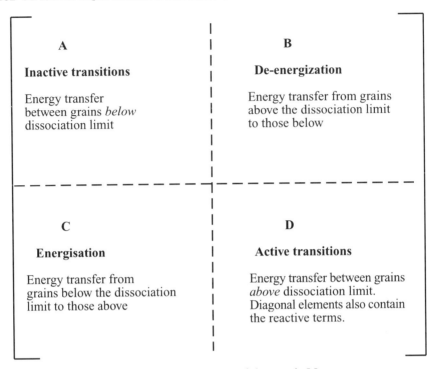

Figure 7.1 Block form of the matrix **M**.

divided into four regions, determined by the location of the grain d, which is the first grain above the dissociation limit:

A: $i < d$, $j < d$ contains 'inactive' transitions, i.e. those involving levels which are exclusively below the dissociation limit.

B: $i < d$, $j \geq d$ contains *de-energisation* collisions from states above the dissociation limit to states below.

C: $i \geq d$, $j < d$ contains *energisation* collisions from states below d to states at or above d.

D: $i \geq d$, $j \geq d$ contains active transitions between states above the limit.

The diagonal elements contain $\omega(P_{ii} - 1)$ which is the net rate of collisional energy transfer from grain i, P_{ii} being the probability of remaining in the grain. The diagonal elements in region D of the matrix **M** also contain the microcanonical rate coefficients for dissociation, k_i.

The solution of (7.16) is

$$n(t) = \mathbf{U} \exp(\mathbf{\Lambda} t)\mathbf{U}^{-1} n(0) \tag{7.17}$$

where $n(0)$ is the initial population vector, **U** is the right eigenvector matrix of **M** and $\mathbf{\Lambda}$ is the diagonal matrix of eigenvalues of **M** (see Appendix 6).

TIME DEPENDENCE

If we define the vector c as

$$c = \mathbf{U}^{-1}n(0)$$

then we can rewrite (7.17) in the form:

$$n(t) = \sum_i u_i c_i \exp(\lambda_i t) \tag{7.18}$$

where u_i is the ith eigenvector of $\mathbf{M}$, with eigenvalue λ_i. Note that the λ_i are all negative. Thus the population distribution evolves as a sum of first-order decays, defined by the first-order rate coefficients, $|\lambda_i|$. For real systems it is found that there is one eigenvalue, λ_1, which is much smaller in magnitude than the rest: $|\lambda_1| \ll |\lambda_2|, |\lambda_3| \ldots$.[11] Thus, after an initial transient period, during which the terms in $i > 1$ decay, the population is given, to a good approximation, by the exponentially decaying function

$$n(t) \cong u_1 c_1 \exp(\lambda_1 t) \tag{7.19}$$

The contribution of the transient terms has been discussed in detail by Pritchard[11]; for our present purposes they need concern us no further. We now consider the relationship between λ_1 and k_{uni}. The concentration of A at time t is given by

$$[A](t) = \sum_i n_i(t) = c_1 \exp(\lambda_1 t) \sum_i (u_1)_i \tag{7.20}$$

Differentiation with respect to t gives

$$\frac{d[A](t)}{dt} = \lambda_1[A](t) \tag{7.21}$$

so that

$$\lambda_1 = -k_{uni} \tag{7.22}$$

i.e. the eigenvalue of smallest magnitude can be equated to the negative value of the unimolecular rate constant.

An important property of a reacting system during this 'long time' period may be deduced from equation (7.19) by examining the populations of the ith and jth grains:

$$\frac{n_j(t)}{n_i(t)} = \frac{(u_1)_j c_1 \exp(\lambda_1 t)}{(u_1)_i c_1 \exp(\lambda_1 t)} = \frac{(u_1)_j}{(u_1)_i} \tag{7.23}$$

Thus, after the initial transient period, the population distribution reaches a steady shape, which is determined by the interplay of collisional energy transfer and reaction. The form of this distribution will be illustrated by reference to the dissociation of the 2-propyl radical in section 7.6. The steady-state distribution $\boldsymbol{u}_1$ determines the magnitude of k_{uni}, since

$$k_{uni} = \sum_i k_i n_i / \sum_i n_i \qquad (7.24)$$

and hence from (7.19),

$$k_{uni} = \sum_i k_i (\boldsymbol{u}_1)_i / \sum_i (\boldsymbol{u}_1)_i \qquad (7.25)$$

At very high pressure, the pseudo-first-order energy-grained rate constants for energy transfer are much larger than the microcanonical rate constants for reaction so that the master equation reduces, to a good approximation, to the energy-grained form of (7.1) and the steady-state distribution corresponds to the Boltzmann distribution, cf. (7.5). At lower pressures the population is depleted by reaction both above and below the dissociation limit (see section 7.6) and the rate constant, k_{uni}, falls below the high-pressure limit.

7.3 CHEMICAL ACTIVATION

The preceding discussion concerned thermal unimolecular reactions where energisation occurs exclusively as a result of collisional energy transfer. The energy dependence of the rate of energisation is determined by the Boltzmann distribution (7.3), and deviations of the population from the Boltzmann distribution are a direct result of reaction from energised states of the molecule. Distributions far from equilibrium can be generated by chemical activation. Examples of such a process include the reactions (7.26) and (7.27):

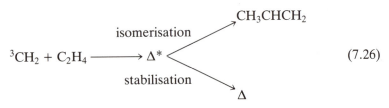

$$(7.26)$$

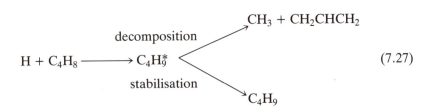

$$(7.27)$$

In reaction (7.26), triplet methylene adds to ethene to form an energised cyclopropane molecule which has sufficient energy to isomerise to form propene. Alternatively, the cyclopropane may be collisionally stabilised. The isomerisation/stabilisation ratio may be measured experimentally, its pressure dependence giving access to energy transfer data (see section 9.1). Similar behaviour is observed for reaction (7.27) which involves addition of hydrogen atoms to butene, resulting in decomposition or stabilisation.

The potential energy surface for both of these reactions is shown schematically in Figure 7.2(a). The chemically activated molecule, C^*, is formed from the reactants via a transition state, $X^{\ddagger}$ and can react to form the products via a lower energy barrier, $Y^{\ddagger}$. Dissociation of C^* to regenerate the reactants via $X^{\ddagger}$ is also feasible, although it is usually slower than the

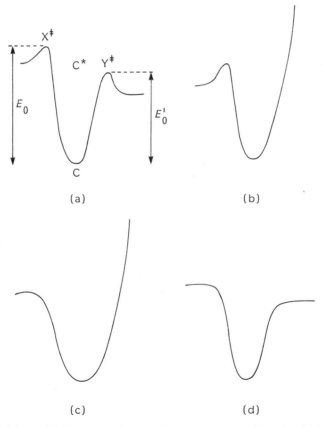

Figure 7.2 Schematic diagrams of potential energy curves for chemical activation reactions: (a) chemical activation via constrained activated complexes; (b) association on a Type I surface; (c) association on a Type II surface; (d) chemical activation via loose transition states.

forward reaction rate via $Y^{\ddagger}$ because of the differences in the barrier heights. An increase in pressure further reduces the rate of the back reaction; it also increases the rate of stabilisation of C^*.

Radical association reactions, e.g.

$$CH_3 + CH_3 \rightarrow C_2H_6 \tag{7.28}$$

and atom addition reaction without a forward dissociation channel, e.g.

$$H + C_2H_4 \rightarrow C_2H_5 \tag{7.29}$$

correspond to the reverse of dissociation reaction and represent a specific form of a chemical activation process. Their potential energy surfaces are represented schematically in Fig. 7.2. Reaction (7.29) occurs on a Type I surface with an energy barrier (Figure 7.2(b)), while reaction (7.28) occurs on a Type II surface where there is no barrier (Figure 7.2(c), cf. Chapter 5).

Chemical activation with a forward dissociation channel can also occur on a Type II surface. An example is the combination of H and C_2H_5:

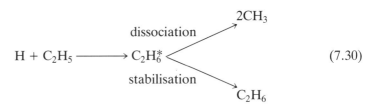

$$ \tag{7.30}$$

This surface is shown schematically in Figure 7.2(d).

Chemical activation may be analysed using a modified strong collision RRKM approach. For the scheme:

$$ \tag{7.31}$$

The rate of formation of A^* in the energy range $E \rightarrow E + dE$ is $v_r(E)$ and $k'_a(E)$ and $k_a(E)$ are the rate constants for dissociation of A^* to form products (with threshold energy E'_0) and to regenerate reactants (with threshold energy E_0) respectively. Here $v_r(E)\,dE$ is equal to the overall rate of forming A^* from $R + R'$, $k_{r,\infty}[R][R']$, multiplied by the fraction, $g(E)\,dE$, of the reactive collisions leading to formation of A^* in this energy range:

$$v_r(E)\,dE = k_{r,\infty}[R][R']g(E)\,dE \tag{7.32}$$

where $k_{r,\infty}$ is the limiting high-pressure rate coefficient for the $R + R'$ reaction. In the limit of infinite pressure ($\beta_c \omega \to \infty$) the rate of the reverse dissociation of A^* is zero so that the rate of forming A^* is equal to the rate of reaction of $R + R'$.

Here $g(E)\,dE$ can be calculated using the principle of detailed balance. Consider a situation where the dissociation process can be ignored in calculating $[A^*]$ and in which an equilibrium is established between $R + R'$ and A. The rate of dissociation from the energised molecules in the range $E \to E + dE$, $v_d(E)\,dE$ is given by

$$v_d(E)\,dE = k_a(E)[A^*(E)]\,dE$$
$$= k_a(E)[A]_e f(E)\,dE \qquad (7.33)$$

where $f(E)\,dE$ is the equilibrium fraction of A molecules in the range $E \to E + dE$ and $[A]_e$ is the equilibrium concentration of A. At equilibrium, the microcanonical rates of dissociation and recombination are equal, i.e. $v_d(E) = v_r(E)$, so that, combining (7.32) and (7.33),

$$g(E)\,dE = \frac{k_a(E)f(E)}{k_{r,\infty}}\,dE \left\{ \frac{[A]}{[R][R']} \right\}_e$$
$$= \frac{k_a(E)f(E)\,dE}{k_{d,\infty}} \qquad (7.34)$$

since $\{[A]/[R][R']\}_e$ is the equilibrium constant, which is equal to $k_{r,\infty}/k_{d,\infty}$, where $k_{d,\infty}$ is the limiting high-pressure rate coefficient for dissociation. Since

$$f(E) = \frac{\rho(E)\exp(-E/kT)}{\displaystyle\int_0^\infty \rho(E)\exp(-E/kT)\,dE} \qquad (7.35)$$

and

$$k_{d,\infty} = \frac{\displaystyle\int_{E_0}^\infty k_a(E)\rho(E)\exp(-E/kT)\,dE}{\displaystyle\int_0^\infty \rho(E)\exp(-E/kT)\,dE} \qquad (7.36)$$

$$g(E)\,dE = \frac{k_a(E)\rho(E)\exp(-E/kT)\,dE}{\displaystyle\int_0^\infty k_a(E)\rho(E)\exp(-E/kT)\,dE} \qquad (7.37)$$

We are now in a position to calculate v_S and v_D, the rates of stabilisation (S) and dissociation (D).

$$v_S = \int_{E_0}^{\infty} \omega \beta_c [A^*(E)] \, dE \tag{7.38}$$

$$= k_r^{\infty}[R][R'] \int_{E_0'}^{\infty} \frac{\omega \beta_c g(E) \, dE}{(k_a'(E) + k_a(E) + \omega \beta_c)} \tag{7.39}$$

and

$$v_D = k_r^{\infty}[R][R'] \int_{E_0'}^{\infty} \frac{k_a'(E) g(E) \, dE}{(k_a'(E) + k_a(E) + \omega \beta_c)} \tag{7.40}$$

If E_0 is significantly greater than E_0', so that $k_a(E) \ll k_a'(E)$ at all accessible energies, then D/S, the ratio of the yields of dissociation to stabilisation of products is given by

$$D/S = \int_{E_0'}^{\infty} \frac{k_a'(E) g(E) \, dE}{(k_a'(E) + \omega \beta_c)} \bigg/ \int_{E_0'}^{\infty} \frac{\omega \beta_c g(E) \, dE}{(k_a'(E) + \omega \beta_c)} \tag{7.41}$$

One major disadvantage of the modified strong collision approach is that it provides an unrealistic treatment of weak collisions. This deficiency is more serious for chemical activation systems than for simple thermal reactions or for single-channel radical combination reactions because several collisions may be needed to deactivate A^* below the dissociation energy, E_0', leading to products. Thus the rate of stabilisation is not equal, simply, to $\omega \beta_c [A^*(E)]$. Such effects can, however, be accommodated in an energy-grained master equation treatment. The reaction is now expressed in the form

$$R + R' \underset{k_i}{\overset{v_i}{\rightleftharpoons}} A_i^* \underset{k_i'}{\overset{}{\rightarrow}} \text{ products}$$
$$\Big\downarrow$$
$$A$$

and the master equation is

$$\frac{dn_i}{dt} = v_i + \omega \sum_j P_{ij} n_j(t) - \omega n_i(t) - (k_i' + k_i) n_i(t) \tag{7.42}$$

which can be compared with (7.15). Two extra terms have been included, v_i, which is the *source term* and is the rate of reaction of $R + R'$ into grain i (cf. $v_r(E) \, dE$ above) and $k_i' n_i(t)$, which is the rate of dissociation of A_i^* to form products. The experimentally accessible time regime corresponds to the steady state, where $dn_i/dt = 0$ and

$$v_i = (\omega + k_i + k_i') n_i - \omega \sum_j P_{ij} n_j \tag{7.43}$$

v_i may be determined using detailed balance (cf. $v(E) \, dE$) which gives

$$v_i = k_r^\infty [R][R'] g_i \tag{7.44}$$

where

$$g_i = \frac{k_i N_i \exp(-E_i/kT)}{\sum_i k_i N_i \exp(-E_i/kT)} \tag{7.45}$$

Equation (7.43) can be expressed in matrix form as

$$v = Qn \tag{7.46}$$

where $Q = \omega(I - P) + k + k'$ using the notation introduced in equation (7.16) with the addition of k', which is a diagonal matrix containing the microcanonical rate constants, k_i', and v is a vector whose elements are the rates v_i.

The rate of forming products, v_D, is given by

$$v_D = \sum_i k_i' n_i \tag{7.47}$$

It is generally convenient to normalise the rate of reaction into A_i^*, v_i, by setting $[R] = [R'] = 1$. With this restriction,

$$k_D = v_D = \sum_i k_i' n_i \tag{7.48}$$

The rate constant for stabilisation, k_s, can be calculated by difference:

$$k_s = \sum v_i - \sum (k_i + k_i') n_i \tag{7.49}$$

This method can also be used to calculate rate constants for association reactions (e.g. 7.28, 7.29) by setting k_i' to zero. The net rate constant for formation of the stabilised adduct is given by k_s.

One problem with this formulation of the master equation is that unnecessarily large matrices are required. Temperatures are generally reasonably low in chemical activation and association studies, so that reactivation of the stabilised molecule A is negligible. Indeed, once A^* has been deactivated a little below the lowest dissociation limit, E_0', the probability of reactivation becomes extremely small. This property can be exploited by introducing an *absorbing barrier*, c, some way below E_0' such that molecules in grains higher than c can be deactivated into c, but upward transitions from c are forbidden. The absorbing barrier thus acts as a *sink*. The introduction of an absorbing barrier allows the number of energy grains, and hence the size of the matrix Q, to be kept to a minimum for a given grain width, δ, and also provides a simple method for calculating the rate of stabilisation, v_s. Here

v_S is simply set equal to the rate of transfer into the absorbing barrier from the higher grains, i.e.

$$v_s = \omega \sum_i P_{ci} n_i \qquad (7.50)$$

for normalised radical concentrations, $k_S = v_S$.

7.4 EQUILIBRATION

Radical recombination reactions (e.g. 7.28) and association reactions (e.g. 7.29) can be treated using the formalism developed above with k_i' set to zero (equation 7.43) and the rate of association calculated via equation (7.49) (see section 7.6 below). Association reactions such as (7.29), which can be studied under pseudo-first-order conditions, can also be analysed using an eigenvalue method.

For the reaction

$$B + C \rightleftharpoons A^* \rightarrow A$$

with $[C] \gg [B]$, the technique involves extending by one element the population vector n originally containing m elements and describing the grained population of A. The $(m + 1)$th element corresponds to the concentration of B. The matrix form of the master equation still takes the form of equation (7.16):

$$\frac{dn}{dt} = M'n \qquad (7.51)$$

where M' is a net transfer matrix describing energy transfer in the adduct, A, its dissociation to form $B + C$ and its formation from the element $i = m + 1$. The top left-hand $(m \times m)$ block of M' is identical to the full matrix contained in equation (7.16). The top m elements of the right-hand column of M' contain the source term elements ϕ_i (cf. v_i, equation 7.44) which operate on n_{m+1} ($\equiv [B]$). The source term is modified from (7.44), since it is now a pseudo-first-order rate constant rather than a rate. In the present context

$$\phi_i = k_{r,\infty}[C]g_i \qquad (7.52)$$

where g_i is defined by equation (7.45) and k_a^∞ is the limiting high-pressure association rate constant. The first m elements of the bottom row of M contain the microcanonical rate constants for dissociation of A^*. Operating on the grained populations of A^* (n_i, $i = 1$ to m) these terms provide the rate of repopulation of B via dissociation of A^*. Finally, the bottom right element $(m + 1, m + 1)$ is $-k_{r,\infty}[C]$; operating on n_{m+1} it gives the rate of loss of B to form A^*.

The full form of $\mathbf{M}'$ is shown below:

$$\mathbf{M}' = \begin{pmatrix} M_{11} & M_{12} & \cdots & M_{1m} & \phi_1 \\ M_{21} & M_{22} & \cdots & M_{2m} & \phi_2 \\ M_{m1} & M_{m2} & \cdots & M_{mm} & \phi_m \\ k_1 & k_2 & \cdots & k_m & -k_{r,\infty}[C] \end{pmatrix} \qquad (7.53)$$

where $M_{ji} = \omega P_{ji} - (\omega + k_i)\delta_{ij}$ and δ_{ij} is the Kronecker delta function.

IRREVERSIBLE REACTION

The rate constant for the irreversible reaction

$$B + C \xrightarrow{k_r} A$$

can be calculated from (7.51) by incorporating an absorbing barrier in the grains corresponding to A. The unimolecular rate constant corresponds to the eigenvalue of smallest magnitude. Alternatively, it can be calculated from the eigenvector n and the net flux into the absorbing barrier (equation 7.50).

REVERSIBLE REACTION

Hanning-Lee et al.[12] used equation (7.51) to model the reaction

$$H + C_2H_4 \rightleftharpoons C_2H_5$$

under reversible conditions. The experiment in question involved the generation of H by laser flash photolysis in the presence of C_2H_4 and at sufficiently high temperatures that C_2H_5 dissociated on the time scale of the reaction. The system evolves exponentially to equilibrium, as in a classical relaxation experiment[13]. Setting $[H] = [B]$, $[C_2H_4] = [C]$, $[C_2H_5] = [A]$, a canonical relaxation analysis gives

$$k_d = k_{rel}/(1 + [C]/K_c) \qquad (7.54a)$$

and

$$k_r = k_{rel}/([C] + K_c) \qquad (7.54b)$$

where k_d and k_r are the dissociation and association rate constants, k_{rel} is the experimental relaxation rate constant, determined from the exponential relaxation to equilibrium and K_c is the equilibrium constant. Experiments of this type enable both kinetic and thermodynamic data to be obtained. A microcanonical analysis, via the master equation, reveals some interesting and novel features in equilibrating systems of this type.

Since the rate of energisation of stabilised A is important in this system, it is not possible to use an absorbing barrier and the full vector n, containing all energy grains of A from $i = 1$ to $i = m$, must be employed. The reaction system is now conservative—there is no population loss—and if the eigenvalues are ranked in order of increasing magnitude, $\lambda_1 = 0$ and u_1 describes the time-independent (Boltzmann) distribution. (It can be shown that c_1 (equation 7.18) is unity). The eigenvalue λ_2 is well separated from the other eigenvalues of larger magnitude and on long (experimental) time scales, the relaxation of n, is now given by

$$n = u_1 + c_2 u_2 \exp(\lambda_2 t) \tag{7.55}$$

The time-dependent population of B ($\equiv$ H) is equal to n_{m+1}:

$$[B]_t = n_{m+1} = (u_1)_{m+1} + c_2(u_2)_{m+1} \exp(\lambda_2 t) \tag{7.56}$$

while

$$[B]_\infty = (u_1)_{m+1} \tag{7.57}$$

the equilibrium population. Thus

$$[B]_t - [B]_\infty = c_2(u_2)_{m+1} \exp(\lambda_2 t) \tag{7.58}$$

The master equation analysis, therefore, leads to the expected exponential relaxation to the equilibrium population, with $k_{rel} = -\lambda_2$. The rate constant k_{rel} and hence λ_2 can be decomposed into k_a and k_d using equation (7.54).

An interesting feature of an equilibrating system is that the shape of the population vector, n, varies during the exponential phase, i.e. after the short-time transients, described by the higher eigenvalues, have decayed away. This contrasts with an irreversible reaction, either dissociation or association, where n, maintains a constant shape (equation 7.23). For a reversible system, we find, by rewriting (7.55),

$$n - u_1 = c_2 u_2 \exp(\lambda_2 t)$$

and, for specific elements,

$$\frac{n_j(t) - (u_1)_j}{n_i(t) - (u_1)_i} = \frac{(u_2)_j}{(u_2)_i} \tag{7.59}$$

Thus it is the *difference* from the equilibrium distribution that both relaxes exponentially and that maintains a constant shape.

This observation raises an interesting question. The population distribution of energised molecules is qualitatively different in reversible and irreversible reactions: time invariant in the latter case, time dependent in the former. The canonical rate constants are determined by these population distributions; does this mean that the canonical rate constants differ under reversible and irreversible conditions?

This question was first raised by Quack[14] for isomerisation reactions and analysed by Hanning-Lee et al.[12] for association/dissociation. They demonstrated numerical equivalence for the reversible and irreversible rate constants from the master equation analysis outlined above. Hanning-Lee *et al.*[12] also examined a three-level Lindemann scheme

$$B + C \rightleftharpoons A^* \rightleftharpoons A$$

and showed that, in the exponential phase, the system could be constructed exactly from the two irreversible schemes:

$$B + C \rightleftharpoons A^* \rightarrow A$$

$$B + C \leftarrow A^* \rightleftharpoons A$$

In the former scheme, the forward flux from A^* to A is limited to those energised molecules generated from $B + C$ and, in the latter, the flux from A^* to $B + C$ is limited to those A^* formed from A. For this simple model, the canonical rate constants under reversible and irreversible conditions are quantitatively and causally linked.

Hanning-Lee *et al.*[12] extended this analysis to an extended distribution in A^* using the strong collision model. Exactly the same conclusions were obtained. It was not feasible to analyse the full master equation similarly, but the numerical equivalence of the rate constants suggests that the same considerations apply.

7.5 ISOMERISATION

Isomerisation reactions

$$A \rightleftharpoons B \tag{7.60}$$

can be analysed using coupled differential equations for the populations of A and B:

$$\frac{dn_i^A}{dt} = \omega \sum_j P_{ij} n_j^A - (\omega + k_i^A) n_i^A + k_l^B n_l^B \tag{7.61}$$

$$\frac{dn_l^B}{dt} = \omega \sum_q P_{lq} n_q^B - (\omega + k_l^B) n_l^B + k_i^A n_i^A \tag{7.62}$$

where k_i^A denotes the microcanonical rate constant for isomerisation from A to B and k_l^B that for isomerisation from B to A from the isoenergetic grain, l, in B.

This system of equations can be transformed into a readily soluble form by generating a single vector, n, by concatenating the energy grains of A and B. The master equation takes the form:

$$\frac{\mathrm{d}n}{\mathrm{d}t} = \mathbf{M}''n \tag{7.63}$$

where n contains the grain populations of isomer A in elements $1 \le i \le m$ and of isomer B in elements $m + 1 \le i \le 2m - h$, where $h\delta = \Delta H_0^0$, the difference in zero-point energies of the isomers, for a grain width δ. Figure 7.3 illustrates the system diagrammatically.

The matrix $\mathbf{M}''$ is given by

$$\mathbf{M}'' = \omega(\mathbf{P} - \mathbf{I}) - \mathbf{k} + \mathbf{k}' \tag{7.64}$$

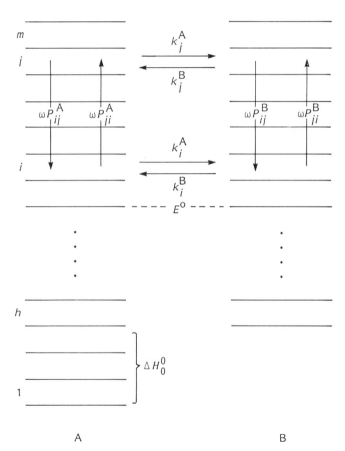

Figure 7.3 Schematic representation of the energy grains and rate processes in a model reversible isomerisation reaction. From Green *et al.*, *J. Chem. Phys.*, **96**, 5896 (1992). Reproduced by permission of the American Institute of Physics.

where $\mathbf{k}$ is a diagonal matrix containing the microcanonical rate constants, k_i^A and k_i^B, and describes the reactive loss from each isomer; $\mathbf{k}'$ contains k_i^A and k_l^B as off-diagonal elements and describes the reactive gain of each isomer. The full form of the matrix is shown in Figure 7.4, while Figure 7.5 shows it in block form. The top left-hand block is of the same form as $\mathbf{M}$ in

$$
\begin{pmatrix}
M_{11}^A & \omega P_{12}^A & \cdots & \omega P_{1m}^A & 0 & \cdots & 0 & 0 \\
\omega P_{21}^A & M_{22}^A & \cdots & \omega P_{2m}^A & \vdots & \ddots & \vdots & \vdots \\
\vdots & \vdots & \ddots & \vdots & 0 & \cdots & k_{y-1}^B & 0 \\
\omega P_{m1}^A & \omega P_{m2}^A & \cdots & M_{mm}^A & 0 & \cdots & 0 & k_y^B \\
0 & \cdots & 0 & 0 & M_{xx}^B & \omega P_{xx+1}^B & \cdots & \omega P_{xy}^B \\
\vdots & \ddots & \vdots & \vdots & \omega P_{x+1x}^B & M_{x+1x+1}^B & \cdots & \omega P_{x+1y}^B \\
0 & \cdots & k_{m-1}^A & 0 & \vdots & \vdots & \ddots & \vdots \\
0 & \cdots & 0 & k_m^A & \omega P_{yx}^B & \omega P_{yx+1}^B & \cdots & M_{yy}^B
\end{pmatrix}
$$

Figure 7.4 Detailed form of the matrix $\mathbf{M}''$ for an isomerisation reaction: $x = m + 1$, $y = 2m - h$, and $M_{ii}^A = (\omega P_{ii}^A - \omega - k_i^A)$.

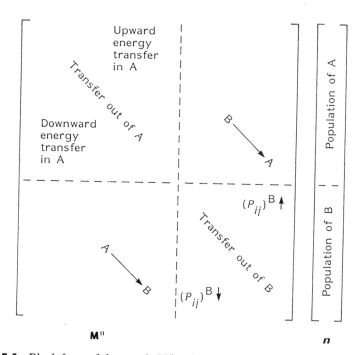

Figure 7.5 Block form of the matrix $\mathbf{M}''$ and vector $\mathbf{n}$ for an isomerisation reaction.

(7.16) and operates on the upper (A) component of n. It describes collisional transfer within isomer A and reactive loss to form B. The bottom right-hand block acts similarly on the B component of n. The top right-hand block describes reactive gain of A from B and the bottom left-hand block reactive gain of B from A.

This form of the master equation can be used to describe both reversible and irreversible isomerisation reactions. For the latter (i.e. for the reaction A $\rightarrow$ B), one of the grains of B below the reaction threshold energy, E^0, is employed as an absorbing barrier. The eigenvalue analysis follows (7.18) *et seq*. and the unimolecular rate constant, k_{uni} is equal to $-\lambda_1$.

The origin of the fall-off in k_{uni} with pressure is somewhat more complex than is the case for a simple dissociation reaction. For the latter k_{uni} decreases as the pressure falls because collisions are less able to maintain the Boltzmann distribution at high energies. This effect also applies in isomerisation reactions, but is augmented by back reaction from the energised grains of B, so that the net forward flux, for a given grain in A, is $(k_i^A n_i - k_l^B n_l)$, where $l = m + i - h$, to ensure equal energies. As the pressure increases the population of B* decreases as energised molecules are collisionally transferred, with increasing efficiency, to the absorbing barrier. In consequence, the reduction in k_{uni} arising from the reverse flux decreases as the pressure is increased.

The forward and reverse rate constants are related by detailed balance:

$$k_i^A N_i^A = k_l^B N_l^B \tag{7.65}$$

As ΔE_0^0 becomes more negative, N_l^B/N_i^A rises as the rovibrational energy in B increases relative to that in A and k_l^B/k_i^A falls. This back reaction effect is, therefore, only significant for endothermic or only slightly exothermic isomerisation reactions. Its origins were first discussed by Lin and Laidler[15] in a strong collision treatment.

Green *et al.*[16] used the master equation model to analyse reversible isomerisation reactions. The absorbing barrier was omitted and the full matrix, $\mathbf{M}''$, and vector n employed. The column sums in $\mathbf{M}''$ are all now zero, so that probability is conserved: there is no reactive loss and the eigenvalue of smallest magnitude is zero. The treatment follows closely that discussed above for reversible association reaction. The system once more relaxes exponentially to equilibrium and k_{rel} can again be decomposed into forward and reverse rate constants via equations equivalent to equations (7.54). The shape of the distribution of energised molecules changes with time even in this exponential regime and it is the difference from the equilibrium distribution that maintains a constant shape. For a simple Lindemann model, Green *et al.*[16] showed that, as described in section 7.4, the equilibrating system can be decomposed into irreversible forward and reverse reactions and that canonical rate constants obtained from an analysis of both reversible and irreversible reactions are equivalent. Numerical equivalence was also established for the full master equation treatment.

7.6 SOLUTION OF THE ENERGY GRAINED MASTER EQUATION

In this section we examine in detail the application of the EGME to two problems, illustrating its usage and detailing some aspects of the techniques involved. We concentrate on the setting up of the matrix equation, rather than on specific methods of solution, which can be found in standard computational texts[17]. Gilbert and Smith[18,19] give more details on one method of solution.

THE DISSOCIATION OF 2-C_3H_7

The spectroscopic data for 2-C_3H_7 are given in Table 7.1. The inverse Laplace transform (ILT) technique (section 6.7) for calculating $k(E)$ from the limiting high-pressure rate constant for the H + C_3H_6 association was employed and the table also contains the required data for C_3H_6 for this part of the analysis.

Table 7.1   Spectroscopic data for the isopropyl radical decomposition reaction[a]

Vibrational frequencies of 2-C_3H_7:
3052.0, 2968.2, 2968.2, 2967.7, 2920.0, 2887.0, 2850.0, 1468.0, 1464.0, 1462.0, 1440.0
1378.0, 1338.0, 1292.0, 1200.0, 1165.0, 1053.8, 921.7, 879.0, 748.1, 369.2, 364.0
External rotational constants of 2-C_3H_7:
$A = 1.528,\ B = 0.2645$
Internal rotational constants of 2-C_3H_7:
5.36053
Barrier height to internal rotation of 2-C_3H_7:
255.0
Vibrational frequencies of C_3H_6:
3091.0, 3022.0, 2991.0, 2973.0, 2952.8, 2931.0 1652.8 1458.5, 1442.5, 1414.0, 1378.0,
1298.0, 1177.5, 1044.7, 990.0, 934.5, 919.0, 912.0, 575.2, 428.0
External rotational constants of C_3H_6:
$A = 1.544,\ B = 0.2905$
Internal rotational constants of C_3H_6:
5.36053
Barrier height to internal rotation of C_3H_6:
699.0

[a]All data are in units of cm^{-1}.

Calculation of Densities of States

Figure 7.6 shows the numbers of states for 2-C_3H_7, N_i, contained in 50 cm^{-1} grains. The N_i were calculated from the SR algorithm (section 4.2) and are based on 1 cm^{-1} cells; 2-C_3H_7 contains two internal rotors which were incorporated in the state count using the technique developed by Robertson and Wardlaw[20].

Calculation of the Collisional Energy Transfer Probabilities, P_{ij}

The P_{ij} were calculated on the basis of the exponential down model (equations (7.10)). The most complex part of the procedure is the calculation of the normalisation constants. This is best achieved by a process of backward substitution, starting with grain, $i = m$. For this grain, all transitions are downward and A_m is calculated straightforwardly from the normalisation condition:

$$A_m \exp\left(-\alpha E_m\right) \sum_{i=0}^{m} \exp\left(\alpha E_i\right) = 1 \qquad (7.66)$$

The procedure corresponds to equating the sum of the mth column of the probability matrix, $\mathbf{P}$, to unity (Table 7.2).

Turning our attention to the $(m - 1)$th column, $P_{m,m-1}$ is an upward

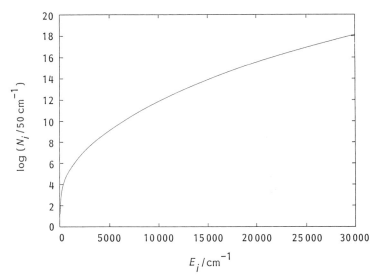

Figure 7.6 Plot of N_i, the number of states in a 50 cm^{-1} grain, versus E_i for 2-C_3H_7.

Table 7.2   Elements of the collisional energy transfer matrix

	$m-4$	$m-3$	$m-2$	$m-1$	m
(a) Matrix **P** before normalisation					
$m-4$	1.00000	0.64741	0.41913	0.27135	0.17567
$m-3$	0.55881	1.00000	0.64741	0.41913	0.27135
$m-2$	0.31222	0.55873	1.00000	0.64741	0.41913
$m-1$	0.17443	0.31215	0.55866	1.00000	0.64741
m	0.09744	0.17436	0.31207	0.55860	1.00000
(b) Matrix **P** after normalisation					
$m-4$	0.24561	0.16115	0.10815	0.07683	0.06194
$m-3$	0.13910	0.24892	0.16704	0.11868	0.09568
$m-2$	0.08056	0.14417	0.25802	0.18331	0.14778
$m-1$	0.04939	0.08838	0.15818	0.28315	0.22827
m	0.03436	0.06148	0.11003	0.19696	0.35259
(c) Matrix $\mathbf{M}/\omega = [\mathbf{P} - \mathbf{I}] - \mathbf{k}/\omega$					
$m-4$	−28.43802	0.16115	0.10815	0.07683	0.06194
$m-3$	0.13910	−29.13115	0.16704	0.11868	0.09568
$m-2$	0.08056	0.14417	−29.83079	0.18331	0.14778
$m-1$	0.04939	0.08838	0.15818	−30.52683	0.22827
m	0.03436	0.06148	0.11003	0.19696	−31.19114

Note: The data refer to 2-C_3H_7, with $T = 720$ K, $[He] = 3.0 \times 10^{17}$ mol cm^{-3}, $\omega = 2.1225 \times 10^9$ s^{-1}, $\alpha^{-1} = 230$ cm^{-1}, $m = 300$ and grain width = 100 cm^{-1}.

transition which may be calculated from the corresponding downward transition which has already been evaluated in summing the mth column:

$$P_{m,m-1} = A_m \exp\left(-(\alpha + \beta)(E_m - E_{m-1})\right)(N_m/N_{m-1}) \qquad (7.67)$$

The rest of the transitions are downward and the summation over them contains only the unknown, A_{m-1}:

$$P_{m,m-1} + A_{m-1} \exp\left(-\alpha E_{m-1}\right) \sum_{i=0}^{m-1} \exp\left(\alpha E_i\right) = 1 \qquad (7.68)$$

The rest of the matrix is calculated similarly and sequentially, evaluating the upward probabilities from detailed balance and previously determined normalisation constants and downward probabilities from the normalisation condition.

Table 7.2 shows the elements in the bottom right-hand corner of **P**, before and after normalisation. The first set of numbers were calculated with the normalisation constant set to unity, so that all the diagonal elements ($P_{ii} = A_i \exp(0)$) are themselves unity. The normalisation procedure described above was then employed to calculate the final elements, as shown for the second set of numbers. The elements in the mth column decrease as

i decreases, because there are no upward transitions. The mth column sum for the five elements shown is only 0.6969—other elements, of smaller i, are populated significantly in energy transfer from the mth grain for $\alpha^{-1} = 230 \text{ cm}^{-1}$.

Figure 7.7 shows a plot of P_{ij} versus energy for 2-C_3H_7. The plot clearly shows the asymmetry arising from detailed balance.

Calculation of k_i

The calculations on the 2-C_3H_7 system were performed as part of an analysis of experimental measurements of the dissociation of 2-C_3H_7 obtained by Seakins et al.[21] using laser flash photolysis/photoionisation mass spectrometry. The measurements were made well into the fall-off region, so that extrapolation to k_∞ was not easy. Instead the k_i were calculated using ILT, based on the association data of Harris and Pitts[22] who obtained

$$k_{r,\infty} = 2.21 \times 10^{-11} \exp\left(-785 \ K/T\right) \text{ cm}^3 \text{ mol}^{-1} \text{s}^{-1}.$$

Seakins et al.[21] applied the ILT technique and transformed k_i, using ΔH_0^0 as a variable parameter. Equation (6.52) becomes, in this case,

$$k_a(E^*)\rho_{C_3H_7}(E^*) =$$
$$A_\infty C'(2\sqrt{\pi}) \int_0^{E-E_\infty-\Delta H_0^0} \rho_{C_3H_6}(E')(E^* - (E' + E_\infty + \Delta H_0^0))^{1/2} \, dE' \quad (7.69)$$

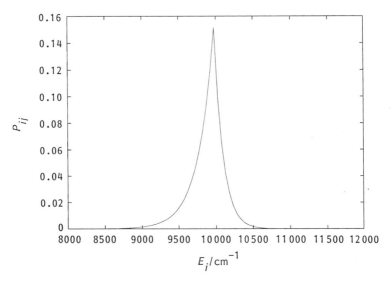

Figure 7.7 The dependence of P_{ij} on the energy of the final state for an initial energy of 10^4 cm^{-1} for 2-C_3H_7.

where $C' = (2\pi M_H M_{C_3H_6}/h^2 M_{C_3H_7})^{3/2}$ and where only $\rho_{C_3H_6}(E^*)$ is needed for the convoluted fragment rovibrational densities of states, since the other fragment is H. Figure 7.8 shows a plot of $\log k_i$ versus E_i for $\Delta H_0^0 = 146.8 \text{ kJ mol}^{-1}$. The third set of numbers in Table 7.2 shows the elements of the matrix $\mathbf{M}/\omega$ for $\omega = 2.1225 \times 10^9 \text{ s}^{-1}$. Under these conditions the matrix is diagonally dominant, but this becomes less evident as ω increases and $\mathbf{M}/\omega \to (\mathbf{P} - \mathbf{I})$.

Reduced Matrix Approximation

In a dissociation reaction, the reactant rovibrational population density is depleted from the Boltzmann distribution except at the high-pressure limit. The degree of depletion decreases as the energy decreases and becomes negligible some way below the dissociation limit. Accordingly, it is not necessary to construct a collision matrix and a population vector that includes all the grains down to $E = 0$; instead, grains only need to be included in the calculation above some cut-off E_c. Below this energy, it is assumed that a Boltzmann distribution pertains exactly[23].

Calculation of $k(p, T)$

We are now in a position to calculate the canonical dissociation rate constant. In the present example this was achieved by determining the

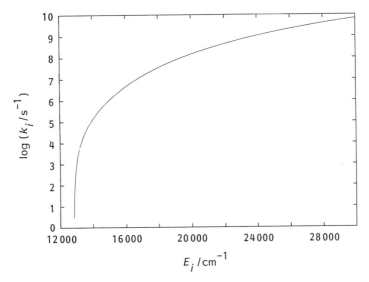

Figure 7.8 The dependence of the microcanonical rate constants for dissociation of 2-C_3H_7 on energy.

eigenvalue of smallest magnitude, although alternative methods can be used. The process is rapid and can be achieved for all 71 conditions of (p, T) studied by Seakins *et al.*[21] for a particular pair of values of α and ΔH_0^0, in a very small amount of processor time. This process can be further speeded up using the diffusion approximation which will be discussed in section 7.7. Robertson *et al.*[23] analysed the data, using this approach, varying the parameters, α and ΔH_0^0, to obtain the best fit. Each change in ΔH_0^0, required a re-evaluation of k_i using the ILT technique, the speed of which was considerably enhanced by using a fast Fourier transform (FFT) algorithm to evaluate the convolution in equation (7.69).

Robertson *et al.*[23] used χ^2:

$$\chi^2 = \sum_l (k_l^{\text{calc}} - k_l^{\text{expt}})^2 \tag{7.70}$$

as their criterion for the best fit, where k_l^{calc} and k_l^{expt} are, respectively, the calculated and experimental rate coefficients and l is an index refer-

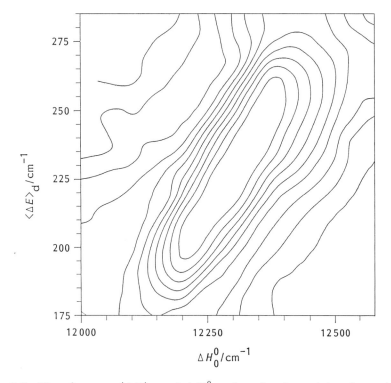

Figure 7.9 The χ^2 versus $\langle \Delta E \rangle_d$ and ΔH_0^0 surface for determining the optimal fitting parameters for the decomposition of 2-C$_3$H$_7$. Reprinted with permission from Robertson *et al.*, *J. Phys. Chem.*, **99**, 13452 (1995). Copyright 1995 American Chemical Society.

ring to the (p, T) conditions. Figure 7.9 shows a contour plot of χ^2 versus $\langle \Delta E \rangle_d$ and ΔH_0^0, showing a strong correlation between these two parameters but also a well-defined minimum. The best fit parameters are $\langle \Delta E \rangle_d = 230 \text{ cm}^{-1}$ and $\Delta H_0^0 = 12\,275 \text{ cm}^{-1}$. The latter corresponds to $\Delta H_{f,298}^0(2\text{-}C_3H_7) = 87.7 \text{ kJ mol}^{-1}$, in excellent agreement with a general analysis by Seakins et al.[24]

THE RECOMBINATION OF METHYL RADICALS

The spectroscopic data for C_2H_6 and CH_3 are given in Table 7.3. The calculation of the densities of states and the collisional transfer probability matrix, M, proceed as for $2\text{-}C_3H_7$. For a recombination reaction, the size of the matrix is reduced by employing an absorbing barrier (section 7.5) which is typically located $10kT$ below the dissociation limit although it is sensible to find its optimal location (highest energy compatible with a converged solution) iteratively. Figure 7.10 shows a plot of log N_i versus energy.

Calculation of k_i

In the calculations used here for illustration, the k_i were obtained by ILT. The procedure adopted for $2\text{-}C_3H_7$ is appropriate, except that now it is necessary to convolute the rovibrational densities of states of the fragments[25]. The k_i are shown on Figure 7.11.

The $CH_3 + CH_3$ system has been widely studied theoretically and an alternative approach would be to calculate k_i from theory and use the master equation as a means of comparing with experiment. The reactions occurs on a Type II surface and a variational approach is necessary. At the simplest level, the experimental data may be analysed using a minimum sum of states approach with a Morse potential function and a variable parameter describing the bond-distance dependence of the frequencies of all the loose

Table 7.3 Spectroscopic data for the methyl radical recombination reaction[a]

Vibrational frequencies of C_2H_6:
2985, 2985, 2969, 2969, 2954, 2896, 1469, 1469, 1468, 1468, 1388, 1379, 1190, 1190, 995, 822, 822
Rotational constants of C_2H_6:
$A = 2.671$, $B = 0.6625$
Internal rotational constant of C_2H_6:
5.342
Vibrational frequencies of CH_3:
3162, 3162, 3044, 1396, 1396, 606
Rotational constants of CH_3:
$A = 4.789$, $B = 9.578$

[a]All data are in units of cm^{-1}.

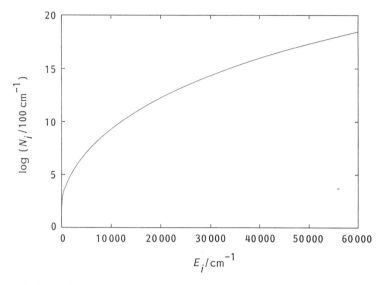

Figure 7.10   Plot of N_i, the number of states in a 100 cm^{-1} grain, versus E_i for C_2H_6.

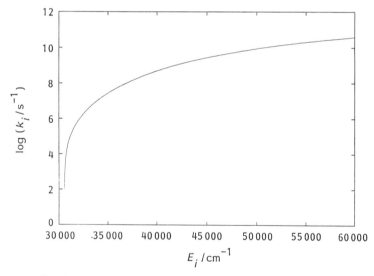

Figure 7.11   The dependence of the microcanonical rate constants for dissociation of C_2H_6 on energy.

vibrations (section 6.2)[26,27]. Brouard, Macpherson and Pilling[28] used a similar method for $CH_3 + H$ (see Chapter 6). At a more realistic level, the statistical adiabatic channel model (SACM) of Quack and Troe[29], or the flexible transition state theory (FTST) of Wardlaw and Marcus[30] could be employed. Both were discussed in Chapter 6. Wardlaw and Wagner[31] used the FTST approach to analyse the experimental data of Slagle et al.[32], although they used a modified strong collision approach rather than a master equation analysis.

Source Term, v

The source term, which describes the rate of population of the energy grains of C_2H_6 by recombination of $CH_3 + CH_3$, is calculated using detailed balance as described in section 7.3:

$$v_i = k_{r,\infty}[CH_3]^2 \left\{ \frac{k_i N_i \exp(-E_i/kT)}{\sum_i k_i N_i \exp(-E_i/kT)} \right\} = k_{r,\infty}[CH_3]^2 g_i \qquad (7.71)$$

Where g_i is given by equation (7.45). Figure 7.12 shows a plot of g_i versus E_i. This function describes the energy dependence of v_i. It is easiest to use normalised rates with $[CH_3]$ set to unity.

Calculation of $k(p, T)$

The matrix form of the equation is given by (7.46), with $\mathbf{k}'$ set to zero. The eigenvalue method cannot be employed because v is second order in CH_3.

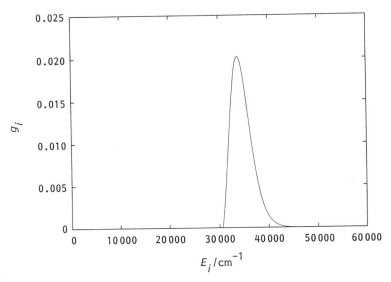

Figure 7.12 The dependence of the source term on energy for $CH_3 + CH_3 \rightarrow C_2H_6$ at 1000 K. Plot of $g_i = k_i N_i \exp(-E_i/kT)/(\Sigma k_i N_i \exp(-E_i/kT))$ versus energy.

The matrix inversion method generates the population vector, n, which can be used to generate the rate constant for recombination, k_r in two ways, following the analysis given in section (7.5):

1. As the difference between the rate of formation and the rate of dissociation of AB* (cf. 7.49),

$$k_r = \sum_i v_i - \sum_i k_i n_i \qquad (7.72)$$

2. Or as the rate of stabilisation into the absorbing barrier (cf. 7.50)

$$k_r = \omega \sum_i P_{ci} n_i \qquad (7.73)$$

These two methods provide a check on programming errors, although at

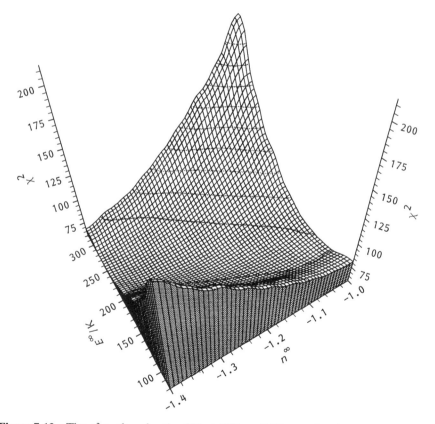

Figure 7.13 The χ^2 surface for the $CH_3 + CH_3 \rightarrow C_2H_6$ reaction for the parameters E_∞ and n_∞. Reprinted with permission from Robertson et al., J. Phys. Chem., **99**, 13452 (1995). Copyright 1995 American Chemical Society.

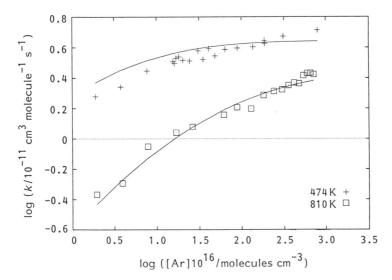

Figure 7.14 Fitted fall-off curves for the $CH_3 + CH_3 \rightarrow C_2H_6$ reaction for temperatures 474 K and 810 K.

very low pressures where $\sum v_i \approx \sum k_i n_i$, numerical errors limit the accuracy of method 1. If one method is to be used, 2 is the method of choice.

The limiting high-pressure rate constant is directly incorporated in the calculation via the ILT. The third-order low-pressure limit may be found by successive decadic reductions of the pressure (and hence of ω) until convergence is obtained in $k_r/[M]$. As always, checks also need to be made on the effect of the grain size, δ, and of the energy of the absorbing barrier.

Robertson et al.[23] analysed a dataset of experimental points of a number of workers using the master equation/ILT approach. They assumed $k_{r,\infty} = A_\infty T^{n_\infty} \exp(-E_\infty/kT)$ and used $\langle \Delta E \rangle_d$, A_∞, n_∞ and E_∞ as variable parameters. Figure 7.13 shows a cut through χ^2 hyperface with $\langle \Delta E \rangle_d$ and A_∞ fixed and n_∞ and E_∞ allowed to vary. Figure 7.14 shows the result of this global fit at $T = 474$ K and 810 K.

7.7 DIFFUSION MODELS OF UNIMOLECULAR REACTIONS

Collisional energy transfer in a unimolecular reaction system is a Markov process[33], i.e. the probability of jumping from state x to state y in no way depends on the previous history of the system prior to its passage to state x. Such a property lies at the basis of our formulation of the master equation, although it was not explicitly stated.[8,11] The master equation is capable of a very detailed description of the reacting system, incorporating, for example, transition rates between individual vibration–rotation states. Such a description is not practically feasible except for the simplest diatomic molecules such as H_2[11]. The transition probability distribution is generally compressed into a simple one-parameter representation based on a stepladder or exponential down model. Such a description is generally perfectly adequate for a unimolecular reaction system; conversely, analysis of experimental results from such systems provides little detailed information on energy transfer. As we shall see in section 9.1, more detailed information is becoming available through laser experiments designed specifically to probe the energy-transfer process.

For the general analysis of reacting systems, the master equation may be an unnecessarily cumbersome and computationally expensive approach. Is there any way of simplifying it while retaining the level of accuracy provided by, say, the exponential down model?

A diffusion process is a Markov process with a continuous path. In liquids, for example, collisions are sufficiently frequent that the motion of a particle is continuous. A diffusion description is appropriate to the evolution of the internal energy of a molecule if collisions are occurring very frequently and with very weak colliders—a rather restrictive constraint. In this section we briefly examine the diffusion approximation and establish rather wider limits for its application.

Green, Robertson and Pilling[34] discussed the derivation of a diffusion equation from the master equation via the Kramers–Moyal expansion[35]. The diffusion equation is derived by truncating the infinite sum in this expansion after the second term:

$$\frac{dn(y, t)}{dt} = \frac{1}{2}\frac{d^2}{dy^2}[\sigma^2(y)n(y, t)] - \frac{d}{dy}[\mu(y)n(y, t)] \qquad (7.74)$$

where $\sigma^2(y)$ is the variance of the population distribution and $\mu(y)$ is the mean. The variance $\sigma^2(y)$ determines how the distribution broadens with time and is related to the diffusion coefficient

$$D(y) = \frac{\sigma^2(y)}{2} \qquad (7.75)$$

while $\mu(y)$ determines how the mean energy changes with time and corresponds to the *drift* in the diffusion equation. Because the Kramers–Moyal infinite sum has been truncated, the diffusion equation is only an approximation to the master equation and the coefficients, $\sigma^2(y)$ and $\mu(y)$, have to be chosen with care if the quality of the approximation is to be high.

Green, Robertson and Pilling[34] examined earlier treatments by Troe[36] and by Nikitin[37] and showed that problems arise from the need to generate a diffusion equation which obeys detailed balance. In the Kramers–Moyal expansion, $\mu(y)$ and $\sigma^2(y)$ are the first and second moments of the energy transfer probability distribution:

$$a_i = \omega \int_0^\infty (y - x)^i P(x|y) \, dx \tag{7.76}$$

where $P(x|y)$ is defined via, for example, equation (7.2), $\mu(y) = a_1$ and $\sigma^2(y) = a_2(y)$. Troe[36] and Nikitin[37] derived their diffusion equations from the truncated Kramers–Moyal expansion, adding some additional approximations so that the population distribution evolved towards the Boltzmann distribution. Green, Robertson and Pilling[34] showed that their approaches led to some inaccuracies, in particular the mean of the energy distribution is poorly described. These authors also demonstrated that a more accurate representation can be obtained if the drift coefficient, $\mu(y)$, is calculated from the Kramers–Moyal expansion via (7.76) with $i = 1$ and then the diffusion coefficient is calculated in such a way that detailed balance is obeyed. These requirements give

$$D(y) = \frac{1}{f(y)} \int_0^y \mu(z) \, dz \tag{7.77}$$

where $f(y)$ is the Boltzmann density at energy y.

Figure 7.15 shows a comparison of the master and diffusion equation

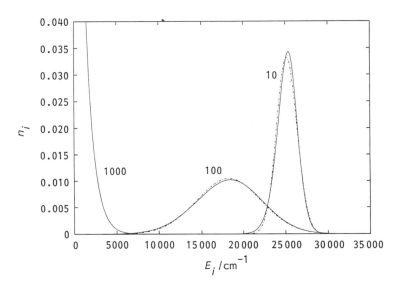

Figure 7.15 Comparison of master (—) and diffusion equation (- · -) population energy distributions for a molecule with an exponentially distributed density of states appropriate for azulene; $p = 100$ Torr, $T = 500$ K, at 10, 100 and 1000 ns. From Green *et al.*, *J. Chem. Phys.*, **100**, 5259 (1994). Reproduced by permission of the American Institute of Physics.

population densities using this approximation for azulene at 100 Torr and 500 K and with an exponential down model with $\langle \Delta E \rangle_d = 270 \ cm^{-1}$. The curves represent the energy distribution after 10, 100 and 1000 ns for an initial distribution such that all the density is in a grain of 25 000 cm^{-1} and show the downward drift and the spreading referred to above. The 1000 ns distribution is narrower than that at 100 ns because the equilibrium distribution has effectively been established. The agreement between the two models is excellent. Green, Robertson and Pilling[34] and later Robertson et al.[23] applied the diffusion model to reacting systems and demonstrated good agreement for typical experimental conditions. Figure 7.16 shows a comparison of fall-off curves for CH_4 dissociation calculated using the master equation and diffusion equation methods.

The main advantage of the diffusion approximation in practical applications is that it considerably simplifies the form of the transfer matrix, $\mathbf{M}$. Except in the simplest of cases, the diffusion equation has to be solved numerically and this is most easily achieved using a finite difference approach.

In a reacting system, the diffusion equation, (7.74), is modified by incorporating the microcanonical reaction term

$$\frac{dn}{dt} = \frac{d^2[D(y)n]}{dy^2} - \frac{d[\mu(y)n]}{dy} - k(y)n \tag{7.78}$$

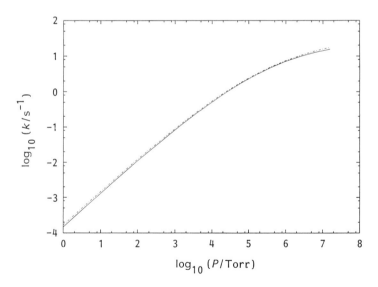

Figure 7.16 Fall-off curves for $CH_4 \rightarrow CH_3 + H$ at 1500 K: $(-)$ master equation solution; $(\cdots)$ diffusion equation solution. From Green et al., *J. Chem. Phys.*, **100**, 5259 (1994). Reproduced by permission of the American Institute of Physics.

The finite difference method approximates the continuous derivatives in y by difference quotients over a small interval, δ:

$$\frac{d}{dy}(\mu(y)n(y)) \cong \frac{\mu(y + \delta)n(y + \delta) - \mu(y - \delta)n(y - \delta)}{2\delta} \quad (7.79)$$

$$\frac{d^2}{dy^2}(D(y)n(y)) \cong \frac{D(y + \delta)n(y + \delta) - 2D(y)n(y) + D(y - \delta)n(y - \delta)}{\delta^2}$$

$$(7.80)$$

Collecting terms, the time-dependent equation becomes

$$\frac{dn}{dt} = \left(\frac{D(y + \delta)}{\delta^2} - \frac{\mu(y + \delta)}{2\delta}\right)n(y + \delta) - \left(\frac{2D(y)}{\delta^2} + k(y)\right)n(y)$$

$$+ \left(\frac{D(y - \delta)}{\delta^2} - \frac{\mu(y - \delta)}{2\delta}\right)n(y - \delta) \quad (7.81)$$

This equation is of the same form as the discretised version of the master equation, with δ equal to the grain width. Equation (7.81) can be expressed in the form

$$\frac{dn_i}{dt} = L_{i,i+1}n_{i+1} - (L_{i,i} + k_i)n_i, + L_{i,i-1}n_{i-1} \quad (7.82)$$

where the $L_{i,j}$ terms are suitably averaged forms of the terms in equation (7.81) and the n_i are the grain populations. Equation (7.82) recognises that both μ and D are proportional to the collisional frequency, ω (see equations 7.76 and 7.77). We now have an equation of the form

$$\frac{dn}{dt} = [\mathbf{L} - \mathbf{k}]\mathbf{n} = \mathbf{M}'''\mathbf{n} \quad (7.83)$$

where $\mathbf{L}$ is a tridiagonal matrix. In effect, the master equation with an exponential down collisional transfer model (or indeed any collisional transfer model), has been transformed into an *equivalent* stepladder model, with collisional energy transfer only between adjacent grains. The equation can be solved in exactly the same way as previously, but the solution is much faster because of the tridiagonal nature of $\mathbf{L}$.

Robertson *et al.*[23] applied the technique to the decomposition of 2-C_3H_7. The term $\mu(y)$ is calculated from a discretised version of (7.76):

$$\mu_i = \omega\sum_j(E_j - E_i)P_{ji} \quad (7.84)$$

and

$$D_i = f_i^{-1}\sum_{j=1}^{i}\mu_j \quad (7.85)$$

Figures 7.17 and 7.18 show plots of μ_i/ω and D_i/ω versus energy; μ_i/ω corresponds to the average energy transferred, $\langle \Delta E \rangle$, from energy E. It is positive at low energies, because upward transfer predominates, and becomes negative at higher energies. At the mean thermal energy for 2-C_3H_7

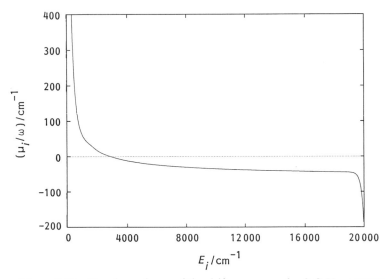

Figure 7.17 The dependence of the drift on energy for 2-C_3H_7 at 720 K.

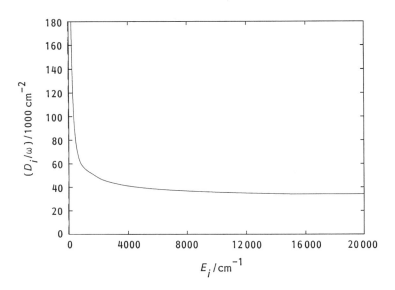

Figure 7.18 The dependence of the diffusion coefficient on energy for 2-C_3H_7 at 720 K.

Table 7.4 Elements of the diffusion collisional energy transfer matrix

	$m-4$	$m-3$	$m-2$	$m-1$	m
Matrix $\mathbf{M}'''/\omega = [\mathbf{L} - \mathbf{k}]/\omega$					
$m-4$	−34.91075	3.86224	0.00000	0.00000	0.00000
$m-3$	3.33368	−35.52717	3.80617	0.00000	0.00000
$m-2$	0.00000	3.28486	−36.09258	3.70553	0.00000
$m-1$	0.00000	0.00000	3.19761	−36.55705	3.52511
m	0.00000	0.00000	0.00000	3.04154	−34.06884

Notes: The data refer to 2-C_3H_7 for the same conditions as those specified in Table 7.2. The above data should be compared with set (c) of Table 7.2.

μ_i/ω is zero. The rapid decrease at high energies is associated with the upper cut-off at $i = m$ and is purely a computational artefact that is of no consequence provided that m has been placed at a sufficiently high energy. Here D_i/ω decreases with energy, approaching a high-energy asymptote.

Table 7.4 shows the highest elements in the probability matrix $\mathbf{M}'''$ for comparison with Table 7.2. Equations (7.81) and (7.82) give

$$L_{i,i-1} = D_{i-1}/\delta^2 - \mu_{i-1}/2\delta \tag{7.86a}$$

$$L_{i,i} = 2D_i/\delta^2 \tag{7.86b}$$

$$L_{i,i+1} = D_{i+1}/\delta^2 - \mu_{i+1}/2\delta \tag{7.86c}$$

Note that normalisation is not required in this case, the elements being directly determined for equations (7.76) and (7.77), with a suitably normalised form of P_{ij}.

REFERENCES

1. I. Oref and D. C. Tardy, *Chem Rev*, **90**, 1407 (1990).
2. K. F. Lim and R. G. Gilbert, *J Phys Chem*, **94**, 77 (1990).
3. G. Lendvay and G. C. Schatz, *J Phys Chem*, **98**, 6530 (1994).
4. J. N. Murrell and S. D. Bosanac, *Introduction to Atomic and Molecular collisions*, Wiley, (1989).
5. D. C. Clary and G. J. Kroes, *Advances in Chemical Kinetics and Dynamics*, JAI Press, (1995).
6. M. Quack and J. Troe in *Gas Kinetics and Energy Transfer*, a specialist periodical report, Vol. 2, Chemical Soc., London, (1977).
7. K. F. Lim and R. G. Gilbert, *J. Chem. Phys.*, **84**, 6129 (1986); **92**, 1819 (1990).
8. A. P. Penner and W. Forst, *J. Chem. Phys.*, **67**, 5296 (1977).
9. G. Z. Whitten and B. S. Rabinovitch, *J. Chem. Phys.*, **38**, 2466 (1963).
10. S. E. Stein and B. S. Rabinovitch, *J. Chem. Phys.*, **60**, 908, (1974).
11. H. O. Pritchard, *The Quantum Theory of Unimolecular Reactions*, Cambridge University Press, Cambridge (1984).

12. M. A. Hanning-Lee, N. J. B. Green, M. J. Pilling and S. H. Robertson, *J. Phys. Chem.*, **97**, 860 (1993).
13. M. Eigen, G. Kurtze and K. Tamm, *Ber. Bunsenges. Phys. Chem.*, **87**, 103 (1983).
14. M. Quack, *Ber. Bunsenges. Phys. Chem.*, **88**, 94 (1984).
15. M. C. Lin and K. J. Laidler, *Trans. Faraday Soc.*, **64**, 94 (1968).
16. N. J. B. Green, P. J. Marchant, M. J. Perona, M. J. Pilling and S. H. Robertson, *J. Chem. Phys.*, **96** 5896 (1992).
17. W. H. Press, B. P. Flannery, S. A. Teukolsky and W. H. Vetterling, *Numerical Recipes*, 2nd edn, Cambridge University Press (1992).
18. R. G. Gilbert and S. C. Smith, *Theory of Unimolecular and Recombination Reactions*, Blackwell Scientific Publications, Oxford (1990).
19. R. G. Gilbert, S. C. Smith and M. J. T. Jordan, UNIMOL Program Suite (calculation of fall-off curves for unimolecular and recombination reactions).
20. S. H. Robertson and D. Wardlaw, *Chem. Phys. Letters*, **199**, 391 (1992).
21. P. W. Seakins, S. H. Robertson, M. J. Pilling, I. R. Slagle, G. W. Gmurczyk, A. Bencsura, D. Gutman and W. Tsang, *J. Phys. Chem.*, **97** 4450 (1993).
22. G. W. Harris and J. N. Pitts Jr, *J. Chem. Phys.*, **77**, 3994 (1982).
23. S. H. Robertson, M. J. Pilling, D. L. Baulch and N. J. B. Green, *J. Phys. Chem.* **99**, 13452 (1995).
24. P. W. Seakins, M. J. Pilling, J. T. Niiranen, D. Gutman and L. Krasnoperov, *J. Phys. Chem.*, **96**, 9847 (1992).
25. J. W. Davies, N. J. B. Green and M. J. Pilling, *Chem. Phys. Letters*, **126**, 373 (1986).
26. J. W. Davies, D. Phil thesis, Oxford University (1989).
27. J. W. Davies, N. J. B. Green and M. J. Pilling, *J. C. S. Faraday Trans.*, **87**, 2317 (1991).
28. M. Brouard, M. T. Macpherson and M. J. Pilling, *J. Phys. Chem.*, **93**, 4047 (1989).
29. M. Quack and J. Troe, *Ber. Bunsenges. Phys. Chem.*, **78**, 241 (1974).
30. D. Wardlaw and R. A. Marcus, *J. Chem. Phys.*, **90**, 5383 (1986).
31. D. Wardlaw and A. F. Wagner, *J. Phys. Chem.*, **92**, 2462 (1988).
32. I. R. Slagle, D. Gutman, J. W. Davies and M. J. Pilling, *J. Phys. Chem.*, **92**, 2455 (1988).
33. G. R. Grimmett and D. R. Stirzaker, *Probability and Random Processes*, Clarendon Press, Oxford (1982).
34. N. J. B. Green, S. H. Robertson and M. J. Pilling, *J. Chem. Phys.*, **100**, 5259 (1994).
35. J. E. Moyal, *J. R. Stat. Soc.*, **B11**, 150 (1948).
36. J. Troe, *J. Chem. Phys.*, **77**, 3485 (1982).
37. E. E. Nikitin, *Theory of Elementary Atomic and Molecular Processes*, Clarendon Press, Oxford (1974).

8 Approximate Techniques of Unimolecular Reactions

The last five chapters have presented detailed techniques for calculating limiting and pressure-dependent rate constants, $k_0(T)$, $k_\infty(T)$ and $k(p, T)$. Many of the procedures are computationally complex and there are frequently occasions on which a rapid, less precise calculation is appropriate, for example in planning the feasibility of an experimental measurement. In addition, it is important to recognise that the output of a master equation calculation of $k(p, T)$ is a set of numerical estimates at specified pressures and temperatures. Simulations of complex kinetic processes, especially in combustion, where both the pressure and temperature can vary during the simulation, are most easily carried out with analytical representations of $k(p, T)$, although look-up tables have been used[1]. Troe[2-4] recognised these requirements and developed a series of formulae for $k(p, T)$ and the high- and low-pressure limits, based on comparisons with more detailed calculations for a range of reactions and empirical fitting. These formulae and their application are briefly reviewed in this chapter.

The calculation of $k_0(T)$, based on threshold energies and the relationship between collisional activation and deactivation processes and the energy-dependent densities of states, is outlined in section 8.1. Troe[2,3] recognised that since, at low pressures, unimolecular reaction is dominated by reacting molecules at or just above the threshold energy, a zeroth-order approximation can be obtained based on equation (3.23) and the density of states at the threshold. He then proceeded to refine this estimate to allow for the contributions of states at higher energies, rotational and internal rotational energy and a range of other effects, such as anharmonicity. The corrections are based on physical insight and empiricism and the Whitten–Rabinovitch[5] formula plays a central role. Provided care is exercised, the approach can provide estimates of k_0 accurate to a factor of approximately 2, with the largest inaccuracies arising from the rotational and weak collision corrections.

The canonical methods described in Chapter 6 for calculating k_∞ are re-examined briefly in section 8.2, while in section 8.3 fall-off calculations are outlined. The latter is the most important section of the chapter because the methods have been widely used to estimate $k(p, T)$, to fit experimental data and as a representation in simulations of complex reaction schemes.

8.1   THE LOW-PRESSURE LIMIT

The limiting low-pressure rate constant for dissociation is given by

$$k_0(T) = \int_{E_0}^{\infty} dk_1(E^*) \tag{8.1}$$

Setting

$$dk_1(E^*) = k_{-1}\rho(E^*)\exp(-E^*/kT)\,dE^*/Q_{vr} \tag{8.2}$$

and

$$k_{-1} = \beta_c Z \tag{8.3}$$

where β_c is the collision efficiency parameter and Q_{vr} is the rovibrational partition function of the reactant, gives

$$k_0(T) = \frac{\beta_c Z}{Q_{vr}}\int_{E_0}^{\infty}\rho(E^*)\exp(-E^*/kT)\,dE^*. \tag{8.4}$$

Strictly, $\rho(E^*)$ is the rovibrational density of states, but Troe[3] developed an approximate representation of $k_0(T)$ by restricting his attention firstly to vibrational excitation and by replacing the energy-dependent density of states by its value at the threshold, $\rho_v(E_0)$. Thus

$$k_0(T) \approx kT\beta_c Z\rho_v(E_0)\exp(-E_0/kT)/Q_v = k_{0,app}(T) \tag{8.5}$$

where $k_{0,app}(T)$ represents an approximation to the true rate constant, $k_0(T)$. The density $\rho_v(E_0)$ is determined using the Whitten–Rabinovitch[5] approximation (equation 4.37).

Troe[3] then proceeded to improve the estimate of $k_0(T)$ by incorporating a sequence of empirical correction factors:

$$k_0(T) = k_{0,app}(T)F_{anh}F_E F_{rot}F_{rotint}F_{corr} \tag{8.6}$$

where F_{anh} corrects the harmonic densities of states for the effects of anharmonicity, F_E corrects $\rho(E_0)$ for the energy dependence of the density of states, F_{rot} and F_{rotint} allow for the incorporation of rotational and internal rotational effects into what is hitherto a purely vibrational model and F_{corr} is a final correction factor, generally set to unity, which allows for any error which might arise from the neglect of coupling between the other factors. The expressions developed by Troe[3] are given below:

1. F_{anh} is primarily influenced by those vibrations that are 'lost' in the

dissociation. Troe proposed

$$F_{anh} = \left[\frac{(s-1)}{(s-3/2)} \right]^m \tag{8.7}$$

where s is the total number of oscillators and m the number lost on dissociation. Thus, for $CH_4 \rightarrow CH_3 + H$, $s = 9$, $m = 3$ and $F_{anh} = 1.21$.

2. Comparing equations (8.4) and (8.5) gives

$$F_E = \frac{1}{kT} \int_{E_0}^{\infty} (\rho_v(E^*)/\rho_v(E_0)) \exp\left(-(E^* - E_0)/kT\right) dE^* \tag{8.8}$$

which Troe[3], via the Whitten–Rabinovitch[5] approximation, writes as

$$F_E = \sum_{i=0}^{s-1} \left[\frac{\Gamma(s)}{\Gamma(s-i)} \right] \frac{1}{(f_1(E_0))^i} \tag{8.9}$$

where

$$f_1(E_0) = [E_0 + a(E_0)E_z]/kT \tag{8.10}$$

E_z being the zero point energy and $a(E_0)$ the Whitten–Rabinovitch[5] parameter. The sum converges rapidly, and even for large species it is not necessary to include more than the first few terms; F_E increases with both temperature and s.

3. The external rotational correction, F_{rot}, provides a major source of uncertainty, especially at low temperatures. The major difficulty arises from the effects of rotation on the threshold energy. The maximum correction factor can be estimated from the Whitten–Rabinovitch approximation:

$$F_{rot,max} = \left[\frac{\Gamma(s)}{\Gamma(s+3/2)} \right] f_1(E_0)^{3/2} \tag{8.11}$$

for a non-linear molecule[5].

This correction factor, which increases with decreasing T and can become very large (Figure 8.1), assumes that there is no rotational contribution to the reaction barrier, i.e. that $E_0(J) = E_0(J = 0)$, so that rotation can contribute fully to reaction. Centrifugal barriers arise, of course, reducing both the effectiveness of rotation and F_{rot}. Troe based his analysis on an approximation introduced by Waage and Rabinovitch[6], obtaining

$$F_{rot} = F_{rot,max} \left\{ \frac{(I^{\ddagger}/I)}{(I^{\ddagger}/I) - 1 + F_{rot,max}} \right\} \tag{8.12}$$

where $I^{\ddagger}$ and I are the moments of inertia at the transition state and the equilibrium geometries respectively. As $I^{\ddagger}/I \rightarrow 1$, $F_{rot} \rightarrow 1$, because rotation is ineffective in overcoming the reaction barrier while as $I^{\ddagger} \rightarrow \infty$, $F_{rot} \rightarrow F_{rot,max}$.

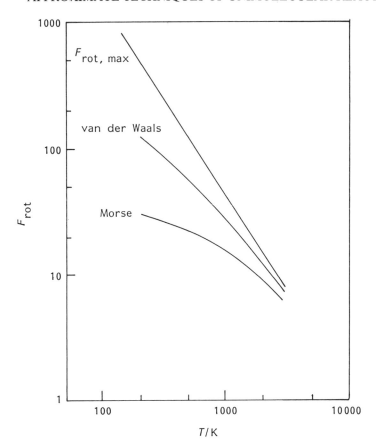

Figure 8.1 The rotational correction factor F_{rot} as a function of T. Reprinted with permission from J. Troe *et al.*, *J. Phys. Chem.*, **83**, 118 (1979). Copyright 1979 American Chemical Society.

The difficulty lies in estimating $I^{\ddagger}/I$ and hence F_{rot} for a Type II potential where there is no maximum and the location of the transition state is energy dependent. The correction depends on the shape of the potential energy function. For a van der Waals' potential ($V \propto -r^{-6}$), Troe showed that the effective ratio, $I^{\ddagger}/I$, is given by

$$I^{\ddagger}/I = \Gamma(2/3)(2C^*/kTr_e^6)^{1/3} \qquad (8.13)$$

for a linear molecule and

$$\frac{I^{\ddagger}}{I} = \left\{ \frac{12}{\pi kT} \frac{ABC}{B_{eff}^3} \frac{C^*}{r_e^6} \right\}^{1/2} \qquad (8.14)$$

for a non-linear molecule. Here Γ is the gamma function, A, B, C are the rotational constants for the molecule and B_{eff} is the effective rotational constant for a quasi-diatomic molecule representation (e.g. $B_{eff} = (B + C)/2$ for a near prolate top), r_e is the equilibrium internuclear separation and C^* is the coefficient in the r^{-6} potential. To a first approximation, $E_0 r_e^6 < C^* < 2E_0 r_e^6$. More complex, approximate expressions can be derived for Morse potentials (Figure 8.1).

Troe[2] had earlier derived more empirical and less transparent expressions for F_{rot}. For a Type II surface based on a van der Waals' potential, he derived

$$F_{rot} = F_{rot,max} \left\{ \frac{2.15(E_0/kT)^{1/3}}{2.15(E_0/kT)^{1/3} - 1 + f_1(E_0)/(s + 1/2)} \right\} \quad (8.15)$$

4. For a molecule with s oscillators and one internal rotor with a barrier V_0 to internal rotation,

$$F_{rotint} = \frac{\Gamma(s)}{\Gamma(s + 1/2)} f_1(E_0)^{1/2} f_3(V_0)[1 - \exp(-E_0/sV_0)] \quad (8.16)$$

where

$$f_3(V_0) = \left\{ [1 - \exp(-kT/V_0)]^{1.2} \right.$$

$$\left. + \frac{\exp(-1.2kT/V_0)}{(2\pi I_m kT/\hbar^2)^{1/2}(1 - \exp[-(n^2\hbar^2 V_0/2I_m(kT)^2)^{1/2}])} \right\}^{-1}$$

$$(8.17)$$

I_m is the reduced moment of inertia for the internal rotation and n is the number of equivalent minima associated with the motion. For $V_0 \to 0$, the correction factor reduces to that for a free internal rotation:

$$\lim_{V_0 \to 0} \{F_{rotint}\} = \frac{\Gamma(s)}{\Gamma(s + 1/2)} f_1(E_0)^{1/2} \quad (8.18)$$

while for $V_0 \to \infty$ the torsional limit is obtained:

$$\lim_{V_0 \to \infty} \{F_{rotint}\} = \frac{f_1(E_0)kT}{sh\nu_{tors}}[1 - \exp(-h\nu_{tors}/kT)] \quad (8.19)$$

where ν_{tors} is the frequency of the torsional vibration.

5. Only centrifugal barriers have been recognised thus far for reactions occurring on Type II potentials, whereas we saw in Chapter 6 that additional maxima arise in the effective potentials because of the significant changes in the frequencies of the transitional modes (see Figure 6.4). Troe[3] added an additional correction factor, which can be considered as a component of F_{corr}, to account for these zero point barrier effects:

$$F_{E_z} = \exp(-[E_0(J = 0) - \Delta H_0^0]/kT) \quad (8.20)$$

where $E_0(J = 0)$ is the effective barrier height on the zero point potential, relative to the zero point energy of the reactant molecule, and ΔH_0^0 is the zero Kelvin enthalpy change on dissociation. Here $F(E_z)$ depends on the shape of the potential and Troe derived an empirical approximation based on the Morse potential parameter, β_M, and the 'loosening' parameter, α, which describes the dependence of the transitional mode frequencies on bond length (section 6.2):

$$E_0(J = 0) - \Delta H_0^0 = E_{zt}(6.2\alpha/(\alpha - \beta_M)) \qquad (8.21)$$

where E_{zt} is the zero point energy in the *transitional* modes.

Finally it is necessary to interpret β_c, the collisional efficiency parameter. Troe[2] has shown that β_c is related to $\langle \Delta E \rangle_{all}$, the average energy transferred in a single collison via the equation

$$\frac{\beta_c}{1 - \beta_c^{1/2}} = \frac{-\langle \Delta E \rangle_{all}}{F_E kT} \qquad (8.22)$$

The major uncertainties in the calculations are F_{rot}, F_{rotint} and β_c. The expressions for the two correction factors were derived semiempirically by fitting the functional form and the numerical parameters to detailed numerical calculations of $k_0(T)$ for a range of systems. Despite the lack of knowledge of some of the parameters needed, e.g. V_0, the approach is remarkably successful, β_c being usually employed as the final fitting parameter; sensible values for β_c and $\langle \Delta E \rangle_{all}$ are generally obtained.

Table 8.1 shows calculations by Troe of $k_{0,app}$ and the correction factors

Table 8.1   Low-pressure limit rate constants calculated by Troe's method

Reaction $E_0/\text{kJ mol}^{-1}$	T/K	$\left\{\dfrac{k_{0,app}}{\beta_c \exp(-E_0/kT)}\right\}$ $/10^{-9}\,\text{cm}^3\,\text{mol}^{-1}\,\text{s}^{-1}$	F_{anh}	F_E	F_{rot}	F_{E_z}
ClNO → Cl + NO	200	9.7	1.72	1.02	63.7	1.0
155.5	300	13.0		1.03	46.18	
	400	15.5		1.04	35.68	
	1000	16.2		1.10	12.96	
NO$_2$ → O + NO	200	8.58	2.37	1.01	52.94	0.96
300.6	300	13.8		1.02	45.28	0.97
	400	19.0		1.02	39.59	0.98
	1000	34.5		1.11	21.42	0.99
H$_2$O → H + HO	200	1.87	2.37	1.01	111.54	0.98
493.7	300	2.98		1.01	95.16	1.0
	1000	11.8		1.08	45.22	
	2000	19.2		1.06	22.94	

for three reactions as a function of temperature. The third column shows values for $k_{0,\text{app}}/[\beta_c \exp(-E_0/kT)]$ calculated from equation (8.5). For these triatomic molecules, the values are very similar but would, of course, be larger for larger molecules. The correction factors are all fairly close to unity except for F_{rot} which was estimated using equations based on the Morse potential rather than the simpler van der Waals' equations given above for illustrative purposes. The values show the substantial enhancement of the zero pressure rate constant by rotational effects.

Troe[3] estimated strong collision rate constants from

$$k_0^{\text{SC}} = (k_{0,\text{app}}/\beta_c) F_{\text{anh}} F_E F_{\text{rot}} F_{\text{rotint}} F_{\text{corr}} \tag{8.23}$$

and compared the values with experimental measurements of recombination or dissociation rate constants to obtain estimates of β_c. For ClNO, for example, the latter ranged from 0.28 (200 K) to 0.12 (1000 K). Patrick and Golden[7] used a similar approach for a range of atmospherically significant reactions. The procedure provides easy access to rough estimates of k_0, but the uncertainties, especially in F_{rot}, substantially limit the effectiveness of the approach for all but the most approximate of applications.

Master equation methods are now much more readily applied, thanks to the programs available such as that of Gilbert, Smith and Jordan[8], than they were when Troe[2,3] developed the approximations. Despite their greater computational requirements, they deserve to be more widely applied in estimates of k_0. Since the population density depends not only on the rate of collisional excitation but also on the microcanonical rate constants for dissociation, any complications associated with centrifugal barriers, and variational transition states still have to be addressed via the techniques described in Chapter 6.

8.2   THE HIGH-PRESSURE LIMIT

The limiting high-pressure rate constant, k_∞, can be calculated using canonical transition state theory (CTST). For reactions occurring on Type I surfaces, with constrained transition states, the procedure is straightforward and reduces to one of estimating transition state parameters (Chapter 3). Approximate methods for calculating k_∞ were described in Chapter 6 for reactions occurring on Type II surfaces, namely the maximum free energy and canonical flexible transition state methods. Neither has a closed form, analytic solution and both rely on a minimisation procedure. The flexible transition state approach provides upper and lower limits on k_∞ via the angular term Θ (see Figure 6.13). Forst[9] has also developed a logarithmic interpolation procedure to describe the bond distance dependence of the transitional modes which can be used to estimate k_0, k_∞ and $k_a(E, J)$. Forst

Table 8.2 Comparison of high-pressure association rate constants for different models

Reaction	T/K	$k_{r,\infty}/10^{-11}\,\text{cm}^3\,\text{mol}^{-1}\,\text{s}^{-1}$			α/β_M
		SACM	MFE	Equation(8.24)	
$CH_3 + CH_3$	300	5.1	4.7	5.5	1.0
	1300	4.8	6.2	5.7	1.0
$O + NO$	300	1.8	2.8	1.3	1.3
	2100	2.0	2.8	1.8	1.3
$Cl + NO$	300	6.2	7.0	6.1	0.8
	2100	5.2	6.4	5.7	0.8

employed a hyperbolic switching function rather that the exponential form used by Quack and Troe[10].

If an approximate closed form is needed for k_∞ then Troe[11] has provided a representation based on his interpolation parameter α/β_M. The representation of k_∞ is very similar to that derived by Robertson, Wagner and Wardlaw[12] from canonical flexible transition state theory. For an association reaction $(B + C \rightarrow A)$ on a Type II surface, Troe obtained (cf. equation 6.38)

$$k_\infty = g_e \frac{kT}{h}\left(\frac{h^2}{2\pi\mu kT}\right)^{3/2} \frac{Q^*_{\text{cent}} F^*_{AM} Q^*_m}{Q_r(A)Q_r(B)} \exp\left(-\Delta E_0/kT\right) \qquad (8.24)$$

where Q^*_{cent} is the pseudo-partition function associated with the centrifugal barriers:

$$Q^*_{\text{cent}} = \sum_J (2J + 1) \exp\left\{(E_0(J) - E_0)/kT\right\}, \qquad (8.25)$$

Here Q^*_m is the partition function for the transitional modes that correlate with rotations in the products and F^*_{AM} corrects Q^*_m for the rotational character of the transitional modes. Troe provided closed-form expressions for all these terms, as a function of α/β_M, which may be found in reference 11. Table 8.2 compares the results with detailed statistical adiabatic channel calculations (SACM) and maximum free energy (MFE) calculations. The results are very satisfactory.

8.3   PRESSURE DEPENDENCE—ANALYTICAL REPRESENTATION OF THE FALL-OFF CURVE

Troe[4,13] has also made a considerable contribution to the description of unimolecular reactions in the fall-off region. His major aim was to generate a formalism that is easily represented and calculated, without recourse to detailed master equation calculations, and yet which is more realistic than Lindemann–Hinshelwood theory. He used the idea of reduced fall-off curves, representing the pressure scale in terms of the dimensionless quantity $P_r(= k_0[M]/k_\infty)$. The effects of the energy dependence of the microcanonical rate constant, $k_a(E^*)$, and of weak collisions is to 'broaden' the fall-off curve and so Troe and co-workers[4,13] suggested a reduced representation of the form

$$k_{uni}/k_\infty = F_{LH}(P_r)F^{SC}(P_r)F^{WC}(P_r) \tag{8.26}$$

where $F_{LH}(P_r)$ is the Lindemann–Hinshelwood representation of the reduced unimolecular rate constant and is given by

$$F_{LH}(P_r) = \frac{P_r}{1 + P_r} \tag{8.27}$$

The term $F^{SC}(P_r)$ is a strong-collision broadening factor, accounting for the effects of the energy dependence of the microcanonical rate constants and $F^{WC}(P_r)$ is a weak-collision broadening factor. The reduced form of the representation means that exactly the same expression can be used for association reactions.

Troe[4] first tried to express the form of $F^{SC}(P_r)$ in terms of Kassel integrals[14]. The integrals are related to two parameters S and B which, in the original classical theory, were equated to the number of effective oscillators and the reduced threshold energy, E_0/kT, respectively. To account for quantum effects, S and B were replaced by S_K and B_K defined by

$$S_K = 1 + U_{vib}^\ddagger/kT = 1 - \frac{1}{T}\frac{\partial \ln Q_v^\ddagger}{\partial(1/T)} \tag{8.28}$$

where $Q_v^\ddagger$ is the partition function for the transition state, and $U_{vib}^\ddagger$ is the average molecular vibrational energy of the transition state. Thus S_K is related directly to k_∞ via $Q_v^\ddagger$. Troe[4] obtained B_K in analytic form via approximations based on the Whitten–Rabinovitch[5] correction $a(E_0)$ (see section 4.4):

$$B_K = \frac{(S_K - 1)f_1(E_0)}{s - 1} \tag{8.29}$$

where s is the total number of oscillators and $f_1(E_0)$ is defined in equation (8.10). Troe[4] then related the reduced broadening term, F^{SC}, to B_K and S_K by comparison with detailed RRKM calculations, using the reduced Kassel integrals as a guide to the functional form to use. The analysis was carried out for 20 molecules to ensure that the final representation was widely applicable.

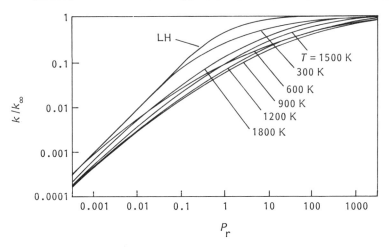

Figure 8.2 The reduced fall-off curves for cycloheptatriene isomerisation as a function of temperature, calculated using RRKM theory assuming strong collisions (Reproduced with permission).

Figure 8.2 shows an example of the reduced fall-off curves calculated using RRKM theory for cycloheptatriene isomerisation as a function of temperature, assuming strong collisions. The curves are compared with the Lindemann–Hinshelwood curve for the same values of k_0 and k_∞ and show substantial broadening resulting from the energy dependence of $k_a(E^*)$. Troe[4] represented F^{SC} via its central value, $F_{cent}^{SC} = F^{SC}(P_r = 1)$, and a width term. The empirical expressions obtained for F_{cent}^{SC} are

$$-\log F_{cent}^{SC} = \frac{(1.06 \log S_K)^{2.2}}{(1 + C_1 S_K^{C_2})} \tag{8.30}$$

where C_1 and C_2 are given by the equations below:

$$C_1 = 0.10 \exp (2.5 B_T^{-1} - 0.22 B_T - 6. \times 10^{-10} B_T^6)$$

$$C_2 = 1.9 + 4.6 \times 10^{-5} B_T^{2.8}$$

$$B_T = \{(S_K - 1)/(s - 1)\}^{0.6} B_K$$

and F^{SC} is then related to F_{cent}^{SC} via the width term

$$\frac{\log (F^{SC})}{\log (F_{cent}^{SC})} = \frac{1}{1 + [\log (P_r)/N^{SC}]^2} \tag{8.31}$$

where N^{SC} is given, for moderate temperatures, by

$$N^{SC} = 0.75 - 1.27 \log (F_{cent}^{SC}) \tag{8.32}$$

At low T, $N^{SC} \approx 1$ while, at higher temperatures, a slightly more complex expression than (8.32) is required[4].

In order to implement this semiempirical, analytic method for calculating reduced fall-off curves with strong-collision broadening included, S_K must be calculated as follows:

$$S_K = 1 + \frac{r}{2} + \sum_{i=1}^{s-r-1} \frac{(hv_i^{\ddagger}/kT)}{\exp(hv_i^{\ddagger}/kT) - 1} \tag{8.33}$$

where s is the total number of internal coordinates and r is the number of free internal rotors; S_K must clearly be compatible with $k_{\infty}(T)$ because of its linkage to $v_i^{\ddagger}$. If information on the transition state cannot be readily obtained, Troe[4] suggested using the equivalent expression for the effective number of oscillators in the ground state, s_{eff}, with

$$S_K = s_{eff} + \Delta S \tag{8.34}$$

and $1 \leqslant \Delta S \leqslant 2$.

Table 8.3 shows calculated values for the various parameters employed in calculating F_{cent}^{SC}, for a range of molecules and for several temperatures. Note the excellent agreement between the exact values ($F_{cent}^{SC}(ex)$) obtained from RRKM calculations and the approximate values ($F_{cent}^{SC}(app)$) obtained from the empirical expressions outlined above. Figure 8.3 compares the equivalent values of F^{SC} over the pressure range for several molecules. The overall shape is represented well, the major errors occurring at low pressures.

Gilbert, Luther and Troe[15] developed a parameterised form of $F^{WC}(P_r)$ to represent broadening arising from weak collision effects. They studied a range of molecular systems, comparing the analytic results, and optimising

Table 8.3 Parameters used in the calculation of F^{SC}

Molecule	T/K	300	500	900	1500
CH$_3$NC	s_{eff}	1.11	2.00	2.74	5.67
	S_K	1.71	2.49	4.10	5.94
	B_K	6.84	8.59	9.93	9.49
	B_T	1.32	2.59	4.64	5.87
	$F_{cent}^{SC}(ex)$	0.95	0.86	0.67	0.53
	$F_{cent}^{SC}(app)$	0.96	0.84	0.63	0.52
C$_7$H$_8$	s_{eff}	3.76	8.08	14.8	21.1
(cyclo-	S_K	4.97	9.43	16.2	22.3
hepatriene)	B_K	19.99	25.3	25.4	21.4
	B_T	5.13	10.3	14.6	15.1
	$F_{cent}^{SC}(ex)$	0.58	0.25	0.12	0.20
	$F_{cent}^{SC}(app)$	0.57	0.29	0.16	0.19

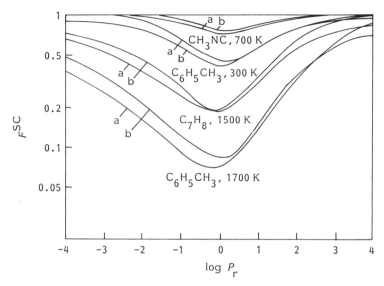

Figure 8.3 Comparison of F_{SC} values for several molecules as a function of pressure: (a) RRKM fall-off curves; (b) simplified fall-off curves (Reproduced with permission).

the parametric form, against detailed master equation calculations with an exponential down model for energy transfer. They again developed the equations via the central value, $F_{cent}^{WC} = F^{WC}(P_r = 1)$:

$$\log(F_{cent}^{WC}) = 0.14\log(\beta_c) \tag{8.35}$$

where β_c is the weak collision efficiency parameter (equation 8.4) which is related to $\langle \Delta E \rangle_{all}$ via equation (8.22). A more detailed form for $F^{WC}(P_r)$, which may be necessary at high temperatures, is given by Gilbert, Luther and Troe[15].

The combined effects of F^{SC} and F^{WC} are contained in composite expressions for k_{uni}/k_∞ (equation 8.26), based on $F_{cent} = F_{cent}^{SC} F_{cent}^{WC}$. At low temperatures,

$$\frac{\log(F)}{\log(F_{cent})} = \frac{1}{1 + (\log(P_r))^2} \tag{8.36}$$

where $F = F^{SC} F^{WC}$. The increased width at higher T requires a more complex expression:

$$\frac{\log(F)}{\log(F_{cent})} = \frac{1}{1 + \{\kappa/(N - 0.14\kappa)\}^2} \tag{8.37}$$

where

$$\kappa = \log(P_r) + c$$

$$c = -0.4 - 0.67 \log(F_{cent})$$

$$N = 0.75 - 1.27 \log(F_{cent})$$

Thus, $k_{uni}([M], T)$ can be represented realistically in terms of equation (8.26) and k_0, k_∞ and F_{cent}, using either simple (equation 8.36) or complex (equation 8.37) expressions to relate F to F_{cent}. Here F_{cent} can be calculated from S_K, B_K and β_c (equations 8.30 and 8.35) and so related to molecular parameters. Alternatively, as is illustrated below, F_{cent} can simply be used, along with k_0 and k_∞, as fitting parameters. In this respect, it has become usual to represent k_0 and k_∞ in modified Arrhenius form (i.e. $AT^n \exp(-E/RT)$) and F_{cent} as

$$F_{cent} = (1 - a) \exp(-T/T^{***}) + a \exp(-T/T^*) + \exp(-T^{**}/T) \quad (8.38)$$

where a, T^*, T^{**} and T^{***} are fitting parameters.

The Troe[13] technique has been widely employed to analyse or represent experimental or theoretical fall-off curves for dissociation and association reactions. The following sections provide some illustrative examples.

1. Experimental measurements are generally made in the fall-off region; extrapolations are needed to estimate k_0 and k_∞ and a representation is needed of $k_{uni}([M], T)$. Lightfoot and Pilling[16] measured rate constants for $H + C_2H_4$, obtaining the data shown in Figure 8.4. The rate constant clearly moves further into the fall-off region as T increases, but the small activation barrier ensures that, under the conditions studied experimentally, k_{uni} increases with T at a given pressure. Lightfoot and Pilling[16] calculated S_K via equation (8.28) using the vibrational frequencies calculated by Hase and Schlagel[17] for the transition state using *ab initio* methods, S_K ranging from 1.89 at 285 K to 3.96 at 604 K.

In order to calculate F_{cent}^{WC}, Lightfoot and Pilling[16] used the low-pressure data of Lee *et al.*[18] on this reaction for a range of diluent gases. They assumed that $\beta_c = 1.0$ for SF_6 as the third body, thus calculating a value of $\beta_c = 0.25$ at 300 K for Helium. $\langle \Delta E \rangle_d$ (section 7.1) was then calculated and was assumed temperature independent, thus allowing $\beta_c(T)$ to be calculated. Work by Gutman and co-workers[19] suggests that $\langle \Delta E \rangle_d$ increases with T for this reaction, although the error introduced in the analysis is negligible (see below). The parameter, β_c, calculated in this way ranges from 0.29 at 285 K to 0.11 at 604 K.

Lightfoot and Pilling[16] then analysed the experimental data at each temperature, using k_0 and k_∞ as variable parameters and equation (8.38) for $F(P_r)$. The fits are shown in Figure 8.4. The values of k_0 and k_∞ obtained were found to be very insensitive to β_c, varying $\log(\beta_c)$ by ± 1 producing

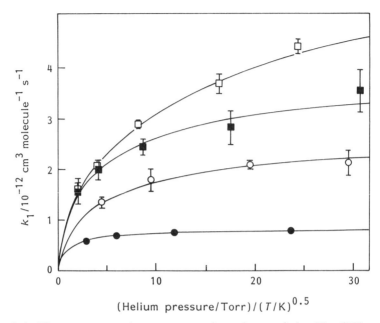

Figure 8.4 The pressure and temperature dependence of the $H + C_2H_4 \rightarrow C_2H_5$ reaction: (●) 285 K; (○) 400 K; (■) 511 K; (□) 604 K. The error bars refer to 95.5% confidence limits, and the curves were obtained from a non-linear least squares fit to the data using the Troe factorisation method. Reprinted with permission from Lightfoot and Pilling, *J. Phys. Chem.*, **91**, 3373 (1987). Copyright 1987 American Chemical Society.

changes in k_0 and k_∞ at 604 K of only ±5% and ±10% respectively. Increasing S_K by one increased k_∞ by 5% at 285 K and 12% at 604 K, but had a dramatic effect on k_0, doubling the value at 285 K.

2. In their analysis of the data for $CH_3 + O_2$, Keiffer, Miscampbell and Pilling[20] utilised the relationship between S_K and k_∞ (equation 8.28) in a global analysis of the data i.e. all the data, over a range of temperatures and pressures, were analysed together.

Neglecting internal rotors, equation (8.28) can be rewritten as

$$S_K = 1 + T\frac{\partial \ln Q_v^\ddagger}{\partial T} \tag{8.39}$$

where $Q_v^\ddagger$ is the vibrational partition function for the transition state. The transition state expression for k_∞ is

$$k_\infty = \frac{kT}{h}\frac{Q^\ddagger}{Q(CH_3)Q(O_2)}\exp\left(-E_r^0/kT\right) \tag{8.40}$$

while a suitable parameterised form for the experimentally determined k_∞ is

$$k_\infty = A_\infty (T/300 \text{ K})^{n_\infty} \exp(-B/T) \qquad (8.41)$$

with $B = E_r^0/k$. Combining $d \ln k_\infty/dT$ from both expressions with equation (8.39), assuming classical translational and rotational partition functions and no temperature dependence of the electronic partition functions or the geometry of the transition state, gives[20]

$$S_K = n_\infty + \frac{5}{2} + T\left[\frac{d \ln Q_v(CH_3)}{dT} + \frac{d \ln Q_v(O_2)}{dT}\right] \qquad (8.42)$$

$$= n_\infty + \frac{5}{2} + \frac{U_v(CH_3)}{kT} + \frac{U_v(O_2)}{kT} \qquad (8.43)$$

where U_v is the average molecular vibrational energy. Similar expressions can be derived if internal rotors are involved.

Equation (8.43) demonstrates the necessary link between S_K and n_∞. The latter is obtained by extrapolation of the experimental data using S_K, and the two must be internally consistent. (This was not, in fact, the case for the procedure adopted by Lightfoot and Pilling[16].) Keiffer, Miscampbell and Pilling[20] fixed $\langle \Delta E \rangle_d$ (initially at 285 cm^{-1} for Ar), set E_r^0 to zero and varied A_0, A_∞, n_0 and n_∞ using the minimum χ^2 as the fitting criterion for a global fit to seven sets of experimental data over a temperature range 298–582 K. They also varied E_r^0 in separate global fits, showing that it must be small (<4 kJ mol^{-1}), while the quality of the fits was insensitive to $\langle \Delta E \rangle_d$.

The optimal expression for k_∞ is

$$k_\infty = [(1.2 \pm 0.2) \times 10^{-12}](T/300 \text{ K})^{1.2 \pm 0.4} \text{ cm}^3 \text{ molecule}^{-1} \text{s}^{-1} \qquad (8.44)$$

corresponding to a small A factor for a reaction occurring on a Type II surface. Cobos et al.[21] reproduced their own, somewhat higher, experimental expression using a very high symmetry number, predicted on the basis of free internal rotation in the transition state. An alternative explanation is a constrained tight transition state which is difficult to rationalise on an attractive potential energy surface. One possible explanation is the orbital rearrangement occurring in forming CH_3O_2 from $CH_3 + O_2$ which is much more substantial than that in forming, say, C_2H_6 from $CH_3 + CH_3$. This rearrangement does not generate an energy barrier, but may lead to a rapid change in the gradient of the potential surface at comparatively short intermolecular separations and to a consequent insensitivity in the location of the transition state to energy.

3. The parametric forms of $k([M], T)$ and of k_0, k_∞ and F_{cent}, have also been employed to provide concise representations of theoretical fall-off curves. A good example is provided by the work of Wagner and Wardlaw[22] on $CH_3 + CH_3$. They employed microcanonical flexible transition state

theory (section 6.3) to calculate $k(E, J)$, using the simple parameterised form of the potential for the reaction which was discussed in Chapter 6. They used α, the interpolation parameter for the CH bond lengths, the H–C–H bond angles and for the conserved frequencies (Chapter 6), as a variable parameter and fitted the data of Slagle et al.[23], using a modified strong collision model with $\langle \Delta E \rangle_d$ as the second fitting parameter. Once the fits to experiment had been performed, the theoretical curves were fitted to equation (8.38) with A_0, A_∞, n_0 and n_∞, E_0, E_∞, a, T^*, T^{**} as variable parameters, giving

$$k_0 = 8.76 \times 10^{-7} T^{-7.03} \exp\left(-1390 \ K/T\right) \text{cm}^6 \text{molecule}^{-2} \text{s}^{-1} \quad (8.45)$$

$$k_\infty = 1.50 \times 10^{-7} T^{-1.18} \exp\left(-329 \ K/T\right) \text{cm}^3 \text{molecule}^{-1} \text{s}^{-1} \quad (8.46)$$

$$F_{\text{cent}} = 0.381 \exp\left(-T/73.2 \ K\right) + 0.619 \exp\left(-T/1180 \ K\right) \quad (8.47)$$

An interesting feature of this fit is the negative temperature dependence of k_∞ which has been referred to in Chapters 6 and 7 in the discussion of master equation fits to the data for this reaction.

4. Perhaps the most important application of parameterised fall-off curves is in simulations of complex kinetic processes in combustion and atmospheric chemistry. Such simulations involve numerical integration of large sets of coupled ordinary differential equations corresponding to the rate equations for each species. Often these equations are embedded in partial differential equations describing transport in non-homogenous systems. Computational efficiency is paramount and since T and p vary in the simulation, especially for combustion systems, simple parametric forms for $k_{\text{uni}}(p, T)$ are essential.

Troe's equations have been widely adopted in compilations of evaluated kinetic data for use in such simulations. Two leading examples are 'Evaluated data for combustion modelling'[24] compiled as part of the European Community Energy Research and Development Programme and 'Kinetic and photochemical data for atmospheric chemistry'[25] compiled under the auspices of the International Union of Pure and Applied Chemistry.

Alternative representations of the reduced fall-off curve have also been developed and Pawlowska, Gardiner and Oref[26] have compared two of these with the Troe formula.

The first, proposed by Oref[27], is more physically based than the others and employs a parameter $J_{3/2}$ defined as

$$J_{3/2} = (k_\infty/k_{3/2} - 1)^2 \quad (8.48)$$

where $k_{3/2}$ is the dissociation rate constant when the reaction order is 3/2 (Figure 8.5). Oref[27] defined J via the Tolman[28] averages over distributions of reacting molecules:

$$J = \langle k_a(E) \rangle^{Tn} \langle k_a(E)^{-1} \rangle^{Tn} \quad (8.49)$$

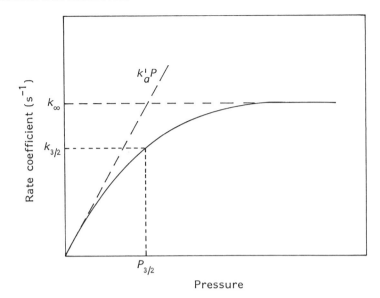

Figure 8.5 The definition of $k_{3/2}$ as the point where the overall order of reaction is $\frac{3}{2}$. Reprinted with permission from I. Oref, *J. Phys. Chem.*, **93**, 3465 (1989). Copyright 1989 American Chemical Society.

where Tn denotes the Tolman[28] average i.e. an average over the distribution function $p(E)$ which is given by

$$p(E) = \frac{\omega f(E) k_a(E)}{k_a(E) + \omega} \tag{8.50}$$

where $f(E)$ is the Boltzmann density and ω is the collision frequency. Defining $J_{3/2}$ for an order of 3/2, Oref[27] showed that,

$$\frac{k}{k_\infty} = \frac{-(1 + P_r) + [(1 + P_r)^2 + 4(J_{3/2} - 1)P_r]^{1/2}}{2(J_{3/2} - 1)} \tag{8.51}$$

and parameterised the temperature dependence of $J_{3/2}$ via

$$J_{3/2}(T) = A_* \exp\left(-B_*(T - T_*)^2\right) \tag{8.52}$$

where A_*, B_* and T_* are fitting parameters. As with the Troe[4] model, $J_{3/2}$ is effectively related to a temperature-dependent F parameter, via equation (8.48), i.e. $F = k_\infty/k_{3/2}$.

The second model, developed empirically, takes the form

$$(k/k_\infty)^a = (P_r)^a + 1 \tag{8.53}$$

in which a is another central parameter. Pawlowska, Gardiner and Oref[26] showed that, using an order of 3/2 to define the centre of the fall-off curve, $a_{3/2}$ is given by

$$a_{3/2} = \log(2)/\log(k_{3/2}/k_\infty)$$
$$= -\log(2)/\log(1 + \sqrt{J}_{3/2}) \tag{8.54}$$

emphasising the relationship between the two models. They also investigated the relationship to F_{cent}, and showed that all three models show fits to calculated fall-off curves for cyclobutane and cyclobutene of comparable quality, with deviations of typically up to 20% although greater discrepancies were also observed. The regions of maximum deviation vary from one model to another; overall, each has no great advantage over the others.

The reduced representation developed by Wang and Frenklach[29] arose from their comparison of $\log(F)/\log(F_{cent})$ for the Troe model and for calculated fall-off curves. They recognised that a Gaussian representation would be more appropriate than the Lorentzian form adopted by Troe (equation 8.36) and proposed

$$\log(F)/\log(F_{cent}) = \exp(-[(\log(P_r) - \alpha)/\sigma]^2) \tag{8.55}$$

where α and σ are shift and width parameters respectively. They compared equations (8.36) and (8.55) with modified strong collision calculations for several decomposition reactions (C_2H_5, C_2H_3, HOCO, CH_4, C_2H_6, C_2H_4, C_3H_6), based on RRKM calculations of $k_a(E)$. They showed that the Gaussian form, obtained by fitting σ and α to the RRKM results, shows far better agreement than the Troe model with F calculated via equation (8.36). The agreement was especially improved at the wings of the $\log(F)/\log(F_{cent})$ curve. The dependence of the fitted α and σ on temperature was found to be quite simple over a wide range of T, showing no extremes of curvature, suggesting that a simple T-dependent form, based on no more than three parameters each, should be appropriate. This Gaussian model, therefore, holds much promise, especially for empirical fitting to experimental and theoretical data for inclusion in complex kinetic simulations. It is not, as yet, widely used.

REFERENCES

1. W. Tsang and R. F. Hampson, *J. Phys. Chem. Ref. Data*, **15**, 1087 (1986).
2. J. Troe, *J. Chem. Phys.*, **66**, 4745 (1977); **66**, 4758 (1977).
3. J. Troe, *J. Phys. Chem.*, **83**, 118 (1979).
4. J. Troe, *Ber. Bunsenges. Phys. Chem.*, **87** 161 (1983).
5. G. Z. Whitten and B. S. Rabinovitch, *J. Chem. Phys.*, **38**, 2466 (1963).
6. E. V. Waage and B. S. Rabinovitch, *Chem. Rev.*, **70**, 377 (1970).
7. R. Patrick and D. M. Golden, *Int. J. Chem. Kinet.*, **15**, 1189 (1983).

8. R. G. Gilbert, S. C. Smith and M. J. T. Jordan, UNIMOL program suite (calculation of fall-off curves for unimolecular and recombination reactions).

9. W. Forst, *J. Phys. Chem.*, **95**, 3612 (1991).

10. M. Quack and J. Troe, *Ber. Bunsenges. Phys. Chem.*, **81**, 329 (1977).

11. J. Troe, *J. Chem. Phys.*, **75**, 226 (1981).

12. S. H. Robertson, A. F. Wagner and D. M. Wardlaw *J. Chem. Phys.* **103**, 291 (1995).

13. J. Troe, *Ber. Bunsenges. Phys. Chem.*, **78**, 478 (1981).

14. L. S. Kassel, *J. Phys. Chem.*, **32**, 225, 1065 (1928).

15. R. G. Gilbert, K. Luther and J. Troe, *Ber. Bunsenges. Phys. Chem.*, **87**, 169 (1983).

16. P. D. Lightfoot and M. J. Pilling, *J. Phys. Chem.*, **91**, 3373 (1987).

17. W. L. Hase and H. B. Schlagel, *J. Phys. Chem.*, **86**, 3901 (1982).

18. J. H. Lee, J. V. Michael, W. A. Payne and L. J. Stief, *J. Chem. Phys.*, **68**, 1817 (1978).

19. Y. Feng, J. T. Niiranen, A. Bencsura, V. D. Knyazev, D. Gutman, W. Tsang, *J. Phys. Chem.*, **97**, 871 (1993).

20. M. Keiffer, A. J. Miscampbell and M. J. Pilling, *J. Chem. Soc. Farad. Trans. II*, **84**, 505 (1988).

21. C. J. Cobos, H. Hippler, K. Luther, A. R. Ravishankara and J. Troe, *J. Phys. Chem.*, **89** 4332 (1985).

22. A. F. Wagner and D. M. Wardlaw, *J. Phys. Chem.*, **92**, 2462 (1988).

23. I. R. Slagle, D. Gutman, J. W. Davies and M. J. Pilling, *J. Phys. Chem.*, **92**, 2455 (1988).

24. D. L. Baulch, C. J. Cobos, R. A. Cox, C. Esser, P. Frank, T. Just, J. A. Kerr, M. J. Pilling, J. Troe, R. W. Walker and J. Warnatz, Evaluated kinetic data for combustion modeling, *J. Phys. Chem. Ref. Data* **21**, 411 (1992).

25. R. Atkinson, D. L. Baulch, R. A. Cox, R. F. Hampson, J. A. Kerr and J. Troe, Evaluated kinetic and photochemical data for atmospheric chemistry supplement IV—IUPAC Subcommittee on Gas Kinetic Data Evaluation for Atmospheric Chemistry, *J. Phys. Chem. Ref. Data*, **21**, 1125 (1992).

26. Z. Pawlowska, W. C. Gardiner and I, Oref, *J. Phys. Chem.*, **97** 5024 (1993).

27. I. Oref, *J. Phys. Chem.*, **93**, 3465 (1989).

28. R. C. Tolman, *Foundations of Statistical Mechanics*, Oxford University Press (1938).

29. H. Wang and M. Frenklach, *Chem. Phys. Lett.*, **205**, 271 (1995).

9 Energy Transfer

This chapter is concerned with two kinds of energy transfer, both of relevance to unimolecular reactions. The first is *intermolecular energy transfer* in collisions. This is the primary step in the energisation of molecules thermally, and a secondary step following other methods of energisation such as photoactivation, overtone pumping or multiphoton absorption. The term may be broadened to include molecular collisions with the vessel walls which may contribute at low pressures. Many studies have now been carried out in which the relative efficiencies of various bath gases have been measured and compared with theoretical predictions. The first part of this chapter gives a brief summary of the theoretical framework used to interpret these experiments and some recent examples of the experimental methods both direct and indirect which have been employed.

The second energy-transfer process of relevance is that of *intramolecular energy transfer*. It is a key assumption of statistical theories of unimolecular reactions, as has been mentioned in earlier chapters, that rapid intramolecular vibrational and rotational energy redistribution occurs in a time which is short compared with that needed for dissociation or isomerisation. Recent experiments to test this assumption are referred to later and generally support this assumption in the majority of cases with randomisation occurring in a time scale of 10^{-11}–10^{-12} s.

9.1 INTERMOLECULAR ENERGY TRANSFER

The RRKM theory originally assumed that energisation and de-energisation processes occurred stepwise with the transfer of relatively large amounts of energy per collision. This assumption and its failure in many experiments involving non-thermal energisation has been discussed in Chapter 3 (section 3.12). It is now customary to incorporate *weak collisions* (energy transferred $\leqslant 2$ kJ mol^{-1}) into routine calculations rather than by crude modifications of the *strong collision* rate constant given by equation (3.17).

Discussion of the theoretical framework of collisional energy transfer is now based upon the treatment of a unimolecular reaction in terms of the master equation. It is therefore necessary to recapitulate some of the equations given earlier in Chapter 7 to which the reader is referred for more detail.

In the master equation, if the energy-dependent population density of molecules (i.e. those in energy states between y and $y + dy$) is represented by $n(y, t)$, then the time dependence of $n(y, t)$ is represented by

$$\frac{dn}{dt}(y, t) = \begin{Bmatrix} \text{rate of collisional energy} \\ \text{transfer into states in the} \\ \text{range y to } y + dy \end{Bmatrix} - \begin{Bmatrix} \text{rate of collisional energy} \\ \text{transfer out of states in the} \\ \text{range } y \text{ to } y + dy \end{Bmatrix}$$

$$= \omega \int_0^\infty P(y|x) n(x, t)\, dx - \omega n(y, t) \tag{7.1}$$

where ω is the collision frequency, $n(x, t)$ is similarly the energy-dependent population density of molecules in energy states between x and $x + dx$ and $P(y|x)$ represents the probability of transfer of molecules into the range y to $y + dy$ from those in states x to $x + dx$.

Equation (7.1) neglects angular momentum considerations and the occurrence of unimolecular reaction which also depopulates state y. The latter is included by incorporating the microcanonical rate constant $k(y)$ in the more complete master equation given by

$$\frac{dn(y, t)}{dt} = \omega \int_0^\infty P(y|x) n(x, t)\, dx - \omega n(y, t) - k(y) n(y, t) \tag{7.6}$$

Solution of the master equation, subsequently described in Chapter 7, requires knowledge of the energy-transfer probabilities, $P(y|x)$. Since all transitions from state x must end in some level y, integrating the probability over all possible y must produce unit probability. This is the normalisation condition.

$$\int_0^\infty P(y|x)\, dy = 1$$

The probability function must also satisfy the detailed balancing principle when the system is at equilibrium, i.e., when the rate of transfer into states (y to $y + dy$) from states (x to $x + dx$) is equal to the rate of transfer in the opposite direction. Under these conditions, i.e. at infinite time, the population densities $n(y, t)$ and $n(x, t)$ may be replaced by the Boltzmann populations $f(y)$ and $f(x)$, e.g.

$$f(y) = \rho(y) \exp(-y/kT)/Q$$

where Q is the total molecular partition function for the active degrees of freedom so that $P(y|x)f(x) = P(x|y)f(y)$ and hence equation (9.1), which is equivalent to (7.3), follows:

$$P(y|x)/P(x|y) = \rho(y)/\rho(x) \exp(-[y - x]/kT) \tag{9.1}$$

To solve the master equation a model is required in order to define the transition probabilities. Various models have been proposed and used to compare with experimental rate constants and also to interpret energy-transfer processes both in the presence and absence of unimolecular reaction. These models differ in the derived quantities related to energy transfer such as the average energy transferred upon collision which may be defined in a number of different ways.

When the molecule is transferred from energy state x into y, the energy lost or gained is given by $(y - x)$ which may be positive if $y > x$ (up transition) or negative if $x > y$ (down transition).

The average energy lost or gained is then given by

$$\langle \Delta E \rangle (x) = \frac{\int_0^\infty (y - x) P(y|x)\, dy}{\int_0^\infty P(y|x)\, dy} \tag{9.2}$$

Because of the normalisation condition the denominator is equal to unity and since no account of sign is taken (9.2) can be written for all collisions as

$$\langle \Delta E \rangle_{all}(x) = \int_0^\infty (y - x) P(y|x)\, dy \tag{9.3}$$

If we wish to consider only *down* transitions, i.e. $y < x$, the relevant expression for average energy transferred is $\langle \Delta E \rangle_d$ given in

$$\langle \Delta E \rangle_d(x) = \frac{\int_0^x (y - x) P(y|x)\, dy}{\int_0^x P(y|x)\, dy} \tag{9.4}$$

According to (9.4) $\langle \Delta E \rangle_d$ is a negative quantity. A related quantity is the square root of the average squared energy transferred in all collisions which may be calculated from

$$\langle \Delta E^2 \rangle_{all}^{1/2}(x) = \left[\int_0^\infty (y - x)^2 P(y|x)\, dy \right]^{1/2} \tag{9.5}$$

This quantity which is always positive, gives a better impression of the width of the distribution of energies transferred[1]. These energy parameters $\langle \Delta E \rangle_d$ and $\langle \Delta E \rangle_{all}$ are related to the collisional efficiency per unit collision β_c which has been defined previously (section 3.12).

Tardy and Rabinovitch[2] have shown that a 'quasi-universal' relationship exists between β_c and $\langle \Delta E \rangle_d / \langle E^+ \rangle$, where $\langle E^+ \rangle$ is the Boltzmann average energy of the reacting molecules. The exact form of the relationship depends upon the functional form of the transition probability. In the case of the exponential model for the transition probability, which is described below, Troe[3,4] has obtained an analytical solution of the master equation which yields

$$\frac{\beta_c}{1 - \beta_c^{1/2}} = -\frac{\langle \Delta E \rangle_{all}}{F_E kT} \tag{9.6}$$

relating β_c to $\langle \Delta E \rangle_{all}$, where F_E is a correction factor for the energy dependence of the density of states above threshold. The quantities $\langle \Delta E \rangle_d$ and $\langle \Delta E \rangle_{all}$ are clearly related to each other, but again the exact relationship depends upon the model under consideration for the transition probability $P(y|x)$. The major interest in various models for $P(y|x)$ is to obtain the most appropriate values to incorporate in an energy-grained master equation. The most widely used model is the *exponential down model* (sometimes referred to as the exponential gap model) in which it is assumed that the energy removed from the molecule is exponentially distributed and the probability of transfer of an amount of energy $|y - x|$ is given by

$$P(y|x) = A(x) \exp\{-\alpha|y - x|\} \qquad y \geqslant x \tag{7.2}$$

in which $\alpha = \langle \Delta E \rangle_d^{-1}$. In the *stepladder model*[5], the energy transfer is assumed to occur in a series of equal steps. The step size is then the average energy transferred per down transition or $\langle \Delta E \rangle_{S.L} = \langle \Delta E \rangle_d$, and the transition probability is a δ function. The relationship between up and down transitions from a given level is given by the normalisation condition and (9.1). For non-thermally activated systems which produce highly energised species it is possible, although not necessary nor generally advisable, to ignore up transitions. For thermally energised species, however, these are more important and the average energy transferred is then the average of all transitions. In thermally energised reactions at low pressures, energy transfer by collision is the second-order rate-determining process.

Less widely used forms of the transition probability are the Poisson and Gaussian models. The *Poisson model* assumes a unique collision encounter time during which the internal energy is randomly distributed, the probability of transferring energy $|y - x|$ is then related to the average energy transferred down $\langle \Delta E \rangle_d$ by

$$P(y|x) = \frac{\langle \Delta E \rangle_d^{|y-x|}}{|y - x|!} \tag{9.7}$$

In the *Gaussian* or *normal distribution model* the corresponding expression is

$$P(y|x) = \exp[-(|y - x| - \langle \Delta E \rangle_d)^2 / 2\sigma^2)] \tag{9.8}$$

where σ is the standard deviation of the mean energy transferred, given by (9.5).

Gilbert and Smith[1] have developed a *biased random walk model* of collisional energy transfer. They argue that the period of a collision (~ 0.1 ps) is significantly longer than a vibrational period. During this time the internal energy of the reactant undergoes a large number of oscillations and the molecule can be considered to undergo a random walk in energy space. Requirements of microscopic reversibility and of the Boltzmann distribution place a bias on the random walk, favouring energy loss over energy gain. They used the Fokker–Planck equation, a standard approach employed in modelling diffusive and dissipative motion in Cartesian space, to develop expressions for $P(y|x)$, obtaining

$$P(y|x) = (4\pi s^2)^{-1/2} \exp \left\{ -\frac{(zs^2 + x - y)^2}{4s^2} \right\} \qquad (9.9)$$

where $z = -\partial \ln f(x)/\partial x$, $f(x)$ is the Boltzmann distribution and $s^2 = D_E t_C$. The symbol t_C represents the collision duration and D_E the diffusion coefficient. Equation (9.9) is a displaced Gaussian in $(x - y)$. The average energy transferred per collision, $\langle \Delta E \rangle$ is equal to $-s^2 z$ and a somewhat more complex representation can be obtained for $\langle \Delta E \rangle_d$. In principle, D_E can be obtained from trajectory studies of collisional energy flow via an autocorrelation function, comparable to those used to relate the Cartesian diffusion coefficient to short-time molecular motion in liquids.

With the insertion of one of these forms for the transition probability into equation (7.6), a solution for the time-dependent population of molecules in energy level y is theoretically possible. The steady-state populations in all levels would be found by setting the analogous expressions for all levels to zero. The solution of such a set of equations by matrix methods and the necessity of using 'energy graining' is the subject of Chapter 7 and will not be repeated here.

Of more concern in the present chapter is the difference produced in the prediction of unimolecular rate behaviour when various weak collision treatments are incorporated in the theoretical equations. A related matter is the extent to which energy transfer parameters themselves may be derived from experimental data and the variety of methods which have been used for this purpose.

EXPERIMENTAL DATA INVOLVING ENERGY TRANSFER

Many extensive compilations of data involving energy transfer have now been published. These may be found in the first edition of this book[6] and in a number of reviews[7–11] including the recent excellent comprehensive review of Oref and Tardy[12]. The following account is intended to provide an overview of the various methods which have been used both directly and indirectly to study energy-transfer processes.

THERMAL ENERGISATION STUDIES

Historically the first attempts to study the involvement of energy transfer in unimolecular reactions were concerned with thermally energised molecules. Since the rate-determining step in a unimolecular reaction at low pressures is the second-order collisional process

$$A + M \rightleftarrows A^* + M \tag{3.1}$$

direct measurements of the rate of reaction in this region in the presence of added inert bath gas molecules M enable the relative collisional efficiencies to be determined. It was shown in Chapter 3 (section 3.12) that the collisional efficiency per unit collision β_c is related to the observed low-pressure rate constant k_{bim} and the strong-collision rate constant for the forward reaction k_{bim}^{sc}, by the equation $k_{bim} = \beta_c k_{bim}^{sc}$.

From values of β_c it is possible to derive the various energy parameters such as $\langle \Delta E \rangle_d$ referred to earlier, although the exact form of the relationship required depends upon the model assumed for the transition probabilities[2].

Some of the most extensive measurements of relative collisional efficiencies of various bath gases were made by Rabinovitch and co-workers for methyl[13-15] and ethyl[16] isocyanides. The latter has been chosen by Gilbert and Smith[1] as an illustrative example of the methods used to apply theoretical treatments to thermal data.

Tardy and Rabinovitch have given illustrative calculations of weak-collision effects for a number of systems[2,5], some results for the isomerisation of methyl isocyanide are shown in Figures 9.1 and 9.2. Figure 9.1 refers to the second-order region (where non-equilibrium effects are at their maximum) and shows the steady-state concentrations in the energy levels below E_0 as a fraction of the equilibrium (high-pressure) concentrations. The curves represent stepladder models with various $\langle \Delta E \rangle$ values, and illustrate the marked depletion of the levels near E_0 which can result from the irreversible transport of molecules to energies above E_0. Similar curves were obtained with exponential, Poisson or Gaussian distributions of transition probability with the same $\langle \Delta E \rangle$ values. The depletion of the upper levels is negligible for step sizes above $8.4\,kJ\,mol^{-1}$ in Figure 9.1, but becomes quite marked for step sizes below about $2.1\,kJ\,mol^{-1}$. The effective collision efficiencies β_c are about 0.9 and 0.2 for $\langle \Delta E \rangle = 8.4$ and $2.1\,kJ\,mol^{-1}$ respectively. The disequilibrium increases as the overall reaction rate increases, and is thus more marked at higher temperatures. The effective collision efficiency is roughly a function of $\langle \Delta E \rangle / T$, so that β_c values of 0.9 and 0.2 are obtained with step sizes of about 16.7 and about $4.2\,kJ\,mol^{-1}$ respectively at a practical experimental temperature of 546 K. This behaviour is very much as might be expected for the average excitation energy $\langle E^+ \rangle$ of the molecules

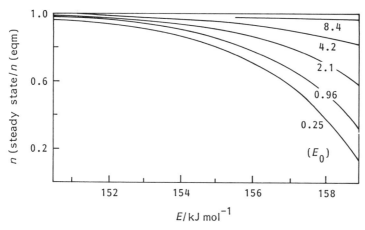

Figure 9.1 Calculated steady-state population distribution[5] of levels below E_0 for thermal methyl isocyanide isomerisation at 273 K in the second-order region, using step ladder models with step sizes ($\langle \Delta E \rangle/\text{kJ mol}^{-1}$) shown on curves. Reproduced by permission of John Wiley and Sons Ltd.

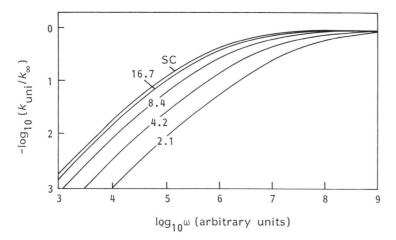

Figure 9.2 Calculated fall-off curves[2] for methyl isocyanide isomerisation at 546 K, using stepladder models with step sizes ($\langle \Delta E \rangle/\text{kJ mol}^{-1}$) shown on curves; SC = strong-collision curve. Reproduced by permission of John Wiley and Sons Ltd.

reacting at this temperature in a strong-collision system, which is 4–13 kJ mol^{-1} depending on the pressure[17]. Tardy and Rabinovitch[2] have given 'quasi-universal' plots of collision efficiency against the ratio $\langle \Delta E \rangle/\langle E^+ \rangle$ at various extents of fall-off, where $\langle E^+ \rangle$ is the average excess energy of the reacting molecules under equilibrium (strong-collision) conditions, and can be estimated by empirical correlation[5].

An important practical aspect of these calculations lies in the derived fall-off curves, some of which are illustrated in Figure 9.2. The illustrated curves refer to stepladder models, but as before, an exponential distribution gave similar results. Reduction of the step size is seen to move the fall-off curve to higher pressures with only a minor change of shape. Again, the effect is only significant for step sizes below $16.7 \, \text{kJ} \, \text{mol}^{-1}$ (the temperature being 546 K). The effect can be large, however; a $\langle \Delta E \rangle$ value of $2.1 \, \text{kJ} \, \text{mol}^{-1}$ gives a curve which is shifted by $\Delta \log \omega \approx 1.4$ and broadened relative to the strong-collision curve.

The experimental results of Rabinovitch and co-workers for the isomerisation of methyl isocyanide in the second-order region in the presence of added bath gases are shown in Figure 9.3.

The relative energisation efficiencies per unit pressure are obtained as the relative slopes of these plots, and are converted to relative efficiencies per collision using the kinetic-theory ratio of collision frequencies. The required

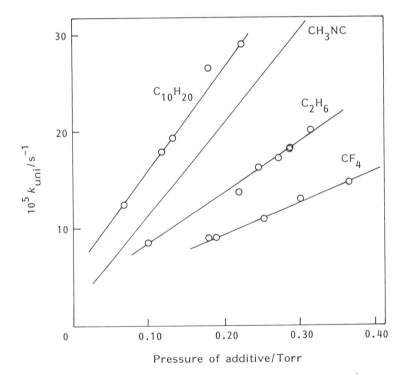

Figure 9.3 Rate data[13,14] for thermal energisation of a constant small pressure of methyl isocyanide in the second-order regions at 280.5 °C by various added gases. The line for CH_3NC summarises many points. Reproduced by permission of John Wiley and Sons Ltd.

expression is

$$\frac{\beta_M}{\beta_N} = \frac{(dk_{uni}/dp_M)}{(dk_{uni}/dp_N)} \times \left(\frac{\sigma_{AN}}{\sigma_{AM}}\right)^2 \times \left(\frac{\mu_{AM}}{\mu_{AN}}\right)^{1/2} \tag{9.10}$$

in which A represents the reactant, M and N are bath gases (one of which is often A in practice), μ represents reduced mass and σ_{ij} is the collision diameter $\frac{1}{2}(\sigma_i + \sigma_j)$ corrected[13,14] for temperature effects. The efficiencies obtained from (9.10) are differential relative efficiencies, as opposed to the integrated values which could be obtained by application of the equation $\beta_c = k_{bim}/k_{bim}^{sc}$ to the mixed reactant-bath system. Integral efficiencies have been measured for over 100 bath gases in the second-order region of methyl isocyanide isomerisation. Values relative to $\beta(\text{MeNC}) = 1.00$ are plotted in Figure 9.4 against the atomicity of the bath gas for noble gases, alkanes, alkenes, alkynes, nitriles, fluorocarbons and other molecules. The plot shows very clearly that the energisation efficiency increases with increasing molecular size and reaches an effectively constant limiting value for molecules containing more than 10–12 atoms, i.e. for C_3 or C_4 and higher

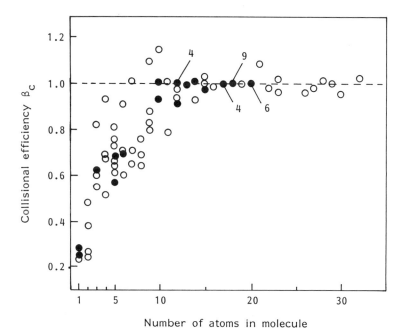

Figure 9.4 Collisional efficiencies β_c for energisation of methyl isocyanide at 280.5 °C by various bath gases relative to $\beta_c(\text{MeNC}) = 1$, using viscosity-derived cross-sections[13]. Filled points represent coincident points for the stated number of bath gases (two if not stated). Reproduced by permission of John Wiley and Sons Ltd.

compounds. It seems very reasonable to assume that the limiting value is unity, and that these sufficiently large molecules behave effectively as strong colliders.

More recent examples of the study of collisional deactivation in thermally energised systems may be found in the reviews cited above. Of special interest are studies in which molecules once activated may react by competing unimolecular channels, as well as being collisionally deactivated. Chow and Wilson[18] first recognised the potential of such systems as a means of probing the mechanism of collisional deactivation.

Marta and co-workers[19] have studied the thermal decomposition of pure 2-methyl oxetane which follows two reaction paths, A and B.

$$\text{2-methyl oxetane} \quad \xrightarrow{k_A} \quad CH_3CH = CH_2 + CH_2O$$
$$\xrightarrow{k_b} \quad C_2H_4 + CH_3CHO$$

Whereas the pressure-dependence of the rate constant for the symmetrical 3-methyl oxetane could equally be explained in terms of strong or weak collisions, the ratio of the observed rate constants k_A and k_B for the two-channel decomposition of 2-methyl oxetane showed the clear failure of the strong collision assumption. The RRKM step ladder model calculations shown in Figure 9.5 for this molecule were calculated from the expression

$$\frac{k_A}{k_B} = \frac{\sum_{E_0^A}^{\infty} k_{Ai} n_i^{ss}}{\sum_{E_0^B}^{\infty} k_{Bi} n_i^{ss}} \tag{9.11}$$

where k_{Ai} and k_{Bi} were calculated for RRKM theory, E_0^A and E_0^B are the critical energies for paths A and B and n_i^{ss} is the steady-state population calculated by the master equation method described in Chapter 7. The results showed best agreement between experiment and the calculated curves assuming

$$\langle \Delta E \rangle_d = 1500 \pm 300 \text{ cm}^{-1} \quad \text{or} \quad (18.0 \pm 3.6) \text{ kJ mol}^{-1} \text{ at 743 K}$$

for the parent molecule. Similar values have been found for other molecules of comparable size acting as deactivating colliders (see refs 7–12). Earlier examples of multichannel decomposition were carried out by Rabinovitch and co-workers[20,21], among others.

The average energy transferred from methyl oxetane molecules by collision with the wall was found to be 2600 cm^{-1} (31.2 kJ mol^{-1}) compared with the value 1500 cm^{-1} given above for gas–gas collisions. This was measured

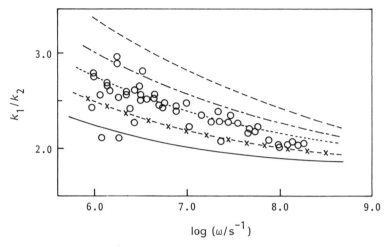

Figure 9.5 Pressure dependence of the ratio of the two reaction pathways for 2-methyl oxetane decomposition[19] at 743 K. ($\bigcirc$) Experimental; ($-$) RRKM stong-collision hypothesis; ($---$) RRKM stepladder model with $\langle \Delta E \rangle_d = 900 \text{ cm}^{-1}$, ($-\cdot-\cdot-$) $\langle \Delta E \rangle_d = 1200 \text{ cm}^{-1}$; ($\cdots$) $\langle \Delta E \rangle_d = 1500 \text{ cm}^{-1}$; ($--\times--$) $\langle \Delta E \rangle_d = 200 \text{ cm}^{-1}$. From L. Zalotai, T. Bérces and F. Mártà, *J. Chem. Soc. Faraday Trans.* **86**, 21 (1990). Reproduced by permission of the Royal Society of Chemistry.

by the variable encounter method[22,23]. In common with previous studies of energy transfer by gas–wall collisions by very low-pressure pyrolysis (VLPP)[24] it was found that $\langle \Delta E \rangle_d$ declines strongly with temperature in the range 800–1100 K.

Both the collision efficiency β_c and $\langle \Delta E \rangle_d$ measured for gas–gas collisions have been found in experimental systems such as these to decrease with increase in temperature, although there are contrary examples such as the ethyl radical given below. The quantitative explanation of this temperature variation at the present is difficult, and a much weaker temperature dependence is observed in experiments where energy-transfer parameters are measured directly as described later[12].

Shock tube studies can in principle be used to obtain energy-transfer parameters at high temperatures and have been used for a number of small molecules such as NO_2, NO and CD_4[24,25]. However, *shock-tube decompositions* of larger molecules are less reliable. An inferred increase in $\langle \Delta E \rangle_d$ with temperature for cyclohexane deactivation by krypton from shock tube studies[26] for example, has been attributed[12] to an incorrect interpretation of the data which was based on the Troe relationship[27] between $\langle \Delta E \rangle_{all}$ and $\langle \Delta E \rangle_d$ which may be inadequate at high temperatures[28].

The thermal decompositions of free radicals are referred to in Chapter 11 where some fall-off characteristics are reported for studies involving simple free radicals. Since these processes are important in high-temperature combustion processes it is necessary to have some knowledge of the energy-transfer parameters over a range of temperatures. A number of studies of radical decompositions in the presence of various bath gases have been carried out by Gutman and co-workers in recent years[29-31].

A typical example of such a study is that carried out on the neopentyl radical decomposition[29]

$$neo - C_5H_{11} \rightarrow CH_3 + C_4H_8 \qquad (9.12)$$

The radical was rapidly generated from pulsed laser photolysis of CCl_4 followed by rapid reaction with neopentane

$$CCl_4 \xrightarrow{193 \text{ nm}} CCl_3 + Cl \qquad (9.13)$$

$$Cl + neo - C_5H_{12} \rightarrow neo - C_5H_{11} + HCl \qquad (9.14)$$

The disappearance of the radical at temperatures between 560 K and 650 K in the presence of He, N_2 or Ar bath gases was monitored in time-resolved photoionisation experiments in a mass spectrometer. The most extensive measurements were made with He bath gas at concentrations $(3-30) \times 10^{16}$ atom cm^{-3}. Under these conditions the decomposition was in its fall-off region and sensitive to collisional deactivators. The rate constants were fitted by means of RRKM theory using a 'best-fit' value of the high-pressure rate constant, and an associated collision efficiency β_c. The $\langle \Delta E \rangle_d$ values for the bath gases used were He (200 cm^{-1}) N_2 (130 cm^{-1}) and Ar (140 cm^{-1}).

This method of derivation of energy transfer parameters, which has been described as a 'modified strong-collision' (MSC)/RRKM method has been applied to other radical decompositions and compared with alternative treatments such as the master equation-based method. For the isopropyl radical[30] the step-down sizes for the same three deactivators were He (130 cm^{-1}) N_2 (129 cm^{-1}) and Ar (130 cm^{-1}) using the MSC/RRKM analysis with similar values from other methods.

More accurate measurements were possible for the ethyl radical decomposition[31] since this is further into its fall-off region at the temperatures (876–1094 K) and pressures (0.8–14.3 Torr) used. In this system, strong evidence was found for an increase of $\langle \Delta E \rangle_d$ with temperature. Here $\langle \Delta E \rangle_d$ was found to increase from 30–40 cm^{-1} at 300 K to 280 cm^{-1} at 1100 K. Between 876 and 1094 K the data is represented by $\langle \Delta E \rangle_d = 0.255 \, T^{1.0(\pm 0.1)}$ cm^{-1}. These results for simple colliding partners are in contrast to the decline of $\langle \Delta E \rangle_d$ with temperature noted for more complex molecules above, and could lead to weak colliders becoming effectively strong colliders in some cases at higher temperatures.

CHEMICAL ACTIVATION

Chemical activation is the term used to describe the preparation of energised molecules by a chemical reaction which itself may in principle be either exothermic or, less frequently, endothermic. The resulting molecule is then energised sufficiently to undergo a subsequent unimolecular reaction. The treatment of such systems by the master equation method has been described in Chapter 7 (section 7.3).

The general treatment of chemical activation systems may be illustrated by considering the formation of a vibrationally excited molecule A* by the association of two radicals according to the following scheme:

$$R + R' \underset{k'_a(E)}{\overset{k_a(E)}{\rightleftharpoons}} A^* \begin{cases} \nearrow & \text{Decomposition products } (D) \\ \searrow & \\ & A \ (S) \end{cases} \quad \omega \tag{9.15}$$

Experimental methods are normally concerned with measuring the ratio of decomposition products to stabilised molecules A (i.e. D/S).

In a strictly monoenergetic system, the experimental ratio D/S is equal to $k_a(E)/\omega$. When there is a spread of energies, the average rate constant $\langle k_a \rangle$ is then given by

$$\langle k_a \rangle = \omega D/S = \omega \frac{\int_{E_0}^{\infty} \{k_a(E)/[k_a(E) + \omega]\} f(E) \, dE}{\int_{E_0}^{\infty} \{\omega/[k_a(E) + \omega]\} f(E) \, dE} \tag{9.16}$$

Equation (9.16) is derived from expressions for the fractions of molecules decomposing or being stabilised according to scheme (9.15) and $f(E)$ is the distribution function of energised molecules. The distribution function is the thermal quantum Boltzmann distribution and the rate constant $k'_a(E)$ is that for the reverse decomposition of energised molecules A*.

Since the rate constant $k'_a(E)$ may be calculated by RRKM methods, equation (9.16) may be used as a test for strong collisions. Strong collisions imply that $\langle k_a \rangle = \omega \cdot D/S$ would be independent of ω. Quite often it is found that plots of $\omega \cdot D/S$ against ω or against S/D show a 'turn-up' at low pressures. This reflects the increased chance of decomposition when weak collisions occur. This behaviour is illustrated in Figure 9.6 for chemically activated 2-pentyl radicals.

Figure 9.6 shows a set of experimental points[32] for the decomposition of

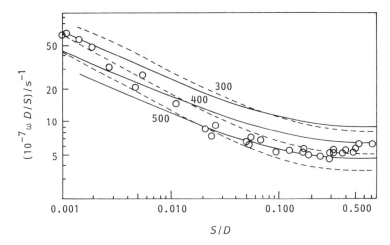

Figure 9.6 Experimental points for chemically activated 2-pentyl radicals compared with theoretical curves of $\omega\, D/S$ versus S/D for stepladder (broken lines) and exponential (solid lines) models with $\langle \Delta E \rangle \equiv 300$, 400 and 500 cm^{-1} (3.60, 4.77 and 5.98 kJ mol^{-1})[32]. The bath gas is mainly hydrogen. Reproduced by permission of John Wiley and Sons Ltd.

2-pentyl radicals formed from H atoms and pent-1-ene in a bath gas which is mainly hydrogen. These results which cover a large range of S/D values show a marked turn-up. Comparison with the theoretical curves shown does not discriminate unambiguously between the stepladder and exponential models, but for either model the $\langle \Delta E \rangle$ values are close to 400 cm$^{-1} \approx$ 4.2 kJ mol^{-1} for collisions with hydrogen bath molecules. This is considerably less than the average excitation energy of the initially formed radicals (about 58.5 kJ mol^{-1}), which accounts for the observation of such pronounced weak-collision characteristics.

Many similar studies have now been carried out and in general the results parallel those from thermal activation experiments in confirming that exponential models appear to fit the experimental data better for deactivators which are atomic or small molecular species, while stepladder models appear to be adequate for large molecular deactivators. The early experimental data has been covered in the reviews of Tardy and Rabinovitch[8] and Quack and Troe[9] while more recent work is covered in the review of Oref and Tardy[12]. The lastnamed review notes that there are a number of discrepancies between step sizes $\langle \Delta E \rangle$ measured in chemical activation experiments and in other types of experiment such as photoactivation. In particular the dependence of $\langle \Delta E \rangle$ upon the excitation energy and upon the size of the deactivating molecule is frequently found to differ between different groups of workers and between different techniques.

PHOTOACTIVATION

When chemical activation is produced by radical recombination and those radicals are generated photochemically it is possible to regard chemical activation as a form of photoactivation. The more obvious examples of photoactivation are single photon activation, multiple photon activation and overtone activation. All these methods are described in more detail in Chapter 11 and only those aspects concerned with energy-transfer studies will be treated here.

Single photon activation of cycloheptatriene and its subsequent isomerisation to toluene was studied by Atkinson and Thrush[33,34]. The absorption produces vibrationally excited ground state molecules and the relative efficiencies of various deactivating gases were determined from the observed quantum yields as a function of pressure (Stern–Volmer plots). The unimolecular rate constants for conversion of cycloheptatriene to toluene were calculated by RRKM theory. The results showed that deactivation of vibrationally excited cycloheptatriene occurs by a multistep process. At an excitation energy of $390–531$ kJ mol^{-1} compared with the critical energy for isomerisation of 214 kJ mol^{-1} the derived average energies removed per collision were found to be: cycloheptatriene 17 ± 5, toluene 11 ± 5, SF$_6$ 5.9 ± 0.9, CO$_2$ 3.8 ± 0.4 and helium 0.6 ± 0.1 (kJ mol^{-1}).

An alternative type of single photon excitation is that employing *overtone excitation*. An example of the use of this technique to study weak collision effects is provided by the work of Baggott[35] on the isomerisation of cyclobutene to buta-1,3-diene. Excitation of the $6 \leftarrow 0$, ($=$C–H) stretching overtone at $16\,603$ cm^{-1} was achieved using an intracavity cw dye laser and the rate constant for isomerisation determined from gas chromatographic measurements of reactant and product pressures after photolysis. The reaction scheme is represented by the following steps:

$$A + h\nu \xrightarrow{k_a} A^*(E^*) \text{ photoactivation} \tag{9.17}$$

$$A^*(E^*) + A \xrightarrow{Z} A + A \text{ collisional deactivation} \tag{9.18}$$

$$A^*(E) \xrightarrow{k(E^*)} B \text{ reaction} \tag{9.19}$$

If the observed rate constant k_{obs} is defined as

$$\frac{-d[A]}{dt} = k_{obs}[A]h\nu$$

a Stern–Volmer analysis leads to the equation (9.20), where ϕ is the quantum yield and Z the rate constant for collisional deactivation of the vibrationally excited reactant.

$$\frac{k_a}{k_{obs}} = \frac{1}{\phi} = 1 + \frac{Z[A]}{k(E^*)} \tag{9.20}$$

Measurements of the reciprocal quantum yield were made as a function of pressure for the bath gases He, Ar, CO_2, CH_4, SF_6 and cyclobutene. The results for all except cyclobutene are shown in Figure 9.7 compared with the predictions of a full collisional master equation calculation, using an exponential down model. It was also possible to derive approximate values for $\langle \Delta E \rangle$ from the collision efficiencies β_c measured from the slopes of the plots in Figure 9.7 and the equation given by Troe[36]. In general the full collisional master equation was found to account satisfactorily for the data shown in Figure 9.7. An exponential model was used for the weak colliders He, Ar, CO_2 and CH_4 and a stepladder model for SF_6 and cyclobutene. The values of $\langle \Delta E \rangle$ obtained from the master equation calculations in this work were lower than those found for the same deactivators by direct methods by other workers[37,38], but the agreement was good considering the big difference in excitation energies involved.

DIRECT PHYSICAL METHODS OF MEASURING ENERGY TRANSFER

The study of energy transfer from excited ground electronic state polyatomic molecules has been historically connected with unimolecular reactions as

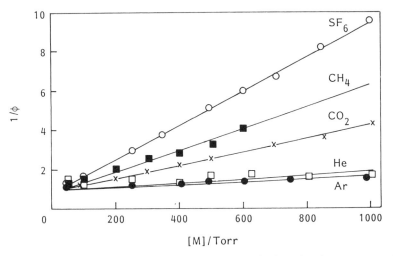

Figure 9.7 Plots of ϕ^{-1} versus pressure for buta-1,3-diene in the presence of the bath gases SF_6(○), CH_4 (■), CO_2 (×), Ar (●) and He (□)[35]. The lines show the behaviour predicted by solution of the collisional master equation: the values for $\langle \Delta E \rangle_d$ giving the best fits to these data are SF_6, 421 cm^{-1}, CH_4 276 cm^{-1}, CO_2, 245 cm^{-1}, Ar, 105 cm^{-1} and He, 105 cm^{-1}. From J. E. Baggott, *Chem. Phys. Lett.*, **119**, 47 (1985). Reproduced by permission of Elsevier Science Publishers BV.

shown by the examples of thermally and chemically activated systems discussed in this chapter so far. It was pointed out by Barker[39] that such studies depend on accurate knowledge of chemical reaction mechanisms and of the values of $k_a(E)$ for the unimolecular reactions involved. To avoid these complications a number of physical methods of measuring energy transfer have been developed. Of these, the most important are time-resolved infrared fluorescence (IRF)[40] and ultraviolet absorption (UVA)[37] which are discussed below, but other methods used have included time-resolved optoacoustics (TROA)[41,42] time-dependent thermal lensing (TDTL)[43] and multiphoton ionisation (MPI)[44].

Experiments using the azulene molecule have been carried out both by IRF[38] and UVA[45] methods since this molecule possesses many advantages for this type of experiment. It has two electronic singlet states $S_1 \sim 700$ nm which can be excited by a tunable dye laser and $S_2 \sim 350$ nm accessible by irradiation with a nitrogen laser. Rapid single photon excitation to either of these states is followed by internal conversion to vibrationally excited ground state molecules. The molecule in either electronic singlet state has negligible fluorescence and probability of triplet state formation, while the quantum yield for chemical reaction to naphthalene is low. The molecule in its excited ground state can be prepared with a very narrow distribution of energy, and subsequent energy transfer to bath gas molecules then permits energy transfer parameters to be determined directly.

Rossi, Pladziewicz and Barker[38] used the infrared fluorescence emission from vibrationally excited azulene and monitored its disappearance with time in the presence of 17 different collider gases. Previous knowledge of the dependence of infrared fluorescence upon the internal energy of the molecule[46] was used to obtain a calibration curve. The infrared fluorescence was found to decay in the presence of the collider gases in an approximately exponential manner with a rate constant k. The average energy transferred in all collisions $\langle \Delta E \rangle_{all}$ was related to the internal energy by use of the calibration curve together with the rate constant k and the Lennard–Jones collision number. Alternatively, $\langle \Delta E \rangle_{all}$ values were obtained by a full collisional master equation. Both methods gave $\langle \Delta E \rangle_{all}$ values which varied almost linearly with the internal energy of the excited molecule. A re-examination of the data by Shi and Barker[47] based upon an improved calibration confirmed the fact that $\langle \Delta E \rangle_{all}$ is nearly directly proportional to the vibrational energy of the excited azulene from 8000 to 33 000 cm^{-1} (96.0 to 395 kJ mol^{-1}).

The work described above[38,47] on the energy dependence of $\langle \Delta E \rangle_{all}$ was carried out at a constant temperature of 300 K. At higher temperature it would be predicted that $\langle \Delta E \rangle_{all}$ would increase in proportion to the increased thermal vibrational energy. Experiments by Barker and Golden[48] have shown that for excited azulene deactivated by unexcited azulene

$\langle \Delta E \rangle_d$ derived from $\langle \Delta E \rangle_{all}$ is independent of temperature over the range 300–630 K, but for deactivation by nitrogen $\langle \Delta E \rangle_d$ has a weak negative temperature dependence. This result may be rationalised in terms of a collision complex[49], but it is in contrast to results on the temperature dependence of $\langle \Delta E \rangle_d$ for other molecules[50,51] and at present no consistent explanation can be given. The infrared fluorescence technique has also been used to study energy transfer from a number of other molecules such as 1,1,2-trifluoroethane[52], benzene[53] and toluene[54].

An alternative method of following collisional relaxation after laser excitation and internal conversion, is to use ultraviolet (UV) absorption of the product molecule. This method has been successfully applied by Troe and co-workers[37,45,55–58] to a number of molecules including the collisional deactivation of toluene produced by isomerisation from excited cycloheptatriene and the collisional deactivation of excited azulene. In similar experiments Nakashima and co-workers[59–61] have studied collisional deactivation of benzene and hexafluorobenzene.

In the experiments on toluene produced from excited cycloheptatriene[37], deactivation of the vibrationally excited ground electronic state toluene was followed by UV absorption monitored at 223 nm. In separate experiments the dependence of absorption coefficient (ε) of the toluene at this wavelength was found to be linearly dependent on the internal excitation energy of the toluene which was varied from 50 to 700 kJ mol^{-1} by shock heating. The variation of absorption coefficient with time was then expressed by equation (9.21) in which t is the time and Z_{LJ} is the calculated Lennard-Jones collision frequency. The time-resolved absorption measurements thus enabled values of the average energies transferred $\langle \Delta E \rangle_{all}$ to be obtained for 60 different bath gases (M).

$$\varepsilon(t) = \langle E_{t=0} \rangle + \langle \Delta E \rangle_{all} Z_{LJ}[M]t \qquad (9.21)$$

The results indicated that the absolute values of $\langle \Delta E \rangle_{all}$ were lower than, but within a factor of about 2 of earlier values from indirect thermal or chemical activation or photoactivation methods. In this work the initial excitation energy of the toluene molecule (623 kJ mol^{-1}) was not varied, but from an analysis of shape of the energy-loss profiles it was concluded that $\langle \Delta E \rangle_{all}$ was independent of energy unlike the energy dependence observed for azulene in the IRF method of Barker and co-workers[38,47].

In a subsequent study of the collisional deactivation of vibrationally excited azulene by UVA; Hippler, Lindemann and Troe[45] found the polar molecules H_2O, NH_3 and CH_3OH to show energy dependence of $\langle \Delta E \rangle_{all}$ varying from a first-order dependence on energy at excitation energies below 10 000 cm^{-1} to half-order dependence around 30 000 cm^{-1}. For the other 20 bath gases studied, the same energy dependence was observed up to $10\,000$ cm^{-1} and then a region between $10\,000$ and $30\,000$ cm^{-1}, where ΔE was virtually independent of energy. In order to study this energy dependence of $\langle \Delta E \rangle_{all}$ at higher energies Damm, Hippler and Troe[62] used UV multiphoton excitation. In this way, energies up to 2–3 photons of UV radiation at $\lambda = 308$ nm were deposited in the azulene molecule, followed by fast internal conversion to the ground electronic state. Under these conditions collisional deactivation competed with a small amount of isomerisation to naphthalene. From a master equation simulation of the process it was shown that the absorption results could be adequately explained at an excitation energy of $66\,000$ cm^{-1} using $\langle \Delta E \rangle_{all}$ values previously determined from UVA absorption measurements at $30\,000$ cm^{-1} without the need for modification. This implies that $\langle \Delta E \rangle_{all}$ is effectively constant at these higher excitation energies.

The IRF and UVA measurements are thus at present in agreement at low energies, but appear to differ at excitation energies above $10\,000$ cm^{-1}. The source of the discrepancy may lie in the calibrations which need to be employed in both methods. In addition the UV multiphoton experiments have a degree of imprecision associated with knowledge of the specific rate constant for the unimolecular isomerisation and the optical pumping range to the excited state involved.

The consequence of an energy-dependent $\langle \Delta E \rangle$ is, however, that collisional energy-transfer processes may become more efficient at higher energies of excitation which could account for differences in the magnitude of $\langle \Delta E \rangle$ values derived from shock tube methods and those calculated by RRKM theory[47].

Direct methods of measuring energy transfer such as IRF and UVA have normally assumed that the initially formed vibrationally excited molecules are produced with a relatively narrow population distribution relative to the excitation energy. Experimental evidence that this is not so is found from the work of Luther and co-workers[63,64] and Oref and co-workers[65,66]. Luther and co-workers[63,64] used a pump-probe multiphoton ionisation technique to produce toluene molecules at high vibrational excitation energy (~ 623 kJ mol^{-1}). Variation of the delay time between the pump laser and the multiphoton ionisation probe laser enabled the vibrational energy distribution to be monitored as a function of time. The results indicated the development of a broad distribution during the relaxation with 'supercollisions', i.e. collisions transferring large amounts of energy having a small but finite probability. Paschutski and Oref[65] studied the sensitised decomposition of cyclohexadiene to benzene and hydrogen by collisions with vibrationally hot 1,3,5-trimethyl-1,1,3,5,5-penta-phenyltrisiloxane produced in a jet. Since the critical energy for the reaction is ~ 178 kJ mol^{-1} the decomposition of cyclohexadiene is a measure of collisions in which energy of this amount or greater is transferred. The extent of decomposition at low pressures of the reactant and collider gases ($\sim$ 20–300 mTorr) was measured by quadrupole mass spectrometry. The results indicated that the probability of transferring amounts of energy $\geqslant 178$ kJ mol^{-1} in a single collision was 5×10^{-7}, i.e. that one collision in 2×10^6 was fruitful. When the results were fitted to an exponential collisional energy transfer probability model the average energy transferred per collision was found to be ~ 13.4 kJ mol^{-1}.

Similar experiments on the collisional activation of cyclobutene by hexafluorobenzene were carried out by Oref and co-workers[66].

The critical energy for conversion of cyclobutene to 1,3-butadiene is 135.5 kJ mol^{-1} and the probability of transferring this amount of energy, or more, in a single collision showed that about one in every 400 collisions of cyclobutene with vibrationally excited hexafluorobenzene (~ 460 kJ mol^{-1}) was fruitful.

The observations of a small proportion of collisions transferring large amounts of vibrational energy have been supported by trajectory calculations carried out by Lendvay and Schatz[67] and are potentially important for many reactive systems.

9.2 INTRAMOLECULAR ENERGY TRANSFER

An important postulate of the RRKM theory of unimolecular reactions is that intramolecular energy transfer, and in particular intramolecular vibrational relaxation (IVR), is sufficiently rapid compared with the rate of reaction, that the rate constant $k(E)$ can be accurately calculated for the microcanonical ensemble of excited molecules. This postulate (discussed in Chapter 3), has been examined by Hase[68,69] who has pointed out the ambiguities in many definitions of IVR and the difficulty of making rigorous tests of this fundamental aspect of RRKM from experimental measurements.

In this section some of the experimental methods used will be discussed. In a comprehensive earlier review of Oref and Rabinovitch[70] it was concluded that little evidence for non-RRKM behaviour exists particularly for polyatomic molecules thermally activated to relatively high levels of excitation. The development of selective excitation techniques such as multiphoton vibrational excitation, overtone excitation and molecular beam experiments have increased the probability of finding evidence for the breakdown of RRKM behaviour particularly for molecules with low threshold energies for reaction. Such molecules might be capable of reacting rapidly before energy randomisation occurred. This was defined by Hase and Bunker[71] as *apparent* non-RRKM behaviour. A different type of non-RRKM behaviour could arise when an inherent weak coupling between the vibrational modes in the molecule is present, often described as a 'bottleneck' to energy flow, this is termed *intrinsic* non-RRKM behaviour.

CHEMICAL ACTIVATION EXPERIMENTS

The method of chemical activation was used in many early experiments designed to test the assumption of rapid vibrational energy transfer. The first experiment was done by Butler and Kistiakowsky[72] who produced chemically activated methyl cyclopropane by two different routes, from addition of singlet methylene to propene or insertion of single methylene into a C–H bond of cyclopropane. Subsequent decomposition of the vibrationally energised methylcyclopropane was found to produce the same ratio of product butenes irrespective of its mode of formation.

In the first experiment to measure the rate of energy randomisation, Rynbrandt and Rabinovitch[73] produced hexafluorobicyclopropyl-d$_2$ which underwent collisional stabilisation or CF$_2$ elimination from either ring as shown in the following scheme:

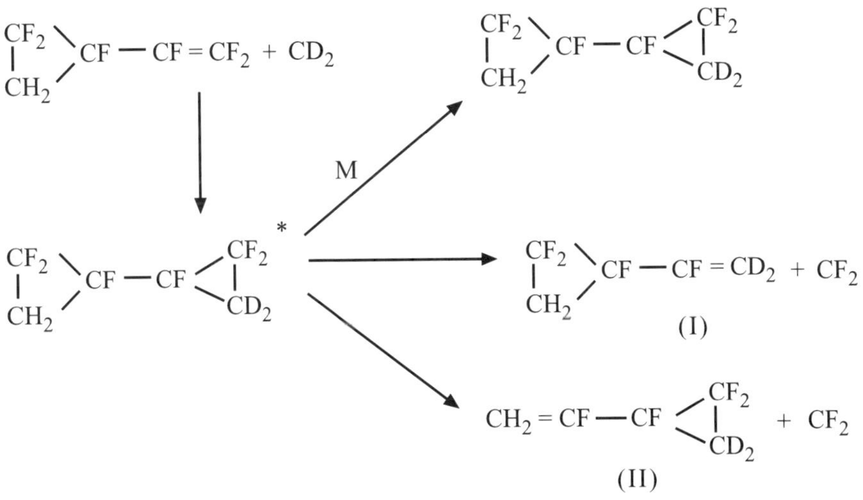

$$(9.22)$$

Measurements of the product ratio I/II by mass spectrometry showed invariance over the pressure range 0.8–310 Torr, indicating complete IVR in times down to 10^{-10} s. Work at higher pressures showed a change in product ratio, enabling an estimate to be made of the relaxation constant for energy exchange of about 10^{12} s^{-1}. Many other chemically activated systems were subsequently studied by Rabinovitch and co-workers[74], Setser and co-workers[75] and others, all yielding relaxation constants of the same order of magnitude.

Chemical activation experiments measure a rate constant which is averaged over the number of collisions, given by

$$k = \omega(D/S) \tag{9.23}$$

where ω is the collision frequency and D/S the ratio of observed decomposition to stabilisation products (see Chapter 11). It is a consequence of this equation that the experimentally observed rate constant at low pressures does not uniquely predict an exponential behaviour of the lifetimes of energised molecules as required by RRKM theory (see Chapter 3—random lifetime assumption). Hase[69] has therefore concluded that such experiments are insensitive to intrinsic non-RRKM behaviour. Despite this reservation, chemical activation is still used to probe the dynamics of intramolecular transfer. Trenwith, Rabinovitch and co-workers[76] have studied intramolecular energy transfer for a homologous series of 2-alkyl radicals vibrationally excited by chemical activation. The motivation for this work was provided by experimental[77] and theoretical[78] studies of the effects of insertion of a central heavy metal atom in a hydrocarbon chain to determine its effect in blocking IVR. The results posed the question as to whether IVR occurs in a 'global' or a 'sequential' manner. In the latter, excitation of a subset of modes of the reactant molecule is envisaged as slowly spreading to the remainder of the modes before randomisation and reaction occurs. Evidence for such sequential relaxation was provided by results showing that the specific rate constant for decomposition of a number of 2-alkyl radicals to give propene was pressure dependent and not independent of pressure, as would be predicted if randomisation occurred throughout the whole radical structure.

PHOTOACTIVATION

The absorption of one or more UV or visible photons by a molecule can raise the molecule to various excited electronic states with a variety of consequences such as dissociation, collisional vibrational de-energisation or rapid internal conversion (see Chapter 11).

Vibrationally excited (S_1) cyclobutanone produced by absorption of UV light of wavelength ≤ 313 nm, has been found to undergo rapid reaction to a mixture of products including cyclopropane. The cyclopropane itself is vibrationally excited and can either decompose to propene or be stabilised to cyclopropane as illustrated in scheme (9.24).

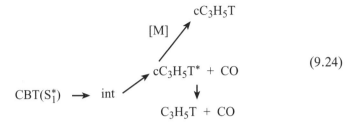

$$CBT(S_1^*) \longrightarrow int \nearrow \begin{array}{c} cC_3H_5T \\ [M] \nearrow \\ cC_3H_5T^* + CO \\ \downarrow \\ C_3H_5T + CO \end{array} \qquad (9.24)$$

This scheme shows the case of tritiated cyclobutane which was studied by Lee and Lee[79] who established that IVR for the intermediate (int) from the parent cyclobutane molecule occurred on a time-scale $\leqslant 10^{-10}$ s. In re-examining the data, Dorer[80] investigated the ratio of propene to stabilised cyclopropane as a function of total pressure and concluded that IVR of the vibrationally excited intermediate cyclopropane as well as that of the intermediate 1,4-biradical arising from cyclobutanone-2-t (CBT) occurs in $< 10^{-10}$ s. The use of lasers to produce photoactivated molecules was stimulated by the search for mode-selective chemistry. It was hoped that vibrational excitation localised in a particular mode of vibration might influence the subsequent chemical behaviour of the molecule and promote one reaction channel in preference to others[81,82] The question of mode specificity is clearly related to the rate of IVR since rapid IVR makes mode specificity impossible to achieve. It is now generally accepted that rapid IVR occurs in all but very few cases. The use of laser excitation has in fact enabled intramolecular relaxation processes to be measured with greater precision than in the more indirect methods such as chemical activation.

MULTIPHOTON EXCITATION

The use of multiphoton infrared absorption[83] to promote unimolecular reactions is a well-established technique (see Chapter 11). Many experiments using multiphoton infrared excitation have been interpreted as supporting the view that IVR is a rapid process (see refs 70, 83 and 84 and refs cited therein). It has been pointed out by several authors[70,83,85], however, that measured product ratios from such experiments do not constitute definite proof of the validity of the rapid IVR assumption for a variety of reasons. Intramolecular vibrational relaxation may be brought about by a mechanism involving long-range collisions for example, which are still operative at very low pressures. The preference of one reaction channel over another may be a consequence of the different populations of reactive intermediates prepared by multiphoton excitation which often differs from the thermal Boltzmann distribution. In a similar way many claims for mode specific behaviour produced by excitation at different frequencies have been shown to be capable of explanation in other ways[86]. The observation that chemical reactivity can depend on vibrational excitation of a reactant in a bimolecular reaction[87] nevertheless continues to prompt speculation that such behaviour may in certain circumstances be found in a unimolecular reaction. Recent attention is now given to small molecules with very weak bonds of which van der Waal's molecules are an extreme example[88].

MOLECULAR BEAMS AND PRODUCT ENERGY DISTRIBUTIONS

The quantitative study of intramolecular vibrational relaxation has been advanced in recent years in experiments conducted under 'collision-free' conditions in molecular beams. These experiments have enabled product translational energy and internal energy distributions to be obtained and compared with predictions based upon RRKM theory[89,90]. Bimolecular reactions occurring via a collision complex may be followed by intramolecular relaxation of energy within the complex. At normal pressures such intramolecular processes may be assisted by intermolecular collisions which complicate the interpretation. At the very low pressures in molecular beams, the translational energies and angular distribution of the products give information on the degree to which energy in the complex is randomised before reaction occurs. Many studies of kinetic energy distributions have been made for ionic decompositions by mass spectrometry[91,92]. While most experiments involving ionic species have shown kinetic energy distributions in agreement with quasi-equilibrium theory (QET) or phase-space theory (see section 11.3), there are some notable exceptions which have been discussed by Lifshitz and co-workers (see references 222 and 229 of Chapter 11). Whether these exceptions are genuine examples of non-RRKM behaviour or are attributable to other causes is at present unclear. The most extensive study of neutral reactions in this way has been made by Lee and co-workers[84]. An example is the work of Farrar and Lee[93] on crossed molecular beams of F atoms and C_2H_4. The chemically activated $C_2H_4F^*$ produced, eliminated an H atom to give C_2H_3F, the recoil energies of which were measured as a function of collision energy. Angular and velocity distributions of the product molecules were measured which indicated that the product distribution was non-statistical.

Unimolecular multiphoton dissociation of SF_6 in a crossed laser-molecular beam was measured by Coggiola and co-workers[94] who determined the translational energy distributions of the fragments. These and many other similar studies in molecular beam experiments[84] have been interpreted as evidence for rapid energy randomisation.

Interpretation of these experiments, however, is difficult and it has been pointed out by Lupo and Quack[83] that the neglect of angular momentum considerations in RRKM formulations employed may have led to false conclusions. It is suggested that product internal state distributions can be better tests of the theory.

It is known that when unimolecular dissociation results in molecular fragments, the potential energy surface is a Type I surface (see Fig 5.1) and coupling between the internal modes of the rigid complex and the exit channel affects the product energy distribution. Two major methods of investigating product energy distributions are those using infrared chemi-luminescence by McDonald and co-workers[95,96] and laser-induced fluores-cence[97–99]. In experiments by McDonald and co-workers[96] on the reaction between fluorine atoms and substituted ethenes C_2H_3X, it was shown by molecular beam-scattering experiments[100] that the substitution of F atoms by reaction (9.25) occurred via a long-lived complex

$$F + C_2H_3X \rightarrow C_2H_3F + X \qquad (9.25)$$

In the series where X = Cl, Br, H and CH_3 the exothermicity of the reaction varies over a wide range and this energy is partitioned among the reaction products. Under conditions where collisions could be ignored the energy partitioned into different vibrational modes of the product C_2H_3F was measured from the observed intensities of spectrally resolved chemilumin-escence bands.

These intensities were compared with calculated ratios on the basis of statistical redistribution. The results showed support for the statistical as-sumption in the case of X = Cl and Br, but non-statistical behaviour for X = H and CH_3. The latter was attributed to exit-channel effects as men-tioned above.

Stephenson and co-workers[98,99] have used pulsed UV/visible lasers to measure the time-resolved induced fluorescence spectra of the products of unimolecular reactions produced by infrared multiple photon excitation at low pressures. Their technique enabled them to measure average transla-tional energy of the products in the CO_2 laser-induced decomposition of C_2F_3Cl:

$$CF_2CFCl \rightarrow CF_2 + CFCl \qquad (9.26)$$

from an assumed translational energy distribution. The technique which also enables measurements to be made of the relative concentrations of products in particular rotational and vibrational energy states to be measured, has been discussed further by Rayner and Hackett[101] who have been able to demonstrate satisfactory agreement between the results of the laser-induced fluorescence and molecular beam methods.

In recent years many detailed experimental studies have been made of the mechanism of intramolecular vibrational relaxation. These include the analysis of non-resonance infrared fluorescence following infrared excitation of the C–H stretching modes of a number of molecules in their ground electronic states. Observation of this fluorescence provides evidence for mixing between the excited vibrational state and various combination and overtone bands of other vibrational modes[102-104]. Experiments of a similar kind have been done to probe IVR in electronically excited states and this work has been reviewed by Smalley[105] and Parmenter[106].

OVERTONE EXCITATION

Unimolecular reactions produced by direct overtone excitation of particular vibrational modes using dye lasers have now been studied for over 20 years. The experimental results and their theoretical interpretation in terms of a local mode description of molecular vibration have been the subject of several reviews[107,108] and are also discussed later in Chapter 11. In contrast with many other methods of excitation this method should yield a narrow distribution of vibrational energy located initially in a particular vibrational mode. In principle, therefore, it should provide an ideal test of the assumption of random intramolecular vibrational relaxation and hence the possibility of mode specific chemistry.

Early experiments by Reddy and Berry[81,82] showed some apparent evidence for non-statistical behaviour. In the case of the allyl isocyanide molecule for example, they interpreted their data as showing that the rate of isomerisation to the corresponding nitrile depended upon the site of excitation. Further experiments on this molecule by Segal and Zare[109] have shown that the variations within a particular overtone band are as significant as the variations between them and can be attributed to different amounts of thermal energy produced by the excitation in different parts of the band. This inhomogeneity can also account for differences in photoisomerisation rate at different temperatures and eliminates the need to invoke non-statistical vibrational relaxation. A similar explanation has been put forward to explain non-adherence to RRKM behaviour in the rates of the isomerisation of cis-1,3,5-hexatriene to 1,3-cyclohexadiene induced by pumping the $4 \leftarrow 0$ and $5 \leftarrow 0$ overtones[110].

Many overtone-induced isomerisation and decomposition reactions can be quite adequately explained by rapid IVR and the absence of mode-specific effects[111]. Recently however, experiments have been designed to probe the dynamics of these reactions in detail using more stringent conditions. An example of such experiments is provided by the work of Zewail and co-workers[112] who used a pico-second pump-probe technique to study the overtone-induced dissociation of hydrogen peroxide:

$$H_2O_2(v_{OH} = 5) \rightarrow 2OH \qquad (9.27)$$

Measurements of the OH radical concentration by laser-induced fluorescence with the probe laser and measurement of the time between pump and probe pulses enabled time-resolved measurements to be made. The resulting non-exponential rise in product formation could be viewed as evidence for intrinsic non-RRKM behaviour in which at short times only a fraction of the available phase space proved accessible to the exit dissociation channel. Similar behaviour has also been observed for the dissociation of van der Waals' clusters of phenol and cresol with benzene[112]. The weak bonding in these clusters and the monitoring of dissociation by a pico-second pump-probe photoionisation mass spectrometry method makes the conditions more favourable for breakdown of the 'rapid IVR before reaction' assumption, and this method is likely to yield even more fruitful results for the study of intramolecular relaxation dynamics in the future.

Crim and co-workers[113] have demonstrated bond-specific effects using photodissociation of overtone-excited HOD. They excited HOD via the third overtone of the OH stretching vibration to produce the state $4\nu_{OH}$. A second photolysis photon, wavelength λ_2, was used to excite the molecule from $4\nu_{OH}$ up to the dissociative $\tilde{A}$ state. A third photon, generated laser-induced fluorescence from the resulting radical fragment and was used to probe the relative yields of OH and OD. Photolysis at $\lambda_2 = 266$ nm or 239.5 nm produced a large excess of OD, but comparable amounts were generated with $\lambda_2 = 218.5$ nm. At the longer wavelengths, the overlap between the initial $4\nu_{OH}$ wavefunction and the continuum wavefunction of the $\tilde{A}$ state occurs predominantly in the exit valley leading to H + OD. Photolysis with $\lambda_2 = 218.5$ nm produces $\tilde{A}$ with energy well above the barrier between the H + OD and D + OH exit valleys and there is good overlap between the initial wave function ($4\nu_{OH}$) and the continuum wave functions for both channels. The experiments demonstrate the occurrence of bond-selective chemistry which can be controlled via the excess energy of the dissociative state.

FLUORESCENCE FROM AROMATIC MOLECULES

Parmenter exploited the fluorescence of aromatic molecules to develop an understanding of IVR[114]. The basis of the experiments is the photo-excitation in a jet of an initial 'isolated' vibration-rotation state(s) in the first excited singlet (S_1) state. The vibration–rotation state is embedded in set of 'dark' states which are optically inaccessible from the lowest vibrational level of the S_0 state because of their low Franck–Condon factors. The density of the dark states increases with the vibrational energy of S_1 and can be probed by simply decreasing the wavelength of the exciting radiation. At low energies the resulting emission is sharp because the initially prepared state is effectively isolated and fluoresces before any IVR occurs. As the wavelength decreases an IVR 'threshold' is observed above which the fluorescence is heavily congested because delocalisation has occurred. The thresholds are remarkably low, e.g. 2570 cm^{-1} for naphthalene and 1200 cm^{-1} for stilbene. The experiments demonstrate IVR on the comparatively long time scale of the fluorescence (> 1 ns). Parmenter probed the time scale further by using oxygen to quench the fluorescence of p-difluorobenzene. By increasing the oxygen pressure to 30 kTorr, he was able to reduce the lifetime from 5 ns to 7 ps. For an excitation wavelength producing little discrete structure at zero oxygen pressure, indicating substantial IVR, a decrease in the lifetime led to an increase in structure as the extent of IVR was reduced. Parmenter referred to this effect as pressure line-narrowing, which contrasts with the more usual broadening.

More recently, the development of ultra-fast techniques have allowed a more direct observation of the time scales of IVR processes in S_1 states of aromatics by Zewail and co-workers[115,116]. Figure 9.8 shows the time dependence of the fluorescence of jet-cooled anthracene as a function of excitation energy. At low energies (< 1200 cm^{-1}) a simple exponential

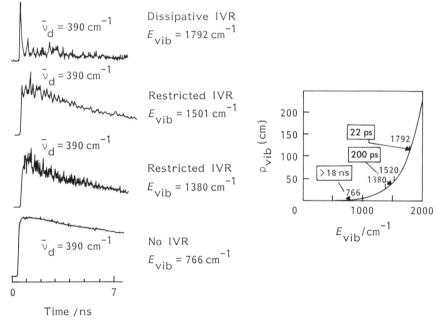

Figure 9.8 Three regimes of IVR: no IVR, restricted IVR, and dissipative IVR observed in a beam of anthracene. The traces on the left show the actual transient observed while detecting emission. The plot on the right shows the increase in density of states as a function of energy and time scale for IVR. From L. R. Khundkar and A. H. Zewail, *Ann. Rev. Phys. Chem.*, **41**, 15 (1990). Reproduced by permission of Annual Reviews Inc.

decay is observed, indicating an absence of vibrational mixing on the time scale of the decay. For energies in the range 1380–1520 cm^{-1}, a much more complex decay is observed with a beat pattern superimposed on top of the decay of the optically active, initially prepared state. In this region, a small number of states (< 10) are coupled and the beats represent the recurrence of the initially prepared state: the time scale of the IVR corresponds to the reciprocal of the beat frequency and the initial state recovers fully because of the low density of states. At higher energies, the densities of states are higher, IVR occurs more rapidly and full recurrence does not take place because of the higher density of coupled states. In this case, a measure of the time scale of IVR is given by the initial, rapid 'dephasing' of the initial state. The inset to Figure 9.8 plots the dependence of the density of states against the vibrational energy and also shows the rapid decrease in the time scale of IVR with $\rho(E_{vib})$. These time scales fall to ~ 20 ps for even modest vibrational excitation.

HIGH-RESOLUTION FTIR SPECTROSCOPY

Quack and coworkers[117,118] have exploited the use of high-resolution infrared spectroscopy, coupled with *ab initio* calculations of potential energy surfaces to study IVR in jet-cooled CHX_3 molecules. They have obtained considerable detail in their descriptions, demonstrating rapid IVR between preferentially coupled modes. In these molecules, there is strong anharmonic coupling, via a Fermi resonance, between the CH stretch and the degenerate bend; the density of the coupled states increases with energy. If a high overtone of the CH stretch is excited, there is an initial, rapid redistribution of energy involving the CH bending modes, on a time scale of ~ 100 fs, provided the coupling is strong. For weak coupling, there is a more classical behaviour similar to the motion of coupled pendula, as the energy moves backwards and forwards between the coupled states (cf. the beating referred to above). On longer time scales (several hundred femtoseconds for strongly coupled systems) there is further delocalisation of the energy, with spreading over the other vibrational modes. A full understanding of IVR is not yet possible, but experiments and theoretical analysis of the type reported by Quack and co-workers are providing an increasingly detailed picture. A critical review may be found in ref. 117.

REFERENCES

1. R. G. Gilbert and S. C. Smith, *Theory of Unimolecular and Recombination Reactions*, Blackwell Scientific Publications, Oxford, p. 220 (1990).

2. D. C. Tardy and B. S. Rabinovitch, *J. Chem. Phys.*, **48**, 1282 (1968).
3. J. Troe, *J. Chem. Phys.*, **66**, 4745 (1977).
4. J. Troe, *J. Chem. Phys.*, **66**, 4758 (1977).
5. D. C. Tardy and B. S. Rabinovitch, *J. Chem. Phys.*, **45**, 3720 (1966).
6. P. J. Robinson and K. A. Holbrook, *'Unimolecular Reactions'*, 1st edn, Wiley, London (1972).
7. P. J. Robinson in *'Reaction Kinetics'*. A Specialist Periodical Report, ed. P. G. Ashmore, The Chemical Society, London, Vol. 1, p. 93 (1973).
8. D. C. Tardy and B. S. Rabinovitch, *Chem. Rev.*, **77**, 369 (1977).
9. M. Quack and J. Troe, Specialist Periodical Reports, *Gas Kinetics and Energy Transfer*, Vol. 2, p. 175 (1977).
10. K. A. Holbrook, *Chem. Soc. Rev.*, **12**, 163 (1983).
11. J. R. Barker, *J. Phys. Chem.*, **88**, 11 (1984).
12. I. Oref and D. C. Tardy, *Chem. Rev.*, **90**, 1407 (1990).
13. S. C. Chan, B. S. Rabinovitch, J. T. Bryant, L. D. Spicer, T. Fujimoto, Y. N. Lin and S. P. Pavlou, *J. Phys. Chem.*, **74**, 3160 (1970).
14. S. C. Chan, J. T. Bryant, L. D. Spicer and B. S. Rabinovitch, *J. Phys. Chem.*, **74**, 2058 (1970).
15. L. D. Spicer and B. S. Rabinovitch, *J. Phys. Chem.*, **74**, 2445 (1970).
16. S. P. Pavlou and B. S. Rabinovitch, *J. Phys. Chem.*, **75**, 1366 (1971).
17. Y. N. Lin, S. C. Chan and B. S. Rabinovitch, *J. Phys. Chem.*, **72**, 1932 (1968); B. S. Rabinovitch, Y. N. Lin, S. C. Chan and K. W. Watkins, *J. Phys. Chem.*, **71**, 3715 (1967); see also refs 13 and 14.
18. N. Chow and D. J. Wilson, *J. Phys. Chem.*, **66**, 342 (1962).
19. L. Zalotai, T. Bérces and F. Márta, *J. Chem. Soc. Farad. Trans.*, **86**, 21 (1990).
20. I. E. Klein and B. S. Rabinovitch, *Chem. Phys.*, **35**, 439 (1978).
21. V. V. Krongauz and B. S. Rabinovitch, *Chem. Phys.*, **67** 201 (1982).
22. D. F. Kelley, B. D. Barton, L. Zalotai and B. S. Rabinovitch, *J. Chem. Phys.*, **71**, 538 (1979).
23. D. F. Kelley, L. Zalotai and B. S. Rabinovitch, *Chem. Phys.*, **46**, 379 (1980).
24. H. Endo, K. Glänzer and J. Troe, *J. Phys. Chem.*, **83**, 2083 (1979).
25. C. C. Chiang, J. A. Barker and G. B. Skinner, *J. Phys. Chem.*, **84**, 939 (1980).
26. J. H. Kiefer and J. N. Shah, *J. Phys. Chem.*, **91**, 3024 (1987).
27. J. Troe, *Ber Bunsenges. Phys. Chem.*, **78**, 478 (1974).
28. D. C. Tardy and B. S. Rabinovitch, *J. Phys. Chem.*, **89**, 2442 (1985).
29. I. R. Slagle, L. Batt, G. W. Gmurczyk, D. Gutman and W. Tsang, *J. Phys. Chem.*, **95**, 7732 (1991).
30. P. W. Seakins, S. H. Robertson, M. J. Pilling, L. R. Slagle, G. W. Gmurczyk, A. Bencsura, D. Gutman and W. Tsang, *J. Phys. Chem.*, **97**, 4450 (1993).
31. Y. Feng, J. T. Niiranen, A. Bencsura, V. D. Knyazev, D. Gutman and W. Tsang, *J. Phys. Chem.*, **97**, 871 (1993).
32. J. H. Georgakakos, B. S. Rabinovitch and E. J. McAlduff, *J. Chem. Phys.*, **52**, 2143 (1970).
33. R. Atkinson and B. A. Thrush, *Proc. Roy. Soc (Lond.)*, **A316**, 123 (1970).
34. R. Atkinson and B. A. Thrush, *Proc. Roy. Soc (Lond.)*, **A316**, 131 (1970).
35. J. E. Baggot, *Chem. Phys. Lett.*, **119**, 47 (1985).
36. J. Troe, *J. Phys. Chem.*, **87**, 1800 (1983).
37. H. Hippler, J. Troe and H. J. Wendelken, *J. Chem. Phys.*, **78**, 5351, 6709, 6718 (1983).
38. M. J. Rossi, J. R. Pladziewicz and J. R. Barker, *J. Chem. Phys.*, **78**, 6695 (1983).

39. J. R. Barker, *J. Phys. Chem.*, **88**, 11 (1984).
40. G. P. Smith and J. R. Barker, *Chem. Phys. Lett.*, **78**, 253 (1981).
41. K. M. Beck and R. J. Gordon, *J. Chem. Phys.*, **87**, 5681 (1987).
42. K. M. Beck, A Ringwelski and R. J. Gordon, *Chem. Phys. Lett.*, **121**, 529 (1985).
43. P. L. Trevor, T. Rothem and J. R. Barker, *Chem. Phys.*, **68**, 341 (1982).
44. H. G. Löhmannsröben and K. Luther, *Chem. Phys. Lett.*, **144**, 473 (1988).
45. H. Hippler, L. Lindemann and J. Troe, *J. Chem. Phys.*, **83**, 3906 (1985).
46. M. J. Rossi and J. R. Barker, *Chem. Phys. Lett.*, **85**, 21 (1982).
47. J. Shi and J. R. Barker, *J. Chem. Phys.*, **88**, 6219 (1988).
48. J. R. Barker and R. E. Golden, *J. Phys. Chem.*, **88**, 1012 (1984).
49. R. C. Bhattacharjee and W. Forst, *Chem. Phys.*, **30**, 217 (1978).
50. V. V. Krongauz, B. S. Rabinovitch and E. Linkaityte-Weiss, *J. Chem. Phys.*, **78**, 5643 (1983).
51. M. Heymann, H. Hippler and J. Troe, *J. Chem. Phys.*, **80**, 1853 (1984).
52. J. M. Zellweger, T. C. Brown and J. R. Barker, *J. Chem. Phys.*, **83**, 6261 (1985).
53. M. L. Yerram, J. Brenner, K. D. King and J. R. Barker, *J. Phys. Chem.*, **94**, 6341 (1990).
54. B. M. Toselli, J. D. Brenner, M. L. Yerram, W. E. Chin, K. D. King and J. R. Barker, *J. Chem. Phys.*, **95**, 176 (1991).
55. H. Hippler, *Ber. Bunsenges. Phys. Chem.*, **89**, 303 (1985).
56. M. Heymann, H. Hippler, H. J. Plach and J. Troe, *J. Chem. Phys.*, **87**, 3867 (1987).
57. H. Hippler, L. Lindemann and J. Troe, *J. Chem. Phys.*, **83**, 3906 (1985).
58. H. Hippler, B. Otto and J. Troe, *Ber. Bunsenges. Phys. Chem.*, **93**, 428 (1989).
59. N. Nakashima and Y. Yoshihara, *J. Chem. Phys.*, **79**, 2727 (1983).
60. T. Ichimura, Y. Mori, N. Nakashima and Y. Yoshihara, *J. Chem. Phys.*, **83**, 117 (1985).
61. T. Ichimura, M. Takahashi and Y. Mori, *Chem. Phys.*, **114**, 111 (1987).
62. M. Damm, H. Hippler and J. Troe, *J. Chem. Phys.*, **88**, 3564 (1988).
63. H. G. Löhmannsröben and K. Luther, *Chem. Phys. Lett.*, **144**, 173 (1988).
64. K. Luther and K. Reihs, *Ber. Bunsenges. Phys. Chem.*, **92**, 442 (1988).
65. A. Pashutski and I. Oref, *J. Phys. Chem.*, **92**, 178 (1988).
66. J. M. Morgulis, S. S. Sapers, C. Steel and I. Oref, *J. Chem. Phys.*, **90**, 923 (1989).
67. G. Lendvay and G. C. Schatz, *J. Phys. Chem.*, **94**, 8864 (1990).
68. W. L. Hase in *'Dynamics of Molecular Collisions'*, Part B, ed. W. H. Miller, Plenum Press, New York, p. 121 (1976).
69. W. L. Hase, *J. Phys. Chem.*, **90**, 365 (1986).
70. I. Oref and B. S. Rabinovitch, *Acc. Chem. Res.*, **12**, 166 (1979).
71. W. L. Hase and D. L. Bunker, *J. Chem. Phys.*, **59**, 4621 (1973).
72. J. N. Butler and G. B. Kistiakowsky, *J. Amer. Chem. Soc.*, **82**, 759 (1960).
73. J. D. Rynbrandt and B. S. Rabinovitch, *J. Phys. Chem.*, **75**, 2164 (1971).
74. B. S. Rabinovitch, J. F. Meagher, K. J. Chao and J. R. Barker, *J. Chem. Phys.*, **60**, 2932 (1974); F. M. Wang and B. S. Rabinovitch, *Can. J. Chem.*, **54**, 943 (1976); A. N. Ko, B. S. Rabinovitch and K. J. Chao, *J. Chem. Phys.*, **66**, 1374 (1977).
75. K. C. Kim and D. W. Setser, *J. Phys. Chem.*, **78**, 2166 (1974); B. E. Holmes and D. W. Setser, *J. Phys. Chem.*, **79**, 1320 (1975).
76. A. B. Trenwith and B. S. Rabinovitch, *J. Chem. Phys.*, **85**, 1696 (1986);

A. B. Trenwith, D. A. Oswald, B. S. Rabinovitch and M. C. Flowers, *J. Phys. Chem.*, **91**, 4398 (1987).

77. P. Rogers, J. I. Selco and F. S. Rowland, *Chem. Phys. Lett.*, **97**, 313 (1983).
78. V. Lopez and R. A. Marcus, *Chem. Phys. Lett.*, **93**, 32 (1982); S. M. Lederman, V. Lopez, G. A. Voth and R. A. Marcus, *Chem. Phys. Lett.*, **124**, 93 (1986).
79. N. E. Lee and E. K. C. Lee, *J. Chem. Phys.*, **50**, 2094 (1969).
80. F. H. Dorer, *J. Phys. Chem.*, **77**, 954 (1973).
81. K. V. Reddy and M. J. Berry, *Chem. Phys. Lett.*, **52**, 111 (1977).
82. K. V. Reddy and M. J. Berry, *Farad. Disc.*, **67**, 188 (1979).
83. D. W. Lupo and M. Quack, *Chem. Rev.*, **87**, 181 (1987).
84. P. A. Schultz, Aa. Sudbø, D. J. Krajnovich, H. S. Kwok, Y. R. Shen and Y. T. Lee, *Ann. Rev. Phys. Chem.*, **30**, 379 (1979).
85. J. D. McDonald, *Ann. Rev. Phys. Chem.*, **30**, 29 (1979).
86. M. N. R. Ashfold and G. Hancock, *Gas Kinetics and Energy Transfer*, The Chemical Society, London, Vol. 4, p. 73 (1981).
87. A. Sinha, M. C. Hsiao and F. F. Crim, *J. Chem. Phys.*, **94**, 4928 (1991).
88. W. H. Green Jr., C. B. Moore and W. F. Polik, *Ann. Rev. Phys. Chem.*, **43**, 591 (1992).
89. S. A. Safron, N. D. Weinstein, D. R. Herschbach and J. C. Tully, *Chem. Phys. Lett.*, **12**, 564 (1972).
90. G. Worry and R. A. Marcus, *J. Chem. Phys.*, **67**, 1636 (1977).
91. E. L. Spotz, W. A. Seitz and J. L. Franklin, *J. Chem. Phys.*, **51**, 5142 (1969).
92. K. C. Kim, J. H. Beynon and R. G. Cooks, *J. Chem. Phys.*, **61**, 1305 (1974).
93. J. M. Farrar and Y. T. Lee, *J. Chem. Phys.*, **65**, 1414 (1976).
94. M. J. Coggiola, P. A. Schultz, Y. T. Lee and Y. R. Shen, *Phys. Rev. Lett.*, **38**, 17 (1977).
95. J. G. Moehlmann and J. D. McDonald, *J. Chem. Phys.*, **59**, 6683 (1973).
96. J. G. Moehlmann, J. T. Gleaves, J. W. Hudgens and J. D. McDonald, *J. Chem. Phys.*, **60**, 4790 (1974).
97. A. Schultz, H. W. Cruse and R. N. Zare, *J. Chem. Phys.*, **57**, 1354 (1972).
98. J. C. Stephenson and D. S. King, *J. Chem. Phys.*, **69**, 1485 (1978).
99. J. C. Stephenson, S. E. Bialkowski and D. S. King, *J. Chem. Phys.*, **72**, 1161 (1980).
100. J. M. Parsons and Y. T. Lee, *J. Chem. Phys.*, **56**, 4658 (1972).
101. D. M. Rayner and P. A. Hackett, *J. Chem. Phys.*, **79**, 5414 (1983).
102. G. M. Stewart and J. D. McDonald, *J. Chem. Phys.*, **75**, 5949 (1981).
103. G. M. Stewart and J. D. McDonald, *J. Chem. Phys.*, **78**, 3907 (1983).
104. H. L. Kim, T. J. Kulp and J. D. McDonald, *J. Chem. Phys.*, **87**, 4376 (1987).
105. R. E. Smalley, *J. Phys. Chem.*, **86**, 3504 (1982).
106. C. S. Parmenter, *J. Phys. Chem.*, **86**, 1743 (1982).
107. F. F. Crim, *Ann. Rev. Phys. Chem.*, **35**, 657 (1984).
108. H. Reisler and C. Wittig, *Ann. Rev. Phys. Chem.*, **37**, 307 (1986).
109. J. Segal and R. N. Zare, *J. Chem. Phys.*, **89**, 5704 (1988).
110. M. C. Chuang and R. N. Zare, *J. Chem. Phys.*, **82**, 4791 (1985).
111. See for example, J. M. Jasinski, J. K. Frisoli and C. B. Moore, *J. Chem. Phys.*, **87**, 3826 (1983) and comments in the reviews cited in references 107 and 108.
112. A. H. Zewail, *Acc. Chem. Res.*, **13**, 360 (1980); N. F. Scherer and A. H. Zewail, *J. Chem. Phys.*, **87**, 97 (1987).
113. R. L. Van der Wal, J. L. Scott, F. F. Crim, K. Weide and R. Schinke, *J. Chem. Phys.*, **94**, 3548 (1991).

114. C. S. Parmenter, *Disc. Faraday Soc.*, **75**, 7 (1983).
115. L. R. Khundkar and A. H. Zewail, *Ann. Rev. Phys. Chem.*, **41**, 15 (1990).
116. P. M. Felker and A. H. Zewail, *J. Chem. Phys.*, **82**, 2975 (1985).
117. M. Quack, *Ann. Rev. Phys. Chem.*, **41**, 839 (1990).
118. R. Marquardt and M. Quack, *J. Chem. Phys.*, **95**, 4854 (1991).

10 Kinetic Isotope Effects in Unimolecular Reactions

The rate of a unimolecular reaction is in general altered by isotopic substitution in the reactant molecule. The study of isotope effects is important since it can often lead to a more detailed understanding of the reaction process than can be obtained by study of the unsubstituted compound alone. Isotopic substitution can have several simultaneous effects, and this complication sometimes produces results which are at first sight curious. For example, a deuterium-substituted compound may isomerise more slowly than the parent compound at high pressures, but more rapidly at low pressures. The general theory of kinetic isotope effects has been adequately presented elsewhere[1] and the present discussion is limited to a review of the principles involved and some detail of their application to unimolecular reactions. In this context the review by Rabinovitch and Setser[2] is useful, and many of the illustrative calculations given later are taken from that

Table 10.1 Terminology used for kinetic isotope effects

Intermolecular	Comparison of rates between two different molecules (e.g. reactions (10.3) and (10.4))
Intramolecular	Comparison of competitive rates of two reactions of one molecule (e.g. reactions (10.14))
Primary	Atom(s) substituted are directly involved in the reaction (e.g. reactions (10.3) and (10.4))
Secondary	Atom(s) substituted are not directly involved in the reaction (e.g. reactions (10.5) and (10.6))[a]
Normal (regular)	$k(\text{light}) > k(\text{heavy})$
Inverse	$k(\text{heavy}) > k(\text{light})$
Statistical-weight	Arises from changes in the distribution of energy levels as opposed to changes in critical energy. Usually predominant in secondary isotope effects
Non-equilibrium	Effects found when transition states are not in equilibrium with reactants, e.g. unimolecular reactions in the fall-off region, or chemically activated systems

[a]More rigorously, the reaction coordinate is orthogonal to all vibrational modes involving the atom(s) substituted; the normal coordinates for these modes have effectively zero weighting in the reaction coordinate.

source. For simplicity the discussion is in terms of hydrogen–deuterium isotope effects, but the principles obviously apply to other systems as well. Some of the terminology used in connection with isotope effects is summarised in Table 10.1.

10.1 GENERAL DISCUSSION OF ISOTOPE EFFECTS

The origin of all kinetic isotope effects lies in the changes in the quantised molecular energy levels which occur when the vibration frequencies and moments of inertia of a molecule are modified by isotopic substitution. The changes are purely mass effects, since isotopic substitution has no effect on the electron distribution or potential-energy surface for a molecule, and hence no effect on the ground-state geometry or the force constants for vibration of the molecule. The changes in moments of inertia are generally less important than those in vibration frequency and will be neglected here. In the case of a simple harmonic oscillator the vibration frequency is given by

$$\nu = \frac{1}{2\pi}\sqrt{\frac{k}{\mu}} \tag{10.1}$$

and isotopic substitution alters the vibration frequency by changing the reduced mass μ with a constant value of the force constant k. A simple illustration of this result is the observation that the stretching vibration frequencies of the bonds X–H and X–D, where X is a relatively heavy group such as Cl, Br, alkyl, etc. are related by $\nu_{XH}/\nu_{XD} \approx \sqrt{2}$ (it will be noted that $\mu_{XH}/\mu_{XD} \approx \frac{1}{2}$ if X is heavy). For the more general case where the molecular distortions must be considered in terms of normal vibrations which are not highly localised (see Chapter 2 and ref. 3), the properties of the isotopic forms are still related in an overall sense by the Teller–Redlich product rule[3]. For the case of a non-linear molecule this is

$$\prod_{i=1}^{3N-6} \frac{\nu_i}{\nu_i'} = \left(\frac{I_A I_B I_C}{I_A' I_B' I_C'}\right)\left(\frac{M}{M'}\right)^{3/2} \prod_{i=1}^{N}\left(\frac{m_i'}{m_i}\right)^{3/2} \tag{10.2}$$

in which I denotes moment of inertia, M molecular weight, m_i the masses of the N atoms in the molecule and primed quantities refer to the isotopically substituted molecule. The product rule is a useful relationship when it is desired to set up consistent models of a molecule and its isotopically substituted form, particularly in the case of transition states where experimental measurement is not possible even in principle[4].

The crucial effect of the changes in vibration frequencies is to modify the vibrational energy levels of the molecule (for example the spacing of the levels is $h\nu$ for a simple harmonic oscillator) and these changes can have essentially two consequences. Firstly, there are *critical energy effects*; the zero-point energies ($\frac{1}{2}h\nu$) of the affected vibrations will change on isotopic substitution, and if these vibrations contribute substantially to the reaction coordinate the critical energy E_0 will be changed. The compounds XH and XD again form a simple example, the relevant energy diagrams being shown in Figure 10.1. The group X is here regarded as essentially a point mass, so that the unimolecular reaction is a pure bond fission and there is no zero-point energy in the transition state. Fission of the X–H or X–D bond thus requires a critical energy which is equal to $D_e - E_z$ and may differ substantially for the two compounds. For R_3C–H and R_3C–D, for example, the zero-point energies are $\frac{1}{2}h\nu_0 \approx 17.2$ and 12.1 kJ mol^{-1} respectively, giving a difference in the critical energies for C–D and C–H fission of

$$E_0(R_3C\text{-}D) - E_0(R_3C\text{-}H) = \Delta E_0 \approx 5.1 \text{ kJ mol}^{-1}$$

This difference would give a rate ratio of

$$k(R_3C\text{-}H)/k(R_3C\text{-}D) = \exp{(\Delta E_0/kT)} = 2.7 \text{ at } 600 \text{ K}$$

which is a substantial difference, and when this effect is present it is frequently predominant. The reaction involving the lighter isotope is generally favoured (as in the above case), and this is known as a *normal kinetic isotope effect* $[k(\text{light}) > k(\text{heavy})]$. The rate ratio will not always be as large as above since the reaction may be such that the hydrogen atom is incompletely removed in the transition state. In this case there will still be some

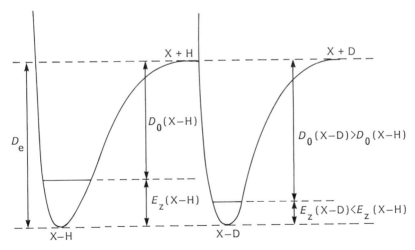

Figure 10.1 Potential energy curves for the molecules X–H and X–D, showing difference in dissociation energy. Reproduced by permission of John Wiley and Sons Ltd.

zero-point energy of the corresponding vibration(s) and the change in E_0 on isotopic substitution will be less than before. If the reaction involves transfer of an atom from one site to another it is even possible for the atom to be bound more tightly in the transition state than in the molecule. In this case the result can be that k(heavy) > k(light), an *inverse kinetic isotope effect*. Inverse effects can also arise for other reasons discussed below.

These examples have referred to the case where the atom to be isotopically substituted actually takes part in the reaction, i.e. the reaction coordinate involves the vibration of which the frequency is altered. Examples would be reactions (10.3) and (10.4).

$$C_2H_5—H \rightarrow C_2H_5 + H \tag{10.3}$$

$$C_2H_5—D \rightarrow C_2H_5 + D \tag{10.4}$$

Such effects are known as *primary kinetic isotope effects*, and a major contribution to such effects is the change in critical energy discussed above. Statistical-weight effects (see below) may also be important, however, and can sometimes predominate.

The second consequence of changes in the energy-level spacing on isotopic substitution is a change in the state densities of reactant and transition state. Even when isotopic substitution occurs at a site remote from the reaction centre the reaction rate may be affected. Reactions (10.5) and (10.6) comprise an example, and any difference in rate is due to a *secondary kinetic isotope effect*.

$$CH_3CH_2—CH_2CH_3 \rightarrow 2CH_3CH_2 \tag{10.5}$$

$$CD_3CH_2—CH_2CD_3 \rightarrow 2CD_3CH_2 \tag{10.6}$$

To a good approximation the critical energy is unaffected by such changes, and secondary isotope effects arise mainly from changes in the quantities $W(E_{vr}^+)$ and $\rho(E^*)$ which (with E_0) principally affect the results of the RRKM calculation. These quantities are certain to change on isotopic substitution because of the changes in energy level spacing; the effects are clearly illustrated in the Whitten–Rabinovitch equations (4.36) and (4.42) and a specific example is given in Table 10.2 and Figure 10.5. Such effects are known as *statistical-weight isotope effects*, since they arise from changes in the numbers of quantum states at various energy levels, i.e. the statistical weights of these levels. Primary isotope effects invariably include statistical-weight effects as well as critical-energy effects, and the statistical-weight effects are not always of smaller magnitude.

Isotopic substitution at sites not directly involved in the reaction can also produce small changes in critical energy, due to small contributions to the reaction coordinate from vibrational modes involving the substituted atoms.

Table 10.2 Calculation of secondary kinetic isotope effects for C_2–C_3 fission in butane and butane-d_6 (reactions (10.5) and (10.6))[a]

	$E^+/\text{kJ mol}^{-1}$				
	0	8	42	168	∞
$W_H(E_{vr}^+)$	1	5.1×10^3	1.6×10^7	6.4×10^{13}	—
$W_D(E_{vr}^+)$	1	1.3×10^4	8.0×10^7	1.1×10^{15}	—
$W_H(E_{vr}^+)/W_D(E_{vr}^+)$	1.00	0.38	0.20	0.060	0.0036
For $E_0 = 146$ $\left\{\begin{array}{l}\rho_H(E^*)\\ \rho_D(E^*)\\ \rho_D(E^*)/\rho_H(E^*)\end{array}\right.$ ($E^* = E_0 + 146$)					
$\rho_H(E^*)$		2.1×10^{10}	4.5×10^{10}	7.9×10^{11}	4.5×10^{15}
$\rho_D(E^*)$		2.9×10^{11}	6.5×10^{11}	1.4×10^{13}	1.4×10^{17}
$\rho_D(E^*)/\rho_H(E^*)$		14	14	18	31
For $E_0 = 356$ ($E^* = E_0 + 356$)					
$\rho_H(E^*)$		5.0×10^{16}	8.1×10^{16}	4.8×10^{17}	1.7×10^{20}
$\rho_D(E^*)$		1.8×10^{18}	2.9×10^{18}	1.9×10^{19}	9.0×10^{21}
$\rho_D(E^*)/\rho_H(E^*)$		36	36	40	53
$k_{aH}(E^*)/k_{aD}(E^*)$ for $\begin{cases} E_0 = 146 \\ E_0 = 356 \end{cases}$	14	5.5	3.5	1.8	1.05
	36	14	7.8	3.1	1.05

[a]Energies are in kJ mol^{-1} and state densities in $(\text{kJ mol}^{-1})^{-1}$. In the published frequency assignment[2] for the C_4-d_6 molecule, the entry 270(2) cm^{-1} should read 207(2) cm^{-1}. In addition, the symmetry numbers were included (as a factor $1/\sigma$ in the rotational partition function) in calculating the entropies of the active degrees of freedom for all the models. With these provisos the above recalculated results are in good agreement with those of ref. 2. The $\rho(E^*)$ values are shown graphically in Figure 10.5.

Isotope effects in such systems are sometimes called secondary effects, although they are not pure secondary effects in the sense used here; the terms 'conventional secondary isotope effects' and 'mixed primary–secondary isotope effects' have been used.

10.2 BASIS OF APPLICATION TO UNIMOLECULAR REACTIONS

The calculation of isotope effects for unimolecular reactions is relatively simple for chemical activation or photochemical activation experiments, in which the energised molecules are produced in a relatively narrow band of energies. There is in practice still a spread of energies which cannot be ignored, but ideal experiments involving monoenergetic excitation would give $k_a(E^*)$ for the isotopic reactions directly. The rate constants are each calculable from RRKM theory (equation 3.16) and the isotopic rate ratio is thus given by,

$$\frac{k_{aH}(E_H^*)}{k_{aD}(E_D^*)} = \frac{W_H(E_{vr}^+)/W_D(E_{vr}^+)}{\rho_H(E_H^*)/\rho_D(E_D^*)} \tag{10.7}$$

in which statistical factors and centrifugal effects have been omitted for simplicity. This equation applies equally to intermolecular isotope effects and (with a simplification described later) to intramolecular isotope effects. The expression becomes relatively simple for a secondary isotope effect, for which a natural comparison can be made for the case where $E_H^* = E_D^*$ and $E_H^+ = E_D^+$. For primary isotope effects the situation is more complicated since E_0 is different for the two compounds, and therefore $E_H^* \neq E_D^*$ and/or $E_H^+ \neq E_D^+$.

In thermal activation experiments, the formulation is complicated even for a secondary isotope effect because of the simultaneous changes in k_a and in the energy distribution function. The fractional concentration $b_H'(E^*)\delta E^*$ of energised molecules in the energy range E^* to $E^* + \delta E^*$ in the steady state is given by

$$b_H'(E^*)\delta E^* = \left(\frac{[A_{H(E^* \to E^* + \delta E^*)}]}{[A]}\right)_{\text{steady state}}$$
$$= \frac{\rho_H(E^*)\exp(-E^*/kT)/Q_{2H}}{1 + k_{aH}(E^*)/k_2[M]}\delta E^* \tag{10.8}$$

and the corresponding quantity $b_D'(E^*)\delta E^*$ is given by a similar equation.

The isotopic rate ratio is thus

$$\frac{(k_{uni})_H}{(k_{uni})_D} = \frac{\int_{E^*=(E_0)_H}^{\infty} b'_H(E^*)k_{aH}(E^*)\delta E^*}{\int_{E^*=(E_0)_D}^{\infty} b'_D(E^*)k_{aD}(E^*)\delta E^*} \tag{10.9}$$

The general equation is clearly very complex, but the high- and low-pressure limits (10.10) and (10.11) are relatively simple and will be useful in the discussion. These equations may be obtained from (10.9) or more simply from (3.21) and (3.24) respectively.

$$(p \to \infty)\ \frac{(k_\infty)_H}{(k_\infty)_D} = \frac{(Q_2^+/Q_2)_H}{(Q_2^+/Q_2)_D} \exp\{[(E_0)_D - (E_0)_H]/kT\} \tag{10.10}$$

$$(p \to 0)\ \frac{(k_{bim})_H}{(k_{bim})_D} = \frac{(Q_2^{*'}/Q_2)_H}{(Q_2^{*'}/Q_2)_D} \exp\{[(E_0)_D - (E_0)_H]/kT\} \tag{10.11}$$

In (10.11) the reasonable approximation has been made that $k_{2H} = k_{2D}$, and it will be recalled that $Q_2^{*'}$ is the partition function for the active degrees of freedom in the energised molecule with the energy zero taken as the ground state of A^+, i.e. the lowest energy an A^* can have.

The type and size of isotope effects in unimolecular reactions can now be discussed. We deal first with the simplest case of secondary effects on $k_a(E^*)$, then primary effects on $k_a(E^*)$ and finally isotope effects on k_{uni} for thermally activated systems.

10.3   SECONDARY KINETIC ISOTOPE EFFECTS ON $k_a(E^*)$ (THEORY AND EXPERIMENT)

The application of (10.7) to calculate secondary isotope effects on $k_a(E^*)$ in monoenergetic systems is fairly straightforward, since it may be assumed that $(E_0)_D = (E_0)_H$ and the values of k_a for the two forms can be compared at the same values of both E^* and E^+. Changes in the moments of inertia of the external (adiabatic) rotations are likely to be small and partially compensating, so the neglect of centrifugal factors will be a reasonable approximation. Deuterium substitution lowers the vibration frequencies and hence the energy-level spacing, and both $W(E_{vr}^+)$ and $\rho(E^*)$ are therefore higher for the deuterio- than for the light molecules. Since $E^* \gg E^+$, however, $\rho(E^*)$ is always increased more than $W(E_{vr}^+)$, and thus $k_{aH}(E^*) > k_{aD}(E^*)$; a normal statistical-weight secondary isotope effect is predicted.[a] The figures

---

[a]It is interesting that statistical-weight isotope effects are normal in chemically activated systems but inverse in non-equilibrium thermally activated systems (see section 10.5)[5].

involved are illustrated by Rabinovitch and Setser's calculations[2] for reactions (10.5) and (10.6), see Table 10.2. The illustrative critical energies of 146 and 356 kJ mol^{-1} were chosen as being typical of C–C fission in an alkyl radical and an alkane molecule respectively. It will be seen that the size of the isotope effect varies markedly with both E_0 and E^+, being largest when E_0 is high and E^+ is low (the difference then being mainly in the $\rho(E^*)$ terms) and smallest when E^+ is large (the maximum compensating effect of the $W(E_{vr}^+)$ terms then being obtained).

The most comprehensive experimental studies of such effects are those of Rabinovitch and co-workers on the decompositions of chemically activated sec-butyl radicals[6]. These radicals were produced by the addition of H or D atoms to butenes or octadeuteriobutenes, e.g. reaction (10.12).

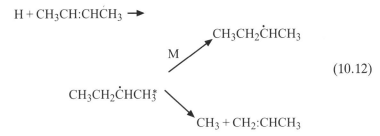

$$H + CH_3CH{:}CHCH_3 \longrightarrow$$

$$\begin{array}{c} \nearrow CH_3CH_2\dot{C}HCH_3 \\ M \\ CH_3CH_2\dot{C}HCH_3^* \searrow \\ CH_3 + CH_2{:}CHCH_3 \end{array}$$

(10.12)

By studying the reactions at two temperatures (195 and 300 K) and at high and low pressures it was possible to produce sec-butyl-d_0, -d_1, -d_8 and -d_9 radicals with a whole range of average energies $\langle E^* \rangle$ and to measure the rates of their unimolecular decomposition to methyl radicals plus propene relative to their rate of collisional stabilization. The technique is discussed in more detail in Chapter 11. As shown there, the excitation is not strictly monoenergetic, but the energy spread in a given experiment is small and the measured rate constants approximate to the values $(k_a(E^*))$ for the appropriate mean energy $\langle E^* \rangle$. The results are presented in Figure 10.2, which shows a large normal isotope effect in favour of the d_0 and d_1 radicals (upper curve) against the d_8 and d_9 radicals (lower curve). The experimental rate constants are in excellent agreement with those calculated from the RRKM theory over a 50-fold range of values. The isotopic rate ratio is best obtained from the smooth-curve representations of the results, since the addition of H to C_4H_8 and C_4D_8 gives radicals with slightly different $\langle E^* \rangle$. The difference arises from the different thermal energies of the two olefins at a given temperature; it is typically 2–4 kJ mol^{-1} at 300 K, which gives a substantial contribution to the experimental rate ratio. The inferred experimental isotopic rate ratio at constant E^* varies from $k_{aH}/k_{aD} \approx 8$ at the lowest energies studied to ≈ 5.5 at the highest energies, and is reproduced very well by RRKM calculations on the basis of a pure secondary statistical-weight effect with $(E_0)_H = (E_0)_D = 138$ kJ mol^{-1}.

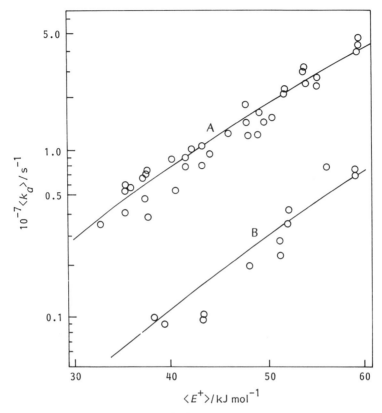

Figure 10.2 Observed rate constants $\langle k_a \rangle$ for decomposition of chemically activated sec-butyl-d_0 and -d_1 (curve A) and -d_8 and -d_9 (curve B) radicals as a function of their excess energy $\langle E^+ \rangle$. Reproduced by permission of John Wiley and Sons Ltd.

Similar results have been obtained for the decompositions of chemically activated n-propyl-d_6 radicals[7] and highly deuteriated sec-hexyl and sec-octyl radicals[8], and for the geometrical and structural isomerisations of cis-1,2-dimethylcyclopropane-d_8[9]. In each case large normal isotope effects were observed and were in reasonable agreement with calculated values.

10.4 PRIMARY KINETIC ISOTOPE EFFECTS ON $k_a(E^*)$ (THEORY AND EXPERIMENT)

For primary isotope effects in monoenergetic systems the appropriate equation is still (10.7), but it is now more difficult to generalise about the nature of the effects because the critical energies $(E_0)_H$ and $(E_0)_D$ are different (usually $(E_0)_H < (E_0)_D$), and different comparisons can arise according to the energy differences involved. Two of the more obvious cases are illustrated in Figures 10.3 and 10.4, which illustrate the energy relationships for the cases where $E_H^* = E_D^*$ and $E_H^+ > E_D^+$ (Figure 10.3) or $E_H^+ = E_D^+$ and $E_H^* < E_D^*$ (Figure 10.4).

The first situation is approximately achieved in chemical activation experiments where a fixed amount of energy E^* is released in the molecule by a reaction which is energetically unaffected by the isotopic substitution. An example would be reaction (10.13):

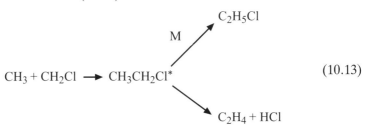

$$CH_3 + CH_2Cl \longrightarrow CH_3CH_2Cl^* \qquad (10.13)$$

and the corresponding perdeuterio case, although it will be seen later that even this simple comparison is in practice confused by differences in the thermal energy distributions of the species involved. Similar remarks apply to the photoactivation of isotopic molecules by light of the same wavelength. The comparison is illustrated in Table 10.3 by some calculated data for reactions (10.3) and (10.4)[2]; these show the salient features, although it would be difficult to study the reactions experimentally in the forward direction. The tabulated rate ratios can be considered to arise from the combination of three effects. First, as noted previously, $\rho_D(E^*)$ is always greater than $\rho_H(E^*)$ (for a common E^*), giving a *normal* statistical-weight contribution. Secondly, for a given E^+ (common to both cases) $W_D(E_{vr}^+)$ would be greater than $W_H(E_{vr}^+)$, tending to give an *inverse* effect. Thirdly, the difference in E_0 makes $E_H^+ > E_D^+$ (by 5.9 kJ mol^{-1} in the present case), and this tends to offset the statistical-weight effect on $W(E_{vr}^+)$. At low E^+ in the present case, the energy difference more than compensates for the statistical-weight effect on $W(E_{vr}^+)$, and the $W(E_{vr}^+)$ terms in (10.7) reinforce the normal isotope effect arising from the $\rho(E^*)$ ratio. At high energies the statistical-weight effect is predominant in the $W(E_{vr}^+)$ ratio, and this offsets the effect of the $\rho(E^*)$ ratio. In all cases, however, a normal isotope effect is predicted.

The second relatively simple case is that in which $E_H^+ = E_D^+$ and therefore $E_D^* > E_H^*$. The isotope effect is still normal, but the rate ratio is generally smaller than in the previous case (see Table 10.4). This is because the effect of a 5.9 kJ mol^{-1} energy difference on $\rho(E^*)$ at relatively high E^* is much less than that of the same energy difference on $W(E_{vr}^+)$ at relatively low E^+.

A practical example of isotope effects in 'monoenergetic' systems is

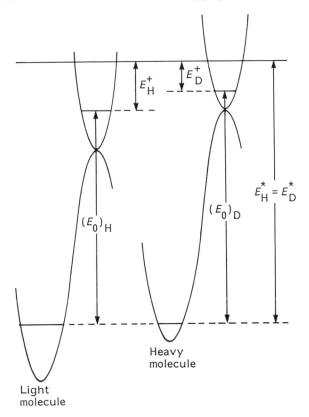

Figure 10.3 Comparative energy diagram for the primary isotope effect $[(E_0)_H < (E_0)_D]$ in monoenergetic systems excited to a common value of E^*. Reproduced by permission of John Wiley and Sons Ltd.

provided by the studies of Dees and Setser[10] on HCl/DCl elimination from chemically activated C_2H_5Cl/C_2D_5Cl and $CH_2ClCH_2Cl/CD_2ClCD_2Cl$, e.g. (10.13). It might be expected that the combination of CD_3 and CD_2Cl radicals would provide $C_2D_5Cl^*$ at the same E^* as the $C_2H_5Cl^*$ produced in (10.13), so that the situation would be analogous to that in Table 10.3. This would be true at the absolute zero, but at the experimental temperature of 298 K the thermal energies in the active degrees of freedom of the energised molecules differ by an amount which is coincidentally similar to the critical energy difference. The result is a situation in which $\langle E_H^+ \rangle \approx \langle E_D^+ \rangle$, similar to that in Table 10.4. The relevant energies are shown in Table 10.5, together with the observed and calculated isotopic rate ratios; normal isotope effects are observed and their magnitudes are in good agreement with the calculated values. The observed effects in this sort of system contain both critical energy and statistical-weight effects, and for this reason

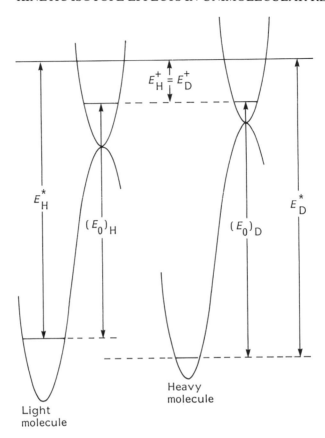

Figure 10.4 Comparative energy diagram for the primary isotope effect in mono-energetic systems excited to a common value of E^+. Reproduced by permission of John Wiley and Sons Ltd.

Table 10.3 Calculated primary isotope effect for C_2H_5-H and C_2H_5-D fission (reactions (10.3) and (10.4)) with $E_H^* = E_D^{*a}$

	Data for $(E_0)_D = 146.0$ $(E_0)_H = 140.6$				Data for $(E_0)_D = 356.0$ $(E_0)_H = 349.8$			
E^*	146	167	230	∞	356	377	439	∞
E_D^+	0	21	84	∞	0	21	84	∞
E_H^+	5.9	26.7	89.5	∞	5.9	26.7	89.5	∞
$\dfrac{k_{a(H)}(E^*)}{k_{a(D)}(E^*)}$	6.1	3.5	2.0	1.4	7.0	3.9	2.2	1.4

[a]Energies in kJ mol^{-1}; for further details see ref. 2.

Table 10.4 Calculated primary isotope effect for C_2H_5-H and C_2H_5-D fission (reactions (10.3) and (10.4)) with $E_H^+ = E_D^+$ and $(E_0)_H = 146.0$ $(E_0)_D = 152.3$ kJ mol^{-1a}

E^+	0	21	84	∞
E_H^+	146	167	230	∞
E_D^+	152.3	173.2	236	∞
$\dfrac{k_{aH}(E_H^*)}{k_{aD}(E_D^*)}$	2.2	1.7	1.6	1.4

[a]Energies in kJ mol^{-1}; for further details see ref. 2.

Table 10.5 Primary isotope effects in the decompositions of chemically activated ethyl chloride and 1,2-dichloroethane molecules[a]

	C_2H_5Cl	C_2D_5Cl	CH_2ClCH_2Cl	CD_2ClCD_2Cl
Critical energy E_0	230	236	230	236
$\langle E^* \rangle$ (calc.)	381	387	369	374
$\langle E^+ \rangle$ (calc.)	151	151	139	138
k_{aH}/k_{aD}(calc.)	3.0		2.8	
k_{aH}/k_{aD}(obs.)	3.3		3.5	

[a]Energies in kJ mol^{-1}; for further details see ref. 10.

are often referred to as 'mixed primary–secondary effects' (cf. section 10.1). Another similar case is the decomposition of the isopropyl-d_6 radical[7].

An interesting simplification occurs for *intramolecular* isotope effects in monoenergetic systems, e.g. reactions (10.14).

$$C_2H_5D^* \nearrow C_2H_4D + H \\ \searrow C_2H_5 + D \qquad (10.14)$$

For such a pair of reactions the $\rho(E^*)$ terms in (10.7) are identical and the rate constant ratio (corrected for the statistical factor of 5) is given simply by

$$k_{aH}(E_H^*)/k_{aD}(E_D^*) = W_H(E_{vr}^+)/W_D(E_{vr}^+)$$

The dominant effect here is the critical energy difference, and the calculated rate constant ratio per bond varies[2] from 4.0 at $E_D^+ = 0$ to 1.4 at $E_D^+ \rightarrow \infty$; again a normal isotope effect is predicted.

Such effects have been measured experimentally for the decompositions of chemically activated ethyl-d_1, -d_2 and -d_3 radicals produced by the addition of H or D atoms to C_2H_4 or *trans*-CHD:CHD[11,12], e.g.:

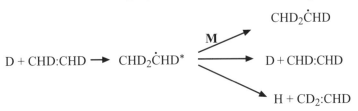

The measured isotopic rate-constant ratios (k_H/k_D per bond) were for ethyl-d_2 10–18 at 195 K and 6–8 at 300 K, and for ethyl-d_3 3.0 at 195 K and 2.0 at 300 K, and were in reasonable agreement with the calculated values. The larger effects for the d_2 case result mainly from the low E^+ values for D-rupture from $CH_2D\dot{C}HD^*$ formed by H-atom addition to CHD:CHD[11].

Intramolecular isotope effects which are largely determined by critical energy differences are particularly important for the decompositions of low-energy ions such as metastable ions which have energies only slightly above their critical energies. Since these ions are generally produced with a narrow range of internal energies, the rate constant ratio is similarly given by the ratio of the sums of states of the respective transition states.

Experimental data on the isotope effects for the decomposition or rearrangement of ions have been reviewed by Derrick and Donchi[13]. These authors have extensively discussed the difficulties in interpreting the sometimes large isotope effects which are observed upon ion abundancies determined by mass spectrometry. Competing processes, isotopic randomisation and isotopically dependent production of ions which affect the energy distribution of the reacting species can complicate the interpretation and obscure pure kinetic isotope effects, particularly for intermolecular comparisons.

The production of isotopically labelled metastable ions with a narrow distribution of energies is frequently achieved by the method of photoion–photoelectron coincidence (PIPECO)[14]. Recent work[15] using a two-laser pump method combined with a reflectron time of flight mass spectrometer has enabled the study of intramolecular isotope effects for various H atom-decay channels of energy-selected benzene cations to be carried out. Measured isotope effects are found to depend strongly on internal energy.

10.5   ISOTOPE EFFECTS IN THERMAL UNIMOLECULAR REACTIONS (THEORY AND EXPERIMENT)

It has already been noted that the relatively simple isotope effects for monoenergetic systems become more complex in thermally activated systems because of changes in the energy distribution of the reacting molecules on isotopic substitution. The general equation (10.9) for calculations in the fall-off region is complicated, but considerable insight can be gained from the limiting high-pressure and low-pressure equations (10.10) and (10.11).

Again the discussion is simplest for a purely secondary (statistical-weight) isotope effect. In this case $(E_0)_H = (E_0)_D$, and with adiabatic rotation factors ignored, the limiting forms (10.10) and (10.11) become (10.15) and (10.16):

$$\frac{(k_\infty)_H}{(k_\infty)_D} = \frac{Q^+_{2H}/Q_{2H}}{Q^+_{2D}/Q_{2D}} \tag{10.15}$$

$$\frac{(k_{bim})_H}{(k_{bim})_D} = \frac{Q^{*'}_{2H}/Q_{2H}}{Q^{*'}_{2D}/Q_{2D}} \tag{10.16}$$

Considering first the low-pressure limit (10.16), it is found that both partition functions Q_2 and $Q^{*'}_2$ are greater for the deuteriated compound, but that the ratio $Q^{*'}_{2D}/Q^{*'}_{2H}$ is considerably greater than the ratio Q_{2D}/Q_{2H}. This is because the density of quantum states increases on deuteriation more at the higher energies important in $Q^{*'}_2$ than at the lower energies appropriate to Q_2. The effect is clearly seen as a divergence of the $\log[\rho(E)]$ plots with increasing energy (Figure 10.5), and its possible numerical magnitude is indicated by the figures for reactions (10.5) and (10.6) at 300 °C with $E_0 = 146 \text{ kJ mol}^{-1}$:

$$\frac{Q_{2D}}{Q_{2H}} = 4.17 \qquad \frac{Q^{*'}_{2D}}{Q^{*'}_{2H}} = 14.5, \qquad \frac{(k_{bim})_H}{(k_{bim})_D} \approx 0.29$$

This effect at low pressures is simply a change in the rate of production of energised molecules, all of which react. The rate of energisation into a small energy range $E^* \to E^* + \delta E^*$ is given by

$$\frac{\delta k_{1(E^* \to E^* + \delta E^*)}}{k_2} = \frac{\rho(E^*) \exp(-E^*/kT)\delta E^*}{Q_2} \tag{3.5}$$

Since $\rho_D(E^*) > \rho_H(E^*)$ and the difference is only partially compensated by the inequality $Q_{2D} > Q_{2H}$, there are more deuteriated molecules than normal molecules excited into any given energy range, and the measured rate of reaction is accordingly greater for the deuterio compound. Thus an inverse isotope effect is predicted as a result of statistical-weight effects for unimolecular reactions in the non-equilibrium situation pertaining at low pressures. Such an effect contrasts with the normal statistical-weight isotope effect found in chemical activation experiments (section 10.3); it was originally predicted by Rabinovitch and co-workers[16] and has subsequently been found experimentally in several systems (see next section). It has been emphasised[17,18] that this is purely a statistical-weight effect and can be very

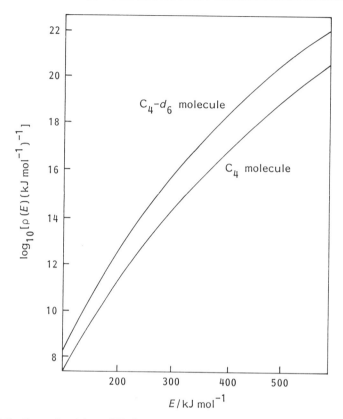

Figure 10.5   State densities $\rho(E)$ for the C_4 and C_4-d_6 molecules by the Whitten–Rabinovitch approximation[2]. Reproduced by permission of John Wiley and Sons Ltd.

much larger than the 'conventional secondary isotope effects', which are usually measured under equilibrium conditions and are due mainly to critical energy differences (cf. section 10.1).

For the high-pressure limit (10.15) may be used. Both Q_2 and Q_2^+ are larger for the deuterio-compound, but this time the effects are roughly compensating. At 300 °C, $Q_{2D}^+/Q_{2H}^+ = 4.8$ for reactions (10.5) and (10.6), giving $(k_\infty)_H/(k_\infty)_D = 0.86$, a small inverse effect compared with the ratio 0.29 at low pressures. It appears that at high pressures the statistical-weight isotope effect may be either normal or inverse but that the rate constant ratio will in any event be close to unity; the experimental results discussed in the next section are in line with this conclusion. The situation can alternatively be discussed in terms of the following equation:

$$k_\infty = \int_{E^*=E_0}^{\infty} k_a(E^*) \frac{\mathrm{d}k_{1(E^*\to E^*+\delta E^*)}}{k_2} \tag{10.17}$$

For any given energy range $E^* \rightarrow E^* + \delta E^*$ it has been seen that $k_{aH}(E^*)/k_{aD}(E^*)$ is substantially greater than unity, while $(\delta k_1/k_2)_H/(\delta k_1/k_2)_D$ is substantially less than unity. The concentration of energised molecules in a given range is higher for the deuterio-compound, but the rate constant for their reaction is lower, resulting in about the same overall rate of reaction.

Finally, when primary isotope effects are present there is usually a general increase in $(k_{uni})_H/(k_{uni})_D$ at all pressures relative to the curve predicted from the statistical-weight effect alone. Values substantially greater than unity may be found at high pressure, and although the ratio will decline as the pressure is reduced, it may or may not invert (i.e. becomes less than unity) at low pressures.

Experimental data for some isotope effects in unimolecular reactions which have been measured over a substantial pressure range are given in Table 10.6. When the measured activation energies for isotopically substituted compounds are the same within experimental error, the results reflect purely statistical-weight effects. This is the case for the extensively studied isomerisations of alkyl isocyanides[17,19]. The difference in critical energy for CH_3NC and CD_3NC is virtually zero, and as predicted there is a weak isotope effect at high pressures $((k_\infty)_H/(k_\infty)_D = 1.1)$ but a large inverse effect at low pressures.

When primary intermolecular effects predominate, k_{light}/k_{heavy} generally falls with pressure as predicted, but complications arise if wall energisation occurs at low pressures as was found for cyclopropane-d_6 at pressures below 0.01 Torr[20].

The primary intramolecular effects for CHD_2CD_2Cl[21] and CHD_2CD_2Br[22] are found to increase as the pressure is decreased. This is understood in terms of equation (10.9) which shows the dependence of the isotopic effect on the distribution functions for the energised species. The transition states for D elimination have a higher average energy than those for H elimination and their concentration therefore falls more rapidly with pressure causing an increase in $k_{(light)}/k_{(heavy)}$ at low pressures. Similar effects have been observed more recently for CH_2DCH_2Cl[23] and CH_2DCH_2Br[24].

Recent experimental work on isotope effects in unimolecular and related reactions has been carried out for a variety of reasons. Termolecular recombination reactions involving small radicals or atoms are capable of treatment in the same theoretical terms employed in this chapter largely for dissociation or isomerisation processes. Alternatively the statistical adiabatic channel method of Quack and Troe[25] or the approximate Troe method[26] may be used. Since these processes are often studied in the fall-off region, interpretation of the results is complicated.

An example of such a study is that by Brouard and Pilling[27] on the reaction between CH_3 and H/D. The $CH_3 + H$ reaction is well into the fall-off region at pressures below 1 atm and temperatures in the range

Table 10.6 Isotope effects in thermally activated unimolecular reactions studied over a range of pressures

Compound	Isotopes involved	Temperature range/°C	ΔE_∞/ kJ mol^{-1}[a]	Temperature/°C	k(light)/k(heavy) at pressures (Torr) in parentheses		Refs
Secondary intermolecular effects							
MeNC-d_1	H/D	—	(0.0)	245	1.0 (10^4)	0.7 (0.1)	34
MeNC-d_3	H/D	180–250	0.0	230	1.1 (10^4)	0.3(0.005)	17, 19
EtNC-d_5	H/D	190–260	0.0	231	1.1 (90)	0.2(0.007)	18
Cyclobutane-d_8	H/D	419–460	3.8	449	1.4 (100)	0.8(0.005)	35
Cyclobutene-d_6	H/D	140–180	4.6	151	1.3 (10)	0.8 (0.01)	36
Primary intermolecular effects							
MeNC-^{13}C	^{12}C/^{13}C	226–243	0.08	226	1.018 (10^3)	1.011(10)[b]	37
Cyclopropane-^{13}C	^{12}C/^{13}C	450–514	0.08	514	1.012 (10^3)	1.003(1)[c,d]	38
Cyclopropane-d_6	H/D	407–514	5.4	510	2.0 (760)	1.0 (0.01)	39, 20
Cyclopropane-t_1	H/T	406–492	1.7	490	1.1 (10^3)	1.0(0.5)[c,e]	e
C$_2$D$_5$Br	H/D	458–692	9.6	553	2.0 (115)	1.8 (6)[f]	22
C$_2$D$_5$Cl	H/D	440–483	10.0	482	2.54 (100)	1.96 (0.1)	40
CH$_3$CD$_2$Cl	H/D	441–482	0.4	482	1.03 (100)	0.80 (0.1)	40
Oxetan-2,2-d_2	H/D	400–485	3.3	459	1.35 (75)	1.22 (0.2)	41
Primary intramolecular effects							
CHD$_2$CD$_2$Cl	H/D	485–716	4.2	559	2.1 (150)	2.4 (0.7)[g]	21
CHD$_2$CD$_2$Br	H/D	425–726	3.8	553	2.1 (200)	2.2 (7)[f,g]	22
CH$_2$DCH$_2$Cl	H/D	397–827	5.1	564	1.8 (25)	2.1 (1.0)[g]	23
CH$_2$DCH$_2$Br	H/D	387–433	4.1	433	1.9 (240)	2.6 (0.2)[g]	24

[a] E_{Arr}(heavy) $- E_{Arr}$(light) at a reasonably high pressure. [b] Observed ratios for CH$_3$NC versus the mixture of ^{13}CH$_3$NC and CH$_3$N^{13}C molecules. [c] Similar trends were also measured at other temperatures. [d] Ratios are those for ^{12}C–^{12}C versus ^{13}C–^{12}C fission derived from the measured rate ratios. [e] No apparent temperature dependence. Ratios are those observed for the whole molecules (C$_3$H$_6$ versus C$_3$H$_5$T), Lindquist and Rollefson[42]; Weston[43]. [f] Results for inhibited reaction. [g] Isotope effect per β-CH or CD bond.

300–600 K and the extrapolation to k_∞ is long and inaccurate. To try to overcome this problem, they studied the reaction between CH_3 and D, which has an isotope exchange channel in addition to the usual association and stabilisation channels:

$$CH_3 + D \underset{k_{aD}(E^*)}{\rightleftharpoons} CH_3D^* \xrightarrow{k_{aH}(E^*)} CH_2D + H$$
$$\Big\downarrow k_2[M]$$
$$CH_3D$$

The threshold energies for the H and D elimination channels differ by 570 cm^{-1} and $k_{aH}(E^*) \gg k_{aD}(E^*)$ at all significant energies. Thus every time D reacts with CH_3 it is removed, either by stabilisation or by H elimination from CH_3D^*; measurement of the rate of the $CH_3 + D$ reaction, therefore, corresponds to measurements of the rate of forming CH_3D^* or $k_{D,\infty}$, the high-pressure limiting rate constant for $CH_3 + D$. This prediction was confirmed experimentally: the rate constant for the $CH_3 + D$ reaction was found to be independent of pressure and gave $k_{D,\infty} \approx 1.65 \times 10^{-10}$ cm^3 mol^{-1}s^{-1} over the temperature range 300–400 K. It is relatively simple to calculate $k_{H,\infty}$ the limiting high-pressure rate constant for $CH_3 + H$, from $k_{D,\infty}$ by the use of CTST. Neglecting vibrational frequency changes and assuming equivalent geometries from CH_4 and CH_3D gives

$$\frac{k_{H,\infty}}{k_{D,\infty}} = \left\{ \frac{m_{CH_4}}{m_{CH_3D}} \right\}^{3/2} \left\{ \frac{m_D}{m_H} \right\}^{1/2} \frac{\mu_{CH_4}}{\mu_{CH_3D}} \approx \sqrt{2}$$

A detailed variational TST treatment of the two reactions, including the vibrational degrees of freedom qualitatively confirms this ratio, giving $(k_{H,\infty}/k_{D,\infty}) = 1.36 - 1.40$, so that, based on the $CH_3 + D$ experiments, $k_{H,\infty}$ is calculated to be $\approx 2.3 \times 10^{-10}$ cm^3 mol^{-1}s^{-1}, over the temperature range 300–400 K.

Problems were encountered, however, in trying to fit the experimental fall-off data at 300 K to this value for $k_{H,\infty}$. Too high a curvature was required and much better fits were obtained with $k_{H,\infty}$ (300 K) $\approx 4.5 \times 10^{-10}$ cm^3 mol^{-1}s^{-1}. The anomaly has not, as yet, been resolved.

Further work on recombination reactions, e.g. OH and OD with NO and NO_2[28] has been carried out primarily because of the importance of such reactions for atmospheric chemistry. Isotope effects can be of great help in establishing reaction mechanisms. Some recent examples include the study of the 1–5 sigmatropic H-shift in 1,3-pentadiene[29], the 1,3-sigmatropic shift in trans 1-methyl-2-(1-tbutylethenyl)cyclopropane[30] and evidence for a stepwise mechanism in the fragmentation of t-butoxide anions[31]. Deuterium and ^{18}O isotope effects calculated from measured mass spectra ion abundancies have been interpreted in favour of a concerted elimination of acetaldehyde from diethoxyxylene radical cations[32].

Isotopic selectivity in infrared multiphoton dissociation of small molecules such as $CDCl_3$ and $CHCl_3$ may be more a consequence of effects on the absorption of radiation such as anharmonicity and 'rotational bottlenecks' than of pure kinetic isotope effects[33]. The usefulness of a study of isotope effects in clarifying the nature of potential energy surfaces is illustrated by recent work by Chickos[44] on the thermolysis of 2,2-dimethyl-1-vinyl cyclobutane. Secondary intermolecular deuterium isotope effects here are interpreted in terms of a relatively flat potential energy surface adjacent to the tetramethylene biradical intermediate. Isotope effects are also frequency used as a check on details of proposed potential energy surfaces when combined with measurements of kinetic energy release in product species, for example in the decomposition of positive ions in the mass spectrometer[45].

Among recent theoretical applications are the use of isotopic ratios in characterisation of transition states for radical decompositions[46] and an important paper by Pollak and Schlier[47] which explains isotopic selectivity found by Carrington and McNab[48] in the decomposition of the D_2H^+ ion in terms of total angular momentum barriers.

REFERENCES

1. J. Bigeleisen and M. Wolfsberg, *Adv. Chem. Phys.*, **1**, 15 (1958); L. Melander, *Isotope Effects on Reaction Rates*, Ronald Press, New York (1960); K. J. Laidler, Chemical Kinetics, 3rd edn. Harper and Row, New York, p. 427; (1987); K. Wiberg, *Chem. Rev.*, **55**, 713 (1955).
2. B. S. Rabinovitch and D. W. Setser, *Advan. Photochem.*, **3**, 1 (1964).
3. E. B. Wilson, J. C. Decius and P. C. Cross, *Molecular Vibrations*, McGraw-Hill, New York, section 8.5 (1955).
4. W. E. Buddenbaum and P. E. Yankwich, *J. Phys. Chem.*, **71**, 3136 (1967).
5. B. S. Rabinovitch and J. H. Current, *Can. J. Chem.*, **40**, 557 (1962).
6. J. W. Simons, B. S. Rabinovitch and R. F. Kubin, *J. Chem. Phys.*, **40**, 3343 (1964), and references cited therein.
7. W. E. Falconer, B. S. Rabinovitch and R. J. Cvetanovic, *J. Chem. Phys.*, **39**, 40 (1963).
8. M. J. Pearson, B. S. Rabinovitch and G. Z. Whitten, *J. Chem. Phys.*, **42**, 2470 (1965).
9. J. W. Simons and B. S. Rabinovitch, *J. Phys. Chem.*, **68**, 1322 (1964).
10. K. Dees and D. W. Setser, *J. Chem. Phys.*, **49**, 1193 (1968), and references cited therein.
11. J. H. Current and B. S. Rabinovitch, *J. Chem. Phys.*, **38**, 783 (1963), and references cited therein.
12. J. H. Current and B. S. Rabinovitch, *J. Chem. Phys.*, **38**, 1967 (1963).
13. P. J. Derrick and K. F. Donchi in *Comprehensive Chemical Kinetics*, **24**, 53 (1983).
14. T. Baer in *Gas Phase Ion Chemistry*, Academic Press, New York, Vol. 1, pp. 153ff. (1979).

15. H. Kuhlewind, A. Kiermeier, H. J. Neusser and E. W. Schlag, *J. Chem. Phys.*, **87**, 6488 (1987).
16. B. S. Rabinovitch, D. W. Setser and F. W. Schneider, *Can. J. Chem.*, **39**, 2609 (1961).
17. F. W. Schneider and B. S. Rabinovitch, *J. Amer. Chem. Soc.*, **85**, 2365 (1963).
18. K. M. Maloney, S. P. Pavlou and B. S. Rabinovitch, *J. Phys. Chem.*, **73**, 2756 (1969).
19. F. J. Fletcher, B. S. Rabinovitch, K. W. Watkins and D. J. Locker, *J. Phys. Chem.*, **70**, 2823 (1966).
20. B. S. Rabinovitch, P. W. Gilderson and A. T. Blades, *J. Amer. Chem. Soc.*, **86**, 2994 (1964).
21. A. T. Blades, P. W. Gilderson and M. G. H. Wallbridge, *Can. J. Chem.*, **40**, 1526 (1962).
22. A. T. Blades, P. W. Gilderson and M. G. H. Wallbridge, *Can. J. Chem.*, **40**, 1533 (1962); A. T. Blades, *Can. J. Chem.*, **36**, 1043 (1958).
23. P. J. Papagiannakopoulos and S. W. Benson, *Int. J. Chem. Kinet.*, **14**, 63 (1982).
24. K-H. Jung, S. H. Kang, C. U. Ro and E. Tschuikow-Roux, *J. Phys. Chem.*, **91**, 2354 (1987).
25. M. Quack and J. Troe, *Ber. Bunsenges. Phys. Chem.*, **78**, 240 (1975).
26. J. Troe, *J. Chem. Phys.*, **66**, 4758 (1977), see also *J. Chem. Phys.*, **75**, 226 (1981).
27. M. Brouard and M. J. Pilling, *J. Phys. Chem.*, **93**, 4047 (1989).
28. I. W. M. Smith and M. D. Williams, *J. Chem. Soc. Farad. Trans 2*, **81**, 1849 (1985).
29. M. J. S. Dewar, E. F. Healey and J. M. Ruiz, *J. Amer. Chem. Soc.*, **110**, 2666 (1988).
30. J. J. Gajewski and M. P. Squicciarini, *J. Amer. Chem. Soc.*, **111**, 6717 (1989).
31. W. Tumas, R. F. Foster, M. J. Pellerite and J. I. Braumann, *J. Amer. Chem. Soc.*, **109**, 961 (1987).
32. C. E. Allison, M. B. Stringer, J. H. Bowie and P. J. Derrick, *J. Amer. Chem. Soc.*, **110**, 6291 (1988).
33. M. L. Azcárate and E. J. Quel, *J. Phys. Chem.*, **93**, 697 (1989).
34. B. S. Rabinovitch, P. W. Gilderson and F. W. Schneider, *J. Amer. Chem. Soc.*, **87**, 158 (1965).
35. R. W. Carr and W. D. Walters, *J. Amer. Chem. Soc.*, **88**, 884 (1966).
36. H. M. Frey and B. M. Pope, *Trans. Farad. Soc.*, **65**, 441 (1969).
37. J. F. Wettaw and L. B. Sims, *J. Phys. Chem.*, **72**, 3440 (1968).
38. L. B. Sims and P. E. Yankwich, *J. Phys. Chem.*, **71**, 3459 (1967).
39. A. T. Blades, *Can. J. Chem.*, **39**, 1401 (1961).
40. R. Jonas and H. Heydtmann, *Ber. Bunsenges, Phys. Chem.*, **82**, 823 (1978).
41. L. Zalotai, Zs Hunyadi-Zoltan, T. Bércès and F. Márta, *Int. J. Chem. Kinet.*, **15**, 505 (1983).
42. R. H. Lindquist and G. K. Rollefson, *J.Chem. Phys.*, **24**, 725 (1956).
43. R. E. Weston, *J. Chem. Phys.*, **26**, 975 (1957).
44. J. S. Chickos, *J. Chem. Soc. Perk. Trans II*, 1109 (1987).
45. D. H. Williams, *Acc. Chem. Res.*, **10**, 280 (1977).
46. M. A. Grela and A. J. Colussi, *Int. J. Chem. Kinet.*, **19**, 869 (1987).
47. E. Pollak and C. Schlier, *Acc. Chem. Res.*, **22**, 223 (1989).
48. A. Carrington and I. R. McNab, *Acc. Chem. Res.*, **22**, 218 (1989).

11 Experimental Data

The last 20 years have seen an enormous increase in the number of published rate constants for unimolecular reactions. In 1972 it was possible to give a fairly comprehensive coverage of thermal unimolecular data in one chapter of the first edition of this book. While some reviews have appeared in the interim[1,2] providing supplementary information, it is no longer possible to give a comprehensive compilation within the scope of the present chapter. It may well be that in future the main sources of primary data will be databases such as those currently produced by the CODATA[3] group and the National Institute of Standards and Technology[4].

Thermal energisation continues to be an important method of obtaining accurate and reliable high-pressure parameters. Section 11.1 provides some recent examples of work both at high pressures and in the low-pressure fall-off region, and a tabulation of some of the data. Further sections deal with bond fission reactions and the decomposition and isomerisation of free radicals. In section 11.2 other methods of energisation such as the use of shock-tubes, photolytic activation, overtone excitation and infrared multi-photon excitation are illustrated. Large amounts of data are now available on ion decompositions and ion–molecule reactions. Some recent examples are given in section 11.3.

Finally, association reactions, often the reverse of unimolecular processes occurring by loose transition states (see Chapter 6) are also increasingly important, especially in connection with processes occurring in the upper atmosphere and the need for reliable data for computer modelling of complex mechanisms. Some data on recombination processes are given in section 11.4.

11.1 THERMAL ENERGISATION

HIGH-PRESSURE DATA

Thermally energised molecules are produced with a Boltzmann distribution of energies, and kinetic studies of their subsequent reactions produce the thermally-averaged rate constant k at a given temperature. Earlier chapters have covered measurements of the microscopic rate constants $k(E)$ related to the specific energy of the energised species (for example chemical activation studies in Chapter 7).

It is possible to estimate the high-pressure Arrhenius parameters A_∞ and E_∞ necessary from the application of RRKM theory. This can be done by the methods illustrated in Chapter 5 for 'tight' transition states and in Chapter 6 for 'loose' transition states. Alternatively, Benson[5] has shown how by concentrating on the enthalpy and entropy changes in forming the transition state from reactants, and using group contributions and reasonable estimates of changes in molecular parameters such as vibration frequencies and moments of inertia, values of A_∞ and E_∞ for reactions of various types can be relatively quickly predicted. It is found for example that simple fissions of complex molecules into two relatively large fragments normally produce A_∞ values in the range 10^{15}–10^{17} s^{-1}. Similarly, considerations of the loss in internal rotation and lowering in bending frequencies associated with the formation of cyclic transition states can be used to estimate A_∞ values for these processes which are often in the range $10^{13.5 \pm 0.5}$ s^{-1}. These methods form a useful means of checking the extent to which an experimental A-factor is considered to be reasonable, or for estimating A and hence deriving a value of the activation energy when experiments have only been carried out over a limited temperature range. They do not, however, replace careful experimental study when this is possible over a range of temperatures and pressures which enables the high-pressure Arrhenius parameters to be obtained with a high degree of precision. Such studies have been made for the decomposition or isomerisation reactions of a number of classes of compound including alicyclic compounds, alkyl halides, esters and heterocyclic compounds. Some recent examples of work on typical compounds in each of these classes are now given and Table 11.1 presents a summary of the experimentally determined Arrhenius parameters. Attention is focused on conclusions relevant to the structures of the transition states involved and to the use of RRKM calculations where appropriate.

In the thermal isomerisation of *cyclopropane* to propene for example, and also for many simple *alkyl-substituted cyclopropanes*, the results can often be rationalised in terms of biradical intermediates. Ring opening to the biradical is followed by hydrogen migration as shown in the basic scheme:

The high-pressure rate constant for formation of product was calculated[5] as $k = K_{ro} k_{Hmig}$, so that

$$A_\infty = \exp\left(\Delta S^0_{ro}/R\right)\{(ekT/h)\exp\left(\Delta S_{Hmig}/R\right)\}$$

Table 11.1 Some Arrhenius parameters for representative examples of various unimolecular reactions at high pressures (Ref. page 374)

Reactant	Product	$\log_{10} A_\infty/$ s^{-1}	$E_\infty/$ $kJ\,mol^{-1}$	Ref.
Cyclopropane and alkyl-substituted cyclopropanes				
Cyclopropane	Propene	15.5	274.4	1
cis-Cyclopropane-d_2	Propene-d_2	15.1	273.6	2
trans-Cyclopropane-d_2	cis-Cyclopropane-d_2	16.1	272.4	
	But-1-ene	14.4	262.8	3
	cis-But-2-ene	14.2	262.3	
Methylcyclopropane	Trans-But-2-ene	14.6	272.8	
	isobutene	14.3	272.4	
	Overall	14.8	260.7	
	cis-Pent-2-ene	13.9	256.9	
	trans-Pent-2-ene	14.0	256.1	
cis-1,2,-Dimethyl-cyclopropane	2-Methylbut-1-ene	13.9	259.0	4
	2-Methylbut-2-ene	14.1	260.7	
	trans compound	15.3	248.5	
	cis-Pent-2-ene	14.4	266.1	
	trans-Pent-2-ene	14.3	263.2	
trans-1,2-Dimethyl-cyclopropane	2-Methylbut-1-ene	13.9	259.0	4
	2-Methylbut-2-ene	14.1	260.7	
Halogenated cyclopropanes				
Fluorocyclopropane	Fluoropropenes	14.6	255.2	5
Chlorocyclopropane	3-Chloropropene	14.8	235.1	6
Bromocyclopropane	Bromopropenes	13.5	197.9	6
1,1-Dichlorocyclopropane	2,3-dichloropropene	15.1	241.8	7
	1,1-dichloropropene	14.5	250	8
cis-1,1-Dichloro-2,3-dimethylcyclopropane	trans-3,4-dichloro-pent-2-ene	13.7	186.6	9
1 Chloro-cis 2,3 dimethyl cyclopropane	HCl+ penta 2,4-diene	13.9	199.6	10
1-Chloro-trans 2,3 dimethyl cyclopropane	HCl + penta 2,4-diene	13.8	190.2	10
	4-chloropent-2-ene	14.6	199.5	
Trimethylsilylcyclopropane	Allyltrimethylsilane	14.3	242.3	11
	cis and trans-propenyl trimethylsilane	14.9	264.8	
Substitution by unsaturated groups				
Vinylcyclopropane	Cyclopentene	13.5	207.5	12

continued overleaf

Table 11.1   (*continued*)

Reactant	Product	$\log_{10} A_\infty/$ s^{-1}	$E_\infty/$ $kJ\,mol^{-1}$	Ref.
cis-1-methyl-2-vinyl cyclopropane	*cis*-hexa-1,4-diene	11.0	130.5	13
1-Ethynyl-2,2,3,3,tetramethyl cyclopropane	4,4,5 trimethyl-1,2,5 hexatriene	12.07	150.3	14
Other reactions of cyclopropanes				
Perfluorocyclopropane	$C_2F_4 + \ddot{C}F_2$	13.3	161.5	15
Perfluoroallylcyclopropane	Perfluoropenta-1,4-diene $+ \ddot{C}F_2$	14.8	178.7	16
Cyclobutane and alkyl-substituted *cyclobutanes*				
Cyclobutane	Ethene	15.6	261.5	17
Methylcyclobutane	Ethene + propene	15.4	256.1	18
cis-1,2-Dimethyl	*trans*-Compound	14.8	251.5	19
Cyclobutane	Propene	15.5	252.7 ⎱	20
	Ethene + but-2-ene	15.6	263.6 ⎰	
trans-1,2-Dimethyl cyclobutane	*cis*-compound	14.6	256.5 ⎱	
	Propene	15.4	257.7 ⎬	19
	Ethene + but-2-ene	15.5	265.3 ⎰	
Chlorocyclobutane	Ethene + vinyl chloride	14.8	255.6 ⎱	21
	HCl + buta 1,3-diene	13.6	233.0 ⎰	
Halogenated cyclobutanes				
Bromocyclobutane	HBr + buta-1,3 diene	13.6	217.6	22
1,1-Difluorocyclobutane	Ethene + vinyl fluoride	15.6	289.7	23
Oxygen-containing substituents				
Cyclobutanone	Ethene + ketene	14.6	217.6	24
	Cyclopropane + CO	14.4	242.7	25
Cyclobutanol	Ethene + CH_3CHO	15.1	251.5	26
	Ethene + $CH_2{=}CH{-}CH_2OH$	15.2	256.2	27
Substitution by unsaturated groups				
Vinyl cyclobutane	Ethene + buta 1,3-diene	14.9	212.2	28
	Cyclohexene	13.9	203.5	
Methylene cyclobutane	Ethene + allene	15.7	264.8	29
Higher cyclic alkanes				
Cyclopentane	Pent-1-ene	16.8	355	30

Table 11.1 (*continued*)

Reactant	Product	$\log_{10} A_\infty/$ s^{-1}	$E_\infty/$ kJ mol^{-1}	Ref.
Methylcyclopentane	+	16.4	341	31
	+ +	16.6	353	31
	CH$_3$ +	16.1	368	31
Cyclohexane	Hex-1-ene	16.9	361	30
Methylcyclohexane	+	16.4	345	31
	+ + +	16.7	360	31
	CH$_3$ +	16.1	368	31
Unsaturated alicyclic compounds				
Cyclopropene	Propyne	13.3	157	32
1-Methylcyclopropene	But-2-yne + buta 1,3 diene	13.5	157	33
3,3-Dimethylcyclopropene	2-Methyl prop-3-yne + 2-methylbuta 1,3 diene	13.3	156	34
Cyclobutene	Buta 1,3-diene	13.4	137.6	35
1-Methylcyclobutene	2-Methyl-buta 1,3-diene	13.8	146.9	36
3-Methylcyclobutene	Penta-1,*trans*-3-diene	13.5	132.2	37
cis-3,4-Dimethylcyclobutene	*cis*,*trans*-Hexa 2,4 diene	13.9	142.3	38
Cyclopentene	Cyclopenta 1,3 diene + H$_2$	13.4	255	39
2,2,5,5 d_4-Cyclohexene	1,3 Dideuterio-cyclohexa 1,3 diene + D$_2$	12.6	257.7	40
Reverse Diels–Alder reactions				
	+ $\|$	15.2	279	41
		15.2	274	42

continued overleaf

Table 11.1   (*continued*)

Reactant	Product	$\log_{10} A_\infty / \text{s}^{-1}$	$E_\infty / \text{kJ mol}^{-1}$	Ref.
		15.2	259.4	43
		15.1	239.7	44
Other reactions				
		10.9	83.3	45
	$+ H_2$	12.5	180.0	46
Polycyclic compounds				
		13.9	161.9	47
		14.4	164.0	47
		13.4	150.6	48
$-F_6$	$-F_6$	13.2 13.7	115.5 119.2	49 50
		14.0	169.8	51
Alkenes and polyenes *cis/trans* **Isomerism** *trans* CHD:CHD *cis*-MeCH:CHD *cis*-But-2-ene	*cis*-CHD:CHD *trans*-MeCH:CHD *trans*-But-2-ene	13.0 13.2 13.8	272.0 256.5 262.8	52 53 54
Cyclisation				
		11.9	125.1	55

Table 11.1 (*continued*)

Reactant	Product	$\log_{10} A_\infty/$ s^{-1}	$E_\infty/$ kJ mol^{-1}	Ref.
1,5-H shift				
CD$_2$ (structure with H)	CD$_2$H (structure)	11.9	151.9	
				56
D / CD$_2$ (structure)	CH$_2$D / CD$_2$ (structure)	11.9	157.7	
H (structure)	(structure)	10.8	136	57
Cope rearrangement				
CD$_2$ CD$_2$ (structure)	CD$_2$–CD$_2$ (structure)	11.1	148.5	58
(structure)	(structure)	10.6	143.1	
				59
	(structure)	10.5	149.4	
Heterocyclic compounds				
3-Membered rings				
(CF$_2$)$_2$O	COF$_2$ + C̈F$_2$	13.7	132.2	60
O—CF$_3$ (structure)	CF$_3$COMe CF$_3$CH$_2$CHO	13.2 14.5	239.7 259.0	61
Me, Me / C—C / Me, Me / O (structure)	Me$_3$CCOMe MeC(=CH$_2$) OCHMe$_2$ ($\rightarrow$ Me$_2$CO + MeCH=CH$_2$) MeC(=CH$_2$)C(OH)Me$_2$	13.6 13.6 11.8	233.4 235.0 209.8	62
Me$_2$C (structure with N‖N)	N$_2$ + [Me$_2$C̈l] $\rightarrow$ MeCH:CH$_2$	14.0	138.9	63
MeCCl N:N (structure)	N$_2$ + [MeC̈Cl]	14.1	130.1	64

continued overleaf

Table 11.1   (*continued*)

Reactant	Product	$\log_{10} A_\infty/$ s^{-1}	$E_\infty/$ $kJ\,mol^{-1}$	Ref.
4-membered rings				
$(CH_2)_3O$	$C_2H_4 + CH_2O$	15.7	263.7	65
		15.4	259.5	66
(oxetane with CH₃)	$CH_3CH{=}CH_2 + CH_2O$	14.5	249.2	67
	$C_2H_4 + CH_3CHO$	15.7	269.8	
(oxetane with two CH₃)	$(CH_3)_2CH{=}CH_2 + CH_2O$	13.5	222.1	68
	$(CH_3)_2CO + C_2H_4$	15.6	270.6	
$\overline{CH_2CH_2CH_2Si(CH_3)_2}$	$C_2H_4 + [CH_2{=}Si(CH_3)_2]$	15.6	261.5	69
	$\rightarrow \overline{CH_2Si(CH_3)_2CH_2Si(CH_3)_2}$			
Larger rings				
(6-membered O ring)	(ring) $+ H_2$	12.7	202.9	70
(6-membered N–H ring)	(ring) $+ H_2$	12.3	186.6	71
(6-membered S ring)	(ring) $+ H_2$	13.2	229.3	72
(6-membered O ring, unsat.)	$+ CH_2O$	14.3	208.1	73
Alkyl halides				
C_2H_5F	$C_2H_4 + HF$	13.7	248.9	74
C_2H_5Cl	$C_2H_4 + HCl$	14.0	244.3	75
		13.8	241.8	76
C_2H_5Br	$C_2H_4 + HBr$	13.5	225.5	77
		12.5	216.3	78
C_2H_5I	$C_2H_4 + HI$	14.1	220.9	79
		13.7	209.2	80
2-Chloropropane	$C_3H_6 + HCl$	13.6	213.8	81
2-Chloro-2-methyl propane	Isobutene $+ HCl$	13.9	193.3	82

Table 11.1 (*continued*)

Reactant	Product	$\log_{10} A_\infty /$ s^{-1}	$E_\infty /$ kJ mol^{-1}	Ref.
Neopentylchloride	$(CH_3)_2C{=}CHCH_3$ $CH_2{=}C(CH_3)\text{-}C_2H_5$ $\Big\}$ + HCl	13.8	258.7	83
Esters Ethyl acetate	$C_2H_4 + CH_3CO_2H$	12.5	200.0	84
Isopropyl acetate	$C_3H_6 + CH_3CO_2H$	13.4	193.7	85
t-Butyl acetate	$iC_4H_8 + CH_3CO_2H$	13.3	169.5	85
Halogeno-esters $CF_3CO_2C_2H_5$	$C_2H_4 + [CF_3CO_2H]$	12.1	184.1	86
$CF_3CO_2CH(CH_3)_2$	$C_3H_6 + [CF_3CO_2H]$	12.7	171.5	86
$ClCH_2CH_2CH_2CO_2CH_3$	$CH_2{=}CHCH_2CO_2Me$ $+CH_3CH{=}CHCO_2Me+HCl$ $\Big\}$	12.95	224.3	87
	$\begin{array}{c} CH_2\!\!-\!\!C{=}O \\ \mid \qquad \mid \\ CH_2 \quad O \\ \diagdown \quad \diagup \\ CH_2 \end{array} \; + CH_3Cl$	12.82	223.0	87

Among the *halogen-substituted cyclopropanes*, the Arrhenius parameters for the isomerisation of monofluorocyclopropane are comparable with those for the alkyl-substituted cyclopropanes and, in view of the strength of the carbon–fluorine bond, it seems reasonable to assume that an H-atom migration mechanism is also applicable in this case. The chloro- and bromo-derivatives isomerise at substantially greater rates, chlorocyclopropane giving only 3-chloropropene and bromocyclopropane giving 3-bromopropene which isomerises further to 1-bromopropene. In view of the much lower activation energies for these isomerisations compared with that of cyclopropane itself, it has been suggested that migration of the halogen atom rather than a hydrogen atom occurs[6].

For the isomerisation of 1,1-dichlorocyclopropane the main product is 2,3-dichloropropene which cannot be formed from the most stable biradical intermediate $\dot{C}Cl_2CH_2\dot{C}H_2$ and is therefore probably produced by a concerted mechanism[7]. A slower process producing 1,1-dichloropropene probably occurs from the above biradical by hydrogen migration[8]. The thermal isomerisation of *cis*-1,1-dichloro-2,3-dimethylcyclopropane is a thousand

times faster than that of 1,1-dichlorocyclopropane, and produces stereospe-
cifically *trans*-3,4-dichloropent-2-ene with no geometrical isomerisation of
the reactant[9]. It has been concluded that the results support a concerted
mechanism.

Substitution of cyclopropane by unsaturated groups such as vinyl[10] produce
stabilisation of the biradical intermediates and consequent lowering of
activation energies. Ring opening can also occur by 1,5 hydrogen shift as
shown by the conversion of 1-methyl-2-vinyl cyclopropane to *cis*-hexa-1,4-
diene[11]. Recent examples of this type of process include triply-bonded
substituents as in *cis*-ethynyl-2-methyl cyclopropane[12] which is converted to
1,2,5-hexatriene and also 1-ethynyl-2,2,3,3-tetramethyl cyclopropane[13]
which forms 4,4,5-trimethyl-hexa-1,2,5-triene (reaction 11.1).

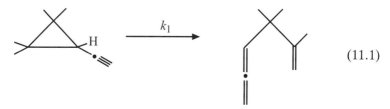

$$\log_{10}(k_1/s^{-1}) = 12.07 \pm 0.07 - 150.3 \pm 0.6 \ (kJ \ mol^{-1})/2.303\mathbf{R}T$$

The latter, however, undergoes a Cope-type rearrangement to produce
5,6-dimethyl-5-hepten-1-yne (reaction 11.2).

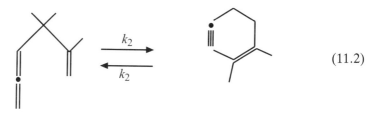

$$\log_{10}(k_2/s^{-1}) = (10.98 \pm 0.81) - (146 \pm 8) \ (kJ \ mol^{-1})/2.303\mathbf{R}T$$

$$\log_{10}(k_{-2}/s^{-1}) = (11.00 \pm 0.80) - (148 \pm 8) \ (kJ \ mol^{-1})/2.303\mathbf{R}T$$

The low *A*-factors associated with this process and its reverse are consistent
with a cyclic transition state as previously postulated for this type of
reaction.

The possibility of other types of process involving substituted cyclopropanes is shown by recent work on silyl cyclopropanes[14].

Silyl cyclopropane isomerises about 10 times faster than simple alkyl
substituted cyclopropanes. In trimethylsilylcyclopropane there is evidence
that the $Si(CH_3)_3$ group migrates, as the reaction produces mainly allyl-

trimethylsilane rather than other products which could also arise from H-migration[14].

Other types of reaction involving cyclopropane derivatives are (a) the rearrangements of methylene cyclopropane[15] and 2,3 gem-difluoromethylene cyclopropane[16], (b) the conversion of cyclopropylamine to N-propylidene cyclopropylamine[17] and (c) the reactions of a number of fully fluorinated compounds. The last category is exemplified by cis-1,2 bis-(trifluoromethyl)-1,2,3,3-tetrafluorocyclopropane[18] which in addition to undergoing unimolecular isomerisation to the $trans$-isomer gives a difluorocarbene elimination as in (reaction 11.3):

$$\qquad\qquad (11.3)$$

with the Arrhenius parameters $\log(A_\infty/\mathrm{s}^{-1}) = 15.4$, $E_\infty/\mathrm{kJ\,mol}^{-1} = 195$. In connection with a series of studies on cyclopropene and its derivatives, Walsh and co-workers[19] have shown that substitution by the $Si(CH_3)_3$ group in the 1-position lowers the A-factor for the major channel in the isomerisation of 3,3-dimethyl-1-trimethylsilyl-cyclopropene to give 3-methyl, 1-trimethylsilyl butyne. Here H-migration appears to be more important than $Si(CH_3)_3$ migration. The interconversion of cyclopropene, allene and propyne is the subject of an important paper by Lifshitz and co-workers[20] discussed in section 11.2.

The thermolysis of cyclobutane and its derivatives has also played an important historical role in the development and application of unimolecular theory. The parent compound decomposes to two molecules of ethene in the temperature range 693–741 K. A concerted mechanism is difficult to reconcile with the high A-factor and is also forbidden on grounds of conservation of orbital symmetry.

The tetramethylene biradical and its analogues appear to explain the observed Arrhenius parameters for cyclobutane and simple substituted cyclobutanes, although the validity of thermochemical estimates has been questioned[21].

Relevant to the biradical mechanism are the relative rates of rotation, cleavage and closure of tetramethylene biradicals which have been directly measured[22] for deuteriated tetramethylene radicals generated by the thermolysis of cis-3,4 d_2-3,4,5,6 tetrahydropyridazine at 439 °C. From the results of these experiments Dervan and Santilli found that at this temperature, rotation is much faster than closure or cleavage and is slowed down by increasing methyl substitution at the biradical centre.

Thermolysis of vinyl cyclobutane[23] shows two reaction paths, one being a ring expansion to cyclohexene analogous to the ring expansion of vinyl cyclopropane, as well as the expected decomposition to ethene and buta-1,3-diene.

The thermal ring-opening reactions of higher cycloalkanes are important in thermal cracking and combustion processes and King and co-workers have reviewed the primary processes which involve ring opening to 1,5 biradicals. Recent work[24,25] has shown that substituted cyclopentanes and cyclohexanes yield a variety of reaction products, thus methyl cyclopentane may produce a variety of open-chain isomers such as hexa-1-ene, and hexa-2-ene from an intermediate C_1, C_2 biradical (11.4):

$$(11.4)$$

and analogous products from C_2C_3 and C_3C_4 biradicals. Bond fission to methyl and cyclopentyl radicals was found to be relatively unimportant at temperatures from 861 to 1218 K. With ethynyl cyclopentane the predominant processes were ring opening at C_1-C_2 to hepta-1-ene-6-yne and hepta-1-yne-3-ene (11.5), but also a propargyl to allene isomerisation (11.6):

$$(11.5)$$

$$(11.6)$$

From very low-pressure pyrolysis (VLPP) measurements the overall rate constant for disappearance of reactant can be obtained and compared with RRKM-calculated values assuming contributions from the possible reaction channels. This is illustrated by Figure 11.1 showing the results of Brown and King for methyl cyclopentane. These workers assigned A-factors and critical energies by analogy with processes in open-chain compounds in performing the RRKM calculations. A slight adjustment of the assumed A_∞ and E_∞ was needed to produce agreement with the observed fall-off curves. The result-

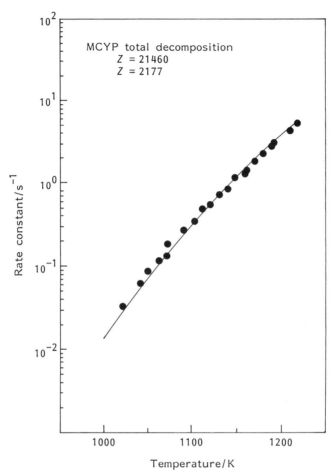

Figure 11.1 Unimolecular rate constant as a function of temperature for the VLPP of methylcyclopentane (from ref. 25) Z = reactor collision number (for explanation see D. M. Golden, G. N. Spokes and S. W. Benson, *Angew. Chem. Int. Ed. (Engl.)*, **12**, 534 [1973]). Points are experimental, solid curve is overall RRKM simulation. From T. L. Brown and K. D. King, *Int. J. Chem. Kinet.*, **21**, 257 (1989), Fig 1. Reproduced by permission of John Wiley & Sons, Inc.

ing high-pressure values A_∞ and E_∞ consistent with the experimental data are given in Table 11.1.

While cyclopropenes are believed to isomerise via biradical or possibly carbene intermediates[19], the unimolecular isomerisation of cyclobutene to buta-1,3-diene may involve a concerted transition state[26]. Higher cyclic alkenes display alternative unimolecular paths; for example cyclohexene decomposes either by H_2 elimination or by the preferred reverse Diels–Alder reaction to ethene and buta-1,3-diene.

There are numerous examples of unimolecular reactions involving poly-cyclic compounds. Release of strain energy often favours the molecular path compared with the radical chain mechanisms involved in straight-chain hydrocarbon decomposition. Some reactions of polycyclic molecules are clearly related to reactions of the cyclopropane, cyclobutane or cyclobutene ring, for example the isomerisation of bicyclo[3.2.0]hept-1-ene (reaction 11.7)

$$(11.7)$$

occurs with Arrhenius parameters very similar to those for the isomerisation of cyclobutene to buta-1,3-diene.

In other cases different possible unimolecular paths may occur and no generalisations are possible. Some examples are given in Table 11.1.

Alkenes and polyenes undergo a variety of unimolecular reactions, most of which are isomerisations typified by the simple *cis-trans*-geometrical isomerisations and molecular rearrangements such as cyclisations and the Cope rearrangement. The high-pressure Arrhenius parameters for a number of such reactions are given in Table 11.1. The simplest *cis-trans-alkene isomerisation* is that of *trans*-1,2-dideuterioethylene[27]. The high-pressure A-factor of $10^{13.0}$ s^{-1} is 'normal' and the high-pressure activation energy of 272 kJ mol^{-1} is close to the value which can be calculated from the potential-energy curves for the ground electronic states of the two isomers. The data for other *cis-trans* isomerisations are in the main quite similar, and it seems likely that the low Arrhenius parameters originally observed in some cases and attributed to non-adiabatic reaction mechanisms were due to surface effects or free-radical reactions.

A number of conjugated polyenes undergo *cyclisation reactions*, an example of such a reaction being (11.8).

$$(11.8)$$

The fact that hexa-1, *cis*-3,5-triene cyclises to cyclohexa-1,3-diene, whereas hexa-1, *trans*-3,5-triene under similar conditions does not cyclise, is taken as evidence that the reaction proceeds through a cyclic transition state rather than a linear biradical intermediate[28].

A cyclic complex is also involved in the rearrangement of 1,3-dienes by *1,5-hydrogen shift*, an example being (11.9)[29].

$$(11.9)$$

The Arrhenius parameters are very similar to those for the conversion of *cis*-1-methyl-2-vinylcyclopropane to *cis*-hexa-1,4-diene, which is thought to occur by a similar mechanism[11].

Many 1,5-dienes undergo unimolecular isomerisation by the *Cope rearrangement*. This reaction can be regarded as the 1,3-shift of any allyl group via a 6-centred activated complex illustrated in (11.10) for 3-methylhexa-1,5-diene[30].

$$(11.10)$$

The loss of entropy associated with the formation of this cyclic complex is assumed to be responsible for the low A-factors ($\approx 10^{10}$ s^{-1}) observed for this type of reaction.

Interest in the thermolysis of heterocyclic compounds has increased in recent years and a number of different types of unimolecular reactions have been identified. Among the oxygen-containing ring compounds studied are the oxiranes and oxetanes. The decomposition of oxirane itself is complex[31], while tetrafluoro-oxirane[32] decomposes to give carbonyl fluoride and difluorocarbene. Various methyl-substituted oxiranes have been studied by Flowers and co-workers[33,34] and these show a variety of concurrent unimolecular paths involving both hydrogen and methyl-shifts after ring opening to give a biradical.

The Arrhenius parameters found for decomposition of oxetane[35] to ethene and formaldehyde have been confirmed by subsequent measurements of Bérces and Mártà and co-workers[36].

Symmetrically substituted oxetanes decompose with Arrhenius parameters similar to oxetane itself and in a manner resembling their cyclobutane analogues. These reactions may be interpreted on the basis of a biradical mechanism. Substitution in the 2-position, however, produces lower A-factors and activation energies and this has been interpreted in some cases[37] as involving concerted processes.

Small ring compounds containing silicon have been studied as precursors to silicon intermediates for synthetic purposes and also to further understanding of knowledge of silicon chemistry.

An early example in this field was the thermolysis of 1,1-dimethylsiletane to produce 1,1,3,3-tetramethyldisiletane by dimerisation of the initially produced 1,1-dimethylsilene[38] (reaction 11.11).

$$CH_2 \!-\!\!-\! Si(CH_3)_2$$
$$CH_2 \!-\!\!-\! CH_2 \qquad \longrightarrow \qquad C_2H_4 \;+\; (CH_3)_2Si = CH_2$$

$$(CH_3)_2Si \!-\!\!-\! CH_2$$
$$H_2C \!-\!\!-\! Si(CH_3)_2$$

$$(11.11)$$

The compound 1-methylsiletane undergoes a similar reaction[39] and both are thought to involve biradical formation by carbon–carbon ring rupture analogous to simple cyclobutanes and oxetanes.

Other analogies with alicyclic carbon compounds have been found including ring expansion of a methylene siletane[40]. Further examples are given in a review of silicon intermediates by Davidson[41] and differences between carbon and silicon thermochemistry are treated in a review by Walsh[42].

Extensive studies have been made by Frey and co-workers[43] of the unimolecular reactions of small ring compounds incorporating nitrogen atoms such as alkyl diazirines and chloroalkyl diazirines. These compounds eliminate nitrogen with the involvement of carbene intermediates. A review by Liu[44] covers the thermolysis and photolysis of diazirines and includes comparison with reactions in solution. Other examples of heterocyclic compounds, some involving larger rings, are included in Table 11.1.

Many alkyl halides decompose by genuine unimolecular reactions involving a 4-centred transition state[45]. The reaction can be represented by

$$-\underset{H}{\overset{\beta}{C}}\!-\!\underset{X}{\overset{\alpha}{C}} \quad \xrightarrow{\text{slow}} \quad \underset{H}{C}\!-\!\overset{+}{C}_{X^-} \quad \xrightarrow{\text{fast}} \quad C\!=\!C \;+\; HX$$

$$(11.12)$$

although a semi-ion-pair rather than an ion-pair transition state may be involved[46,47].

O'Neal and Benson[5] have calculated A-factors for these processes from the canonical transition state theory (CTST) expression

$$A_\infty = (ekT/h)\exp(\Delta S^\ddagger/R)$$

The entropies of activation were calculated by assigning frequencies to the transition state and allowing for reaction path degeneracy. Calculated A-factors agreed to within ± 0.3 log units of experimental $\log_{10} A$ values. These are normally in the range 13.5 ± 0.5 (see Table 11.1) indicating 'tight' transition states.

Activation energies vary more widely, and crudely reflect the carbon–halogen bond dissociation energies, thus for simple alkyl halides the activation energies for fluorides are about (250 ± 20) kJ mol^{-1}, for chlorides (210 ± 20) kJ mol^{-1}, for bromides (190 ± 20) kJ mol^{-1} and for iodides (180 ± 20) kJ mol^{-1}, although Maccoll[48] suggested that heterolytic bond dissociation energies rather than homolytic bond dissociation energies give a better correlation with observed activation energies.

There are problems, however, in isolating genuine unimolecular decompositions. While most alkyl fluorides decompose preferentially by a unimolecular elimination of hydrogen fluoride, alkyl chloride decompositions are frequently complicated by the occurrence of heterogeneous reactions unless the reaction vessel is suitably coated[49]. Activation energies for the unimolecular decomposition of substituted monochlorides reflect the type of substitution around the reaction centre. It is found that α-methylation for example[48] produces an increase in rate equivalent to a lowering of activation energy by about 20–24 kJ mol^{-1} per methyl substitution.

Other work relevant to confirmation of the unimolecular nature of many alkyl halide decompositions is that described in an extensive series of papers by Chuchani[50] and co-workers. This work has been concerned largely with the elucidation of anchimeric assistance of charge development in the transition state and other examples of neighbouring-group participation.

All the alkyl monochlorides (except methyl chloride) are thought to decompose thermally by unimolecular mechanisms in suitably coated vessels, the reactions being first order and unaffected by inhibitors or accelerators. The decline of the first-order rate constants with pressure has been observed in a number of cases and compared with unimolecular reaction rate theories. When more than one chlorine atom is present, chain modes of decomposition may predominate. Of the chlorinated ethanes only 1,1-dichloroethane and ethyl chloride decompose by molecular mechanisms. The chain mechanism is found to be favoured if the radical formed by chlorine atom attack on the parent molecule can easily undergo a chain-propagating step producing alkene[51,52]. For example, reaction (11.13) occurs in the chain decomposition of 1,2-dichloroethane,

$$ClCH_2\dot{C}HCl \rightarrow CH_2{=}CHCl + Cl \qquad (11.13)$$
(from 1,2-dichloroethane)

whereas (11.14) cannot occur easily in a single step and 1,1-dichloroethane decomposes by a unimolecular reaction.

$$Cl_2\dot{C}-CH_3 \xrightarrow{\times}\leftrightarrow CH_2{=}CHCl + Cl \qquad (11.14)$$

Similar arguments have been applied to the thermolysis of alkyl bromides, and radicals analogous to those in (11.13) and (11.14) have been referred to as propagating (P) or stopping (S) radicals[53].

Whereas secondary and tertiary alkyl bromides decompose predominantly by unimolecular mechanisms, primary alkyl bromide decompositions usually give mixed chain and unimolecular mechanisms, and information concerning Arrhenius parameters for the unimolecular channel is frequently obtained from experiments fully inhibited by cyclohexene or propene.

Care is needed in the interpretation of experiments involving the use of conventional radical inhibitors and accelerators in these systems. Conflicting evidence has been presented concerning the inhibition or lack of inhibition by propene in continuous wave (CW)-laser initiated, SF_6 sensitised decomposition of 1,1,1,trichloroethane[54,55]. This difference may be due to differences in the laser conditions used in these experiments.

It has been shown that even if propagating radicals can be produced from the parent molecule, a chain reaction is not a necessary consequence. Huybrechts and Hubin[56] found that the molecule CF_2ClCH_3 when pure, decomposes thermally by a unimolecular elimination (principally to HCl). The chain mechanism, feasible in principle, involving Cl atoms and $CF_2Cl\dot{C}H_2$ radicals appears to be maximally inhibited by conversion of Cl atoms into stopping radicals by the minor product CF_3CH_3:

$$Cl + CF_3CH_3 \rightarrow CF_3\dot{C}H_2(S) + HCl \qquad (11.15)$$

A chain reaction is induced, however, by added CCl_4 and also by the product HCl which are able to convert stopping radicals (S) into propagating radicals (P). Similarly the accelerating effect of HCl upon 1,1,1-trichloro-ethane thermolysis[57] was attributed to the reaction

$$Cl_2C-CH_3(S) + HCl \rightarrow CHCl_2-CH_3 + Cl(P) \qquad (11.16)$$

In *alkyl iodide thermolyses* the unimolecular elimination is

$$RCH_2CH_2I \rightarrow RCH{:}CH_2 + HI \qquad (11.17)$$

This is complicated by an iodine-atom-catalysed elimination which may well be a concerted process (reaction 11.18):

$$(11.18)$$

and by the iodine-atom-catalysed reduction by hydrogen iodide

$$\left.\begin{array}{l} I + RCH_2CH_2I \rightarrow RCH_2CH_2 + I_2 \\ RCH_2CH_2 + HI \rightarrow RCH_2CH_3 + I \end{array}\right\} \qquad (11.19)$$

The overall stoicheiometry of alkyl iodide thermolyses is thus represented by

$$2RCH_2CH_2I \rightarrow RCH{=}CH_2 + RCH_2CH_3 + I_2 \qquad (11.20)$$

Benson[58] has shown that the rate-limiting step is usually either (11.17) or (11.18) and is followed by the rapid reduction (11.19). The molecular elimination appears to be rate-controlling for ethyl, isopropyl and t-butyl iodides. For sec-butyl iodide, the iodine-atom-catalysed reaction (11.19) is important as well as (11.18) in determining the rate, although kinetic parameters relating to the molecular elimination have been derived[59] for the limited temperature range 290–330 °C.

Many other examples of work involving alkyl halide decompositions appear elsewhere in this book. Reference has already been made to work on laser-induced decomposition and other examples are to be found in section 11.2. Many studies have now been made of alkyl halides decomposing by alternative unimolecular channels especially at low pressures. A recent example is the molecule CF_3CH_2Cl produced by chemical activation which shows decomposition by three unimolecular channels[60]. The study of intra-molecular isotope effects exemplified by partially deuteriated alkyl halide decompositions has been discussed in Chapter 10.

The thermolysis of *esters* has been the subject of many comprehensive series of investigations including the work of Smith and collaborators[61], Taylor and co-workers[62] and Chuchani and co-workers[63].

The unimolecular decomposition via a 6-centred transition state was first proposed by Hurd and Blunck[64]:

$$(11.21)$$

This produces (when X represents a simple alkyl group) an alkene and a carboxylic acid. This process is possible for esters containing a hydrogen atom attached to C^β. The observed A-factors for this process are generally in the range $10^{11.5\pm1.5}$ s^{-1} and some examples are given in Table 11.1. A more extensive set of data are to be found in the review by Holbrook[65] which illustrates the range of compounds studied in order to vary the pattern of substitution at carbon atoms C^{α}, C^{β} and C^{γ}.

The main conclusions to be drawn from this work indicate that electron withdrawing groups at C^{γ} enhance the rate as they do at C^{β}, whereas electron donating groups at C^{α} assist the polarisation of the C–O bond and increase the rate.

Further work has extended to a wide range of compounds which decompose via similar 6-centred transition states, these include carbamates[66] (X=NR$_2$, where R = alkyl), carbonates[67] (X=OR), chloroformates[68] (X=Cl) and cyanoformates[69] (X=CN).

For these compounds, the acid produced in (11.21) decomposes further so that the overall reaction is

$$\tag{11.22}$$

A complication is the occurrence of a competing four-centred elimination which can produce alkyl amines, alkyl ethers, alkyl chlorides or alkyl cyanides as indicated in

$$\tag{11.23}$$

This process, which has a lower activation energy than (11.22), is found to be surface catalysed.

Another important class of esters is that of allylic esters[70]. Allyl formates undergo a process not possible for allyl esters of other acids, i.e. a six-centred elimination to give alkenes and carbon dioxide. This is similar to the thermolysis of β–γ unsaturated acids and of vinyl alkyl or allyl ethers which also proceed via 6-centred cyclic transition states.

Allyl esters may also undergo Cope rearrangements which can involve 1–3 or 1–5 allyl shifts.

The thermolyses of some *halogenoesters* have been studied by Chuchani and co-workers. In the case of methyl 4-chloro-butyrate[71], HCl elimination occurs to produce an unsaturated ester and a parallel reaction produces methyl chloride and γ-butyrolactone.

$$ClCH_2CH_2CH_2COOCH_3 \rightarrow CH_2{=}CHCH_2COOCH_3 + HCl \quad (11.24)$$

$$
\begin{array}{l}
CH_2 \!-\! CH_2 \\
| \qquad\quad | \\
CH_2 \quad C{=}O \;\; + CH_3Cl \\
\quad\;\diagdown_O\diagup
\end{array}
\qquad\qquad (11.25)
$$

A different mechanism appears to operate in the thermolysis of ethyl 4-chlorobutyrate[72] which gives initially ethene and 4-chlorobutyric acid. The latter is, however, thermally unstable at 400 °C and decomposes by a rapid consecutive reaction to γ-butyrolactone and HCl.

Many other examples of six-centred eliminations are known and A-factors are commonly in the range $10^{11.5\pm1.5}$ s^{-1}. These reactions include molecular retro-ene reactions of unsaturated hydrocarbons: an example is the recently studied shock-tube thermolysis of buta-1,3-diene[73], the decompositions of vinyl ethers[74], acid anhydrides[75], amides[76] and esters in which the carbonyl double bond is replaced by aromatic double bonds[77].

THERMAL UNIMOLECULAR REACTIONS AT LOW PRESSURES

Whereas the main interest in unimolecular reactions in their high-pressure regions centres on the reaction mechanism, the interest in the low-pressure region is concerned with the change in order, the shape and position of the fall-off curve and in favourable cases with the second-order rate constant. The fall-off characteristics such as $p_{\frac{1}{2}}$ and the shape of the fall-off curve have often been used to test various theories of unimolecular reactions. Energy-transfer processes can also be studied by investigating the relative efficiencies of inert gases in restoring the rate constant to its limiting high-pressure value.

Calculations of RRKM have now been carried out for quite a number of unimolecular reactions. The fall-off curve tests both the assumptions inherent in the RRKM theory and the choice of model for the transition state. Partly for this reason, the fall-off curve in itself is not a very sensitive test of any particular model for the transition state. The low-pressure second-order rate constant, on the other hand, is independent of the properties of the transition state and depends only upon the properties of the energised molecule (see section 3.8).

Table 11.2 lists some of the experimental studies which have been made on unimolecular reactions in their fall-off regions. It is clearly useful to be able to predict the pressure at which the unimolecular rate-constant k_{uni} for a given reaction will begin to decline. All the theories predict that the

Table 11.2 Low-pressure studies of unimolecular reactions with molecular product(s) (Ref. page 376)

Reaction	Product(s)	Temperature/°C	Pressure/Torr	$P_{1/2}$/Torr	Kinetic data	Refs
Cyclopropane	Propene	490 ± 2	0.07–84	5	$\log A_\infty/s^{-1} = 15.5$, $E_\infty/kJ\,mol^{-1} = 274.4$	1–3
Methyl cyclopropane	Butenes	440–490	0.06–200	0.2 (490 °C)	$\log A_\infty/s^{-1} = 15.45$, $E_\infty/kJ\,mol^{-1} = 271.9$	2–5
Ethyl cyclopropane	Pentenes	454–484	0.05–84	~0.01 (468 °C)	$\log A_\infty/s^{-1} = 14.40$, $E_\infty/kJ\,mol^{-1} = 257.7$	6
1,1-dichlorocyclopropane	2,3-Dichloropropene	359–424	0.2–80	2 (424 °C)	$\log A_\infty/s^{-1} = 15.1$, $E_\infty/kJ\,mol^{-1} = 241.8$	7
Cyclobutane	Ethene	449	2×10^{-4}–43	0.4	$\log A_\infty/s^{-1} = 15.84$, $E_\infty/kJ\,mol^{-1} = 264.4$	2, 8
		410–500	5×10^{-5}–20	0.2 (450 °C)		9
Methyl cyclobutane	Ethene + propene	400–450	3×10^{-3}–0.45	0.02	$\log A_\infty/s^{-1} = 15.64$, $E_\infty/kJ\,mol^{-1} = 259.4$	10
			10^{-4}–10	0.02		11
Cyclobutene	Buta, 1,3-diene	130–175	0.02–23	0.5 (150 °C)	$\log A_\infty/s^{-1} = 13.25$, $E_\infty/kJ\,mol^{-1} = 136.8$	12–14
1-Methylcyclobutene	2-Methyl-buta 1,3-diene	150–175	1×10^{-3}–1	0.02 (150 °C)	$\log A_\infty/s^{-1} = 13.84$, $E_\infty/kJ\,mol^{-1} = 146.8$	13, 15
3-Methylcyclobutene	Penta-1, trans-3-diene	124–149	0.01–45	0.06	$\log A_\infty/s^{-1} = 13.53$, $E_\infty/kJ\,mol^{-1} = 132.0$	13, 16

Compound	Products	T range	P range	fall-off	Arrhenius parameters	Ref
Ethyl chloride	Ethene + HCl	402–521	0.2–120	2.8 (521 °C)	$\log A_\infty/\text{s}^{-1} = 14.0$, $E_\infty/\text{kJ mol}^{-1} = 244.3$	17
Methyl cyanide → Methyl isocyanide		200–260, 279–282	0.02–10⁴, 5 × 10⁻³–10	65 (230 °C), —	$\log A_\infty/\text{s}^{-1} = 13.6$, $E_\infty/\text{kJ mol}^{-1} = 160.7$	18, 19
1,1,2,2,-Tetrafluorocyclobutane[a]	$\xrightarrow{k_1}$ $2CH_2{=}CF_2$	542.2	0.03–35	—	$\log A^1/\text{s}^{-1} = 15.34$, $E^1_\infty = 292\ \text{kJ mol}^{-1}$	20
	$\xrightarrow{k_2}$ $CH_2{=}CH_2 + CF_2{=}CF_2$	542.2	0.01–22	0.02	$\log A^2_\infty/\text{s}^{-1} = 15.27$, $E^2_\infty = 308\ \text{kJ mol}^{-1}$	21
Oxetane	$CH_2O + C_2H_4$	459	0.07–52.5	0.52	$\log A_\infty/\text{s}^{-1} = 15.42$, $E_\infty/\text{kJ mol}^{-1} = 259.5$	22
				0.90	$\log A_\infty/\text{s}^{-1} = 15.71$, $E_\infty/\text{kJ mol}^{-1} = 263.7$	23
Oxetan-2-one	$C_2H_4 + CO_2$	262–322	0.27–6.0	~0.17	$\log A_\infty/\text{s}^{-1} = 14.86$, $E_\infty/\text{kJ mol}^{-1} = 180.46$ (extrap)	24
4-Methyloxetan-2-one	$C_3H_6 + CO_2$	209–250	0.10–10.0	~0.04 (estd)	$\log A_\infty/\text{s}^{-1} = 14.39$, $E_\infty/\text{kJ mol}^{-1} = 163.4$	25
(cycloheptatrienone)	$CO^* +$ (triene)	799–995	4.7–612	—	$k_0/[\text{Ar}] = 6.4 \times 10^{12}\ \text{cm}^3\ \text{molecule}^{-1}\ \text{s}^{-1}$, $k_\infty = 5.19 \times 10^4\ \text{s}^{-1}$ (877 °C)	26
(cyclopentenone)	$CO^* +$ (diene)	758–1054	5.7–6.11	—	$k_0/[\text{Ar}] = 1.1 \times 10^{12}\ \text{cm}^3\ \text{molecule}^{-1}\ \text{s}^{-1}$, $k_\infty = 1.38 \times 10^5\ \text{s}^{-1}$ (897 °C)	26

continued overleaf

Table 11.2 (*continued*)

Reaction	Product(s)	Temperature /°C	Pressure/Torr	$P_{1/2}$/Torr	Kinetic data	Refs
	CO* +	448–578	25–742	—	$k_0/[\mathrm{Ar}] = 1.1 \times 10^{11}$ cm^3 molecule^{-1} s^{-1} $k_\infty = 5.25 \times 10^4$ s^{-1} (497 °C)	26
Conformational changes Methyl nitrite CH$_3$–O–N=O (syn) ⇌ CH$_3$–O–N=O (anti) syn ⇌ anti		−39.4 to 0.6	149 CH$_3$ONO + 3000 CO$_2$		$\log(A_\infty/\text{s}^{-1})$ 12.87, $E_\infty/\text{kJ mol}^{-1}$ 51.2	27
		−27 to 15.0	27 CH$_3$ONO + 29 He		$E_0/\text{kJ mol}^{-1}$ 40.3	
Ethylnitrite		−53 to 25	24 C$_2$H$_5$ONO + 4800 Ar		$\log(A_\infty/\text{s}^{-1})$ 12.97, $E_\infty/\text{kJ mol}^{-1}$ 47.4	28
syn ⇌ anti			10.8 to 67.4 (C$_2$H$_5$ONO)	40 (−27 °C)	$\log(A_0/\text{s}^{-1})$ 12.20, $E_0/\text{kJ mol}^{-1}$ 46.3	
n-Propyl nitrite syn ⇌ anti		−38 to 24	11 nC$_3$H$_7$ONO + 50 CO$_2$	<30 (−35 °C)	$\log(A_0/\text{s}^{-1})$ 12.04, $E_0/\text{kJ mol}^{-1}$ 49.9	28
neo-Pentyl nitrite anti ⇌ syn		−33 to 26	1.5 neo C$_5$H$_{11}$ONO + 25 He		$\log(A_0/\text{s}^{-1})$ 13.09, $E_0/\text{kJ mol}^{-1}$ 48.1	28

Compound		Temp. range	Mixture		Energy	Ref.
Cyclohexane chair / boat			$1\,cC_6H_{12} + 6$ to $10^3 SF_6$		$E_\infty/\text{kJ mol}^{-1}$ 52.3	29
		-25 to 0	$1\,cC_6H_{12} + 100\,SF_6$		$E_0/\text{kJ mol}^{-1}$ 46.9	29
Tetrahydropyran (THP) chair ⇌ boat		-38 to -22	$1.3\,THP + 5.5$ to $2057\,SF_6$		$E_\infty/\text{kJ mol}^{-1} = 50.6$	30
			$1.3\,THP + 120\,SF_6$	400 (with SF_6)	$E_0/\text{kJ mol}^{-1} = 47.3$	
Cyclohexylfluoride (CHF) 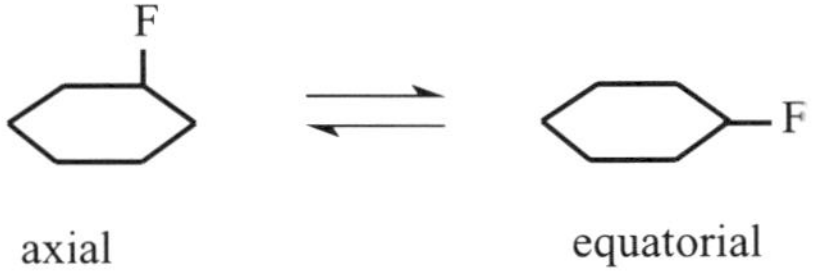		-20.2 to 0	$4.02\,(CHF) +$ $9.96\,CCl_3F +$ $5.6 \rightarrow 2439\,CO_2$	250 (with CO_2) 150 (pure CHF)	$E_\infty/\text{kJ mol}^{-1} = 51.2$ $E_0/\text{kJ mol}^{-1} = 48.9$	31
N,N-dimethylpiperazine (DMPZ) 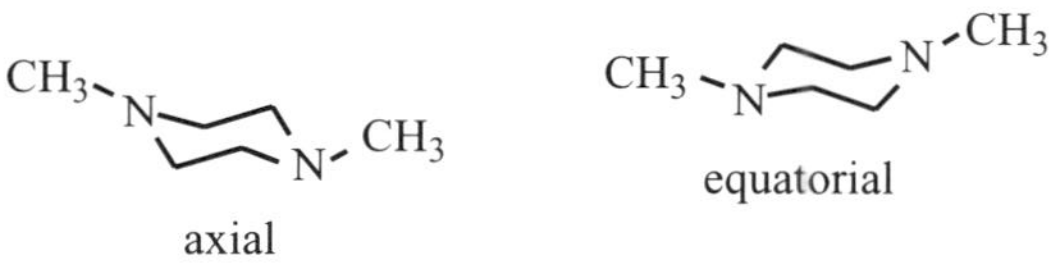		26 to 46	$3\,(DMPZ) +$ 7.6 to $3168\,SF_6$	9 (with SF_6)	$E_\infty/\text{kJ mol}^{-1} = 63.6$ $\log A_\infty/s^{-1} = 14.3$	32

Cyclohexane: chair, boat

Cyclohexylfluoride (CHF):

F (axial) ⇌ F (equatorial)

N,N-dimethylpiperazine (DMPZ):

CH_3–N … N–CH_3 (axial)

CH_3–N … N–CH_3 (equatorial)

---

[a]Fall-off calculations refer to $k_1 + k_2$.

pressure for a given degree of fall-off depends upon the number of modes of vibration involved. From Slater theory, for example, it is seen from equations (2.45) to (2.52) that the fall-off pressure $p_{\frac{1}{2}}$ is proportional to $(E_0/kT)^{-(1/2)(n-1)}$ if one assumes that the terms involving molecular frequencies and also k_2 are approximately constant for different molecules. It then follows that

$$\log p_{\frac{1}{2}} = \text{constant} - \tfrac{1}{2}(n - 1)\log(E_0/kT)$$

Since E_0/kT is usually about 40 at temperatures which are convenient for conventional studies of thermal unimolecular reactions, one would expect an approximate correlation between $\log p_{\frac{1}{2}}$ and $(n - 1)$. In Figure 11.2 some experimental fall-off pressures are plotted as $\log p_{\frac{1}{2}}$ versus the number of atoms (N) in the molecule. The result clearly suggests that $(n - 1)$ is approximately a constant multiple of N. The dangers in assuming Kassel's s to be a constant fraction of the total number of degrees of freedom have been pointed out by Rabinovitch and co-workers[78]. Figure 11.2 shows, however, that the experimental data do conform approximately to this correlation, and this graph is clearly useful in a practical way for predicting the approximate fall-off pressure for any reaction.

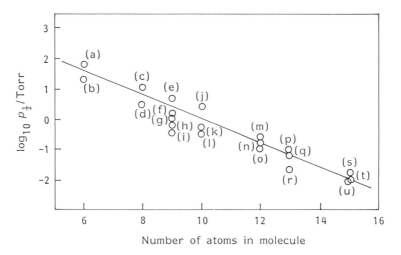

Figure 11.2 Dependence of transition pressure $p_{1/2}$ upon atomicity: (a) methyl isocyanide; (b) CD_3NC; (c) ethane; (d) ethylchloride; (e) cyclopropane; (f) 1,1-dichlorocyclopropane; (g) cyclopropane-d_6; (h) ethyl isocyanide; (i) C_2D_5NC; (j) azomethane-d_6; (k) cyclobutene, (l) cyclobutene-d_6; (m) cyclobutane; (n) methylcyclopropane; (o) cyclobutane-d_8; (p) bicyclo[1:1:1]pentane; (q) 3-methylcyclobutene; (r) 1-methylcyclobutene, (s) methylcyclobutane, (t) 1,1-dimethylcyclopropane, (u) ethylcyclopropane. Reproduced by permission of John Wiley and Sons Ltd.

The experimental study of unimolecular reactions at very low pressures is complicated by the difficulties of measuring the extent of reaction in very small samples. In addition, when the mean free path is greater than the diameter of the reaction vessel, energisation will occur predominantly at the walls rather than in the gas phase; this complication was first treated theoretically by Maloney and Rabinovitch[79], and later developed into a practical method for direct measurement of energy transfer by gas–wall collisions[80]. As a result of these difficulties, the fall-off regions are not easily experimentally accessible for molecules undergoing unimolecular reactions with activation energies from 160 to 250 kJ mol^{-1} at temperatures from 500 to 700 K unless the molecule contains fewer than 10–12 atoms. This fact has restricted the number of examples of unimolecular reactions which have been studied over a wide extent of fall-off. The technique of VLPP[81,82] has, however, allowed the experimentally accessible pressure range to be extended in recent years.

The information obtainable from studies of unimolecular reactions in their fall-off regions depends, as has been shown, on molecular size and therefore how far into the second-order region the study extends. For the larger molecules which decompose into smaller stable molecules quoted in Table 11.2, the principal information is the primary data, the measured or estimated half-pressure ($P_{\frac{1}{2}}$) and the extrapolated high-pressure rate constant (k_∞). Smaller molecules which undergo fission into radicals or atoms can also yield extrapolated values of the second-order rate constant (k_{bim}). Some data of this kind are included in Table 11.3.

The theories used to enable the extrapolation of fall-off data to the high- and low-pressure limits have been dealt with in earlier chapters. Discussion

Table 11.3 Kinetic data in the fall-off region for unimolecular reactions which produce atoms and free radicals (Ref. page 377)

Reactant (+ M)	Product(s) (+ M)	Kinetic data[a]	Ref.
CH_4	CH_3 + H	$\log k_0 = 17.00 - 358.9/\theta$ [M = Ar]	1
CD_4	CD_3 + D	$\log k_0 = 16.32 - 355.2/\theta$ [M = Ar]	2
C_2H_6	$2CH_3$	$\log k_0 = 111.29 - 25.26 \log T - 677.8/\theta$ [M = C_2H_6 or Ar]	3
CH_3Cl	CH_3 + Cl	$\log k_0 = 15.56 - 247/\theta$ [M = Ar]	4
CH_3I	CH_3 + I	$\log k_0 = 15.40 - 178.1/\theta$ [M = Ar]	5
CH_3NH_2	CH_3 + NH_2	$\log k_0 = 13.50 - 147/\theta$ [M = Kr]	6

[a]‘$\log k_0$’ $\equiv \log_{10}(k_0/\text{cm}^3\,\text{mol}^{-1}\,\text{s}^{-1})$, $\theta = 19.147 \times 10^{-3}\,T/\text{K}$.

of the most recent approach based on the approximate treatment of Troe[83] is given in Chapter 8.

Early studies of the fall-off curves of cyclopropanes and cyclobutanes by Wieder and Marcus[84], Lin and Laidler[85], and others showed that curves calculated by RRKM theory on the basis of strong collisions were frequently displaced to lower pressures when compared with experimental data. The effect of weak collisions was at first measured crudely by assuming a constant collision efficiency parameter λ. Subsequent treatments by Tardy and Rabinovitch[86] and by Troe[87,88] have evaluated the efficiency of collisional energy transfer in terms of the collisional efficiency per unit collision β_c and the incremental energy transferred by reactant or added bath gases per collision $\langle \Delta E \rangle$. Many studies of unimolecular reactions in the fall-off region are directed towards measurement of these parameters and are described more fully in Chapter 9.

Since the fitting of the shape and position of the fall-off curve is generally found to be insensitive to most features of the model of the transition state (see Chapter 5), the curve itself is a poor test of the theory once the correct high-pressure Arrhenius parameters have been established. Recent work has therefore often been concerned with the 'fine tuning' of the structure of the transition state by use of supplementary information.

Bérces and Mártà and co-workers[89] have, for example, studied the fall-off curves for oxetane and oxetane-2,2-d_2 in order to probe more deeply the nature of the proposed biradical mechanism.

Bailey and Frey[90] improved the description of the fall-off for 1,1,2,2-tetrafluorocyclobutane at low pressures by measuring the ratio of rate constants for competing reaction channels. It was first pointed out by Chow and Wilson[91] that information enabling distinction between different mechanisms for vibrational energy transfer can be obtained from such measurements. Figure 11.3 shows schematic fall-off curves for two competing channels, channel 2 having the higher activation energy. All other things being equal, the rate constant for the path with the lower activation energy declines first from its infinite pressure value on lowering the pressure. The effect of weak collisions is that more are needed to maintain the equilibrium concentration of energised molecules and maintain a high-pressure rate constant. In the presence of weak collisions, both curves of Figure 11.3(a) are therefore displaced to higher pressures, thus increasing the divergence between them as the pressure decreases. This is reflected in the plot of the ratio of rate constants as a function of pressure in Fig. 11.3(b). This effect was predicted by Chow and Wilson and since confirmed by Waage and Rabinovitch[92] for the structural and geometrical isomerisations of trans-cyclopropane-d_2 and by numerous studies in other systems[93,94]. Complete treatment of multichannel systems must take into account depletion of the concentration of species energised for one channel by the presence of

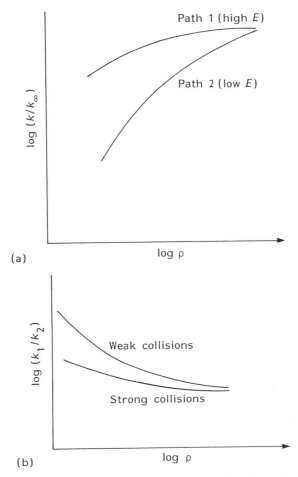

Figure 11.3 (a) Schematic fall-off curves for two reaction channels $(E_1 > E_2)$; (b) ratio of rate constants as a function of pressure for strong and weak collisions.

others, and is most easily done by the master equation method[95] (see Chapter 7).

A study of the thermal decarbonylations of unsaturated cyclic ketones[96] illustrates the use of the shock tube to obtain fall-off data for large molecules at much higher pressures than would be observed with thermal activation. In this work the product CO molecule was monitored by use of a CW CO probe laser which also enabled a study of the CO vibrational distribution. Fall-off curves were fitted by means of the Troe equations[83] permitting evaluation of high- and low-pressure rate constants (Table 11.2). Of particular relevance to unimolecular fall-off behaviour is a series of studies by Bauer and Lazaar[97] and by True and co-workers[98] on conformational changes in suitable molecules.

Alkyl nitrites can exist in syn- and anti-forms with the former generally more stable than the latter. In the case of ethyl, n-propyl and n-butyl nitrites the difference is about 400–800 J mol^{-1}. The rate of interconversion of these forms can be studied at low temperatures in the gas phase and at various pressures from analysis of nuclear magnetic resonance (NMR) line shape. It has been shown that the activation energies for these changes in the above-mentioned molecules are all about 44–48 kJ mol^{-1} and the $P_{\frac{1}{2}}$ values decrease with increasing size of the alkyl group. Since $k(E)$ varies inversely with the density of states $\rho(E)$ (see equation 3.18), the value of $P_{\frac{1}{2}}$ would be expected to decrease in this manner if rapid intramolecular vibrational relaxation occurs and $\rho(E)$ is directly related to molecular size. This postulate has therefore been confirmed for such processes which have low-energy barriers and low densities of states. Similar results have been obtained for ring inversions and some examples are given in Table 11.2. Figure 11.4 shows experimental and calculated fall-off curves for the axial–equatorial proton exchange in N,N-dimethyl piperazine.

These intramolecular isomerisations with low-energy barriers are a particularly sensitive test for the adequacy of conventional RRKM theory. As

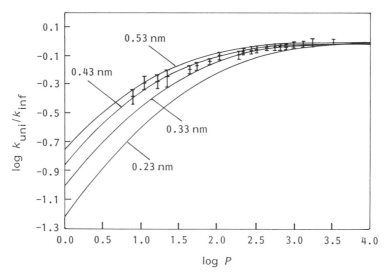

Figure 11.4   Reduced plots of the experimental values of $\log(k_{uni}/k_\infty)$ versus $\log P$ and values (smooth curves) obtained from RRKM calculations. The collision diameter is varied from 0.23 to 0.53 nm. $E_a = 64.9$ kJ mol^{-1}, $E_0 = 61.7$ kJ mol^{-1}, $A = 3.1 \times 10^{14}$ s^{-1}. Reprinted with permission from *J. Phys. Chem.*, **93**, 3993 (1989), Fig. 4a. Copyright 1989 American Chemical Society.

originally developed, the equations of conventional RRKM theory apply only to non-reversible reactions. Lin and Laidler[99] first drew attention to the fact that in an isomerisation reaction the energised product isomer is always sufficiently energised to undergo the back reaction. The isomerisation $AB \to BA$ may therefore be written more explicitly as

$$AB \underset{-1}{\overset{1}{\rightleftharpoons}} AB^* \underset{-2}{\overset{2}{\rightleftharpoons}} BA^* \overset{3}{\to} BA$$

At high pressures BA^* and BA are also in equilibrium. The potential energy profile is shown schematically in Figure 11.5. If the forward reaction is sufficiently exothermic then $k_2 \gg k_{-2}$ and the conventional equations give an adequate treatment.

For reactions which have low endothermicity or are thermally neutral (as mentioned briefly in section 5.5) a more elaborate treatment is necessary involving calculation of both k_2 and k_{-2}. Equations employing the RRKM nomenclature have been given by Lin and Laidler[99]. Bauer[100] has more recently criticised the implicit involvement of a transition state in such a treatment since it is 'inherently a non-observable transient structure' and unnecessary if the process is alternatively treated as a relaxation process. The syn–anti-isomerisation of methyl nitrite is one of two examples chosen as illustrations of the application of the relaxation method. The derived

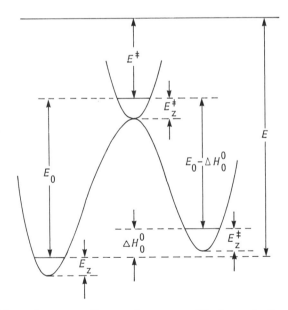

Figure 11.5 Schematic energy diagram for a reversible isomerisation. From M. C. Lin and K. J. Laidler, *Trans. Faraday Soc.*, **64**, 94 (1968). Reproduced by permission of the Royal Society of Chemistry.

fall-off curves differ from those calculated by the modified RRKM procedure and predict slightly lower rate constants, although it is impossible to make a distinction between the two methods of correcting for the back reaction.

BOND FISSION REACTIONS

Unimolecular bond fission reactions producing atoms and/or free radicals often play an important role in complex thermolysis and combustion processes. Knowledge of the relevant high-pressure Arrhenius parameters and fall-off characteristics is therefore essential for computer modelling in such systems.

A much-discussed reaction is the decomposition of ethane into two methyl radicals. Experimental data up to 1980 were reviewed both for the dissociation of ethane and the reverse combination of methyl radicals by Baulch and Duxbury[101]. An early review by Waage and Rabinovitch[102] failed to produce a satisfactory model using RRKM theory and thermodynamic data able to correlate both the dissociation and association data. Since these two reviews there have been further experimental studies notably by Trenwith[103] who carried out an extensive series of experiments in the fall-off region at very low extents of decomposition and measured rate constants based on methane produced in the consecutive reaction steps (11.26) and (11.27):

$$C_2H_6 \rightarrow 2CH_3 \tag{11.26}$$

$$CH_3 + C_2H_6 \rightarrow CH_4 + C_2H_5 \tag{11.27}$$

Skinner, Rogers and Patel[104] have correlated experimental data on ethane decomposition at moderate temperatures (547–717 °C) with shock-tube experiments on ethane in dilute mixtures with inert gas at higher temperatures (727–1027 °C) as well as with room temperature association data. Recent work on the association of methyl radicals is considered in section 11.4.

For the high-pressure rate constant in the range 727–1027 °C the expression of Skinner, Rogers and Patel[104] derived from an RRKM model is

$$\log_{10}(k_1^{\infty}/s^{-1}) = 17.2 - 45\,797/2.303[T/K] \tag{11.28}$$

while that of Baulch and Duxbury[101] is

$$\log_{10}(k_1^{\infty}/s^{-1}) = 16.38 - 44\,010/2.303[T/K] \tag{11.29}$$

The two expressions give identical values for k_1^{∞} at 727 °C. A value approximately three times higher is obtained from the recommended expression[105]

$$k_{\infty} = 1.8 \times 10^{21} T^{-1.24} \exp(-45\,700/T) \text{ s}^{-1} \text{ in the range } 300–2000 \text{ K} \tag{11.30}$$

which is based on the value in

$$k_\infty = 6.0 \times 10^{-11} \text{ cm}^3 \text{ molecule}^{-1} \text{s}^{-1} \text{ (300–2000 K)} \qquad (11.31)$$

for methyl radical recombination and the calculated equilibrium constant.

The high A_∞ value observed for ethane decomposition is characteristic of a loose transition state. Activation energies for bond fissions are normally close to the bond dissociation energies. For a compilation of data on other such processes at high pressures the reader is referred to the first edition of this book and subsequent reviews[1–3].

For small molecules which are generally in their second-order regions at accessible pressures and temperatures, the second-order rate constants are readily measured. Some typical values are given in Table 11.3.

UNIMOLECULAR DECOMPOSITION AND ISOMERISATION OF FREE RADICALS

In common with bond fission processes producing free radicals or atoms, the decompositions of free radicals have often been studied as a part of more complex mechanisms. Decomposition of the ethyl radical, for example, occurs during the thermolysis of ethane[106] and decomposition of the acetyl radical occurs during the photolysis of 3,3-dimethylbutan-2-one[107]. Frequently kinetic data for the decomposition of radicals has been derived from rate measurements relative to the rate of a reference reaction. The decompositions of alkoxy radicals (a), for example, have been measured relative to the rate of their combination with nitric oxide (b)[108]

$$\text{RO} \xrightarrow{k_a} \text{Products}$$

$$\text{RO} + \text{NO} \xrightarrow{k_b} \text{RONO}$$

$$\frac{R_a}{R_b} = \frac{k_a}{k_b[\text{NO}]}$$

This procedure has been criticised by Choo and Benson[109] on the grounds of uncertainties in the rates of some reference reactions.

This criticism notwithstanding, the fall-off curve for the t-butoxy radical was determined by Batt and Robinson[110] relative to the rate of formation of t-butyl nitrite in step (b) and the high-pressure rate constant extrapolated from measurements from 129.3–170.0 °C is expressed by

$$\log_{10} (k_\infty/\text{s}^{-1}) = 14.6 - \frac{66.95 \text{ kJ mol}^{-1}}{2.303RT} \qquad (11.32)$$

The fall-off curves shown in Fig. 11.6 obtained from experiments in the presence of CF_4[108] and subsequently investigated for a series of bath gases[110] agreed well with RRKM calculations[111].

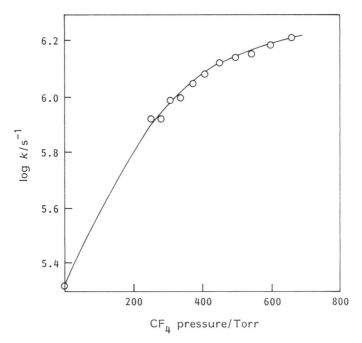

Figure 11.6 Pressure dependence of the rate constant for the unimolecular decomposition of the t-butoxy radical[108]. From L. Batt, *Int. J. Chem. Kinet.*, **11**, 977 (1979). Reproduced by permission of John Wiley & Sons, Inc.

Another radical whose fall-off data have been derived from a complex system is the 1,2-dichloroethyl radical. In the chlorine-catalysed decomposition of 1,2-dichloroethane, the following reaction steps occur:

$$Cl + C_2H_4Cl_2 \rightarrow \dot{C}_2H_3Cl_2 + HCl \tag{a}$$

$$\dot{C}_2H_3Cl_2 \rightarrow C_2H_3Cl_2 + Cl \tag{b}$$

$$\dot{C}_2H_3Cl_2 + Cl_2 \rightarrow C_2H_3Cl_3 + Cl \tag{c}$$

enabling the rate constant for step (b) to be measured relative to that of step (c). The reaction was shown to be in its fall-off region between 8 and 150 Torr at temperatures around 277 °C. Calculations of RRKM by Ashmore, Owen and Robinson[112] involving centrifugal corrections were in good agreement with experimental data[113,114].

Theory and experiment have been combined in a recent study of the decomposition of the methoxy radical by Lin and co-workers[115]. Rate constants have been obtained in two ways, by the thermal decomposition of methyl nitrite from 177 to 427 °C and by shock-wave decomposition of methyl nitrite/argon mixtures at 787–1347 °C. In both cases unimolecular decomposition of methoxy radicals is followed by an H-atom chain decomposition of CH_2O to CO and hydrogen (11.33), (11.34), (11.35):

$$CH_3O + M \xrightarrow{k_1} CH_2O + H + M \qquad (11.33)$$

$$H + CH_2O \rightarrow CHO + H_2 \qquad (11.34)$$

$$CHO + M \rightarrow H + CO + M \qquad (11.35)$$

In the low-temperature experiments the rate constant for step 1 was derived from FTIR measurements of the products and computer modelling of the previously established mechanism. In the shock tube experiments it was derived directly from CO determination by use of a CW CO laser.

Theoretical and experimental values of k_1 are compared on an Arrhenius plot. The calculated curves are for the RRKM results with and without the incorporation of a correction for quantum-mechanical tunnelling.

Tunnelling was considered because *ab initio* molecular orbital calculations revealed the existence of both forward and reverse activation barriers and because of the involvement of the H-atom in motion along the reaction coordinate. Modification of the RRKM calculation involved replacement of the sum of states $W(E_{vr}^+)$ in the equation for the microcanonical rate constant $k(E)$ (equation 3.16) by the quantum-mechanical probability of transmission through the barrier at energy (E_{vr}^+). This modification which is based on the work of Garrett and Truhlar[116] and Miller[117] takes a simple analytical form if an Eckhart potential is assumed[118], the defining parameters for which are obtained from the *ab initio* potential energy surface calculations. Although the data do not permit a clear conclusion on whether tunnelling is important in this case, the predictions from the calculations are of interest. Since tunnelling allows molecules which are collisionally energised below the classical barrier height to react, the rate constant in the second-order region is increased. This is shown by the calculated fall-off curves in Figure 11.7.

It is likely that in future, more work will be directed to the direct determination of rate constants using time-resolved measurements. One way in which this can be done is illustrated by the work of Brouard, Lightfoot and Pilling[119] on the decomposition of the ethyl radical. The equilibrium

$$H + C_2H_4 \rightleftharpoons C_2H_5 \qquad (11.36)$$

has been investigated by excimer laser flash photolysis/resonance fluorescence, enabling direct measurement of both forward and reverse rate constants. An added advantage is that such experiments combined with spectroscopic data permit the more precise determination of heats of formation of radical species. In turn, these may be used to calculate unknown equilibrium constants and thus radical decomposition rate constants from the often more easily measured rate constants of recombination reactions.

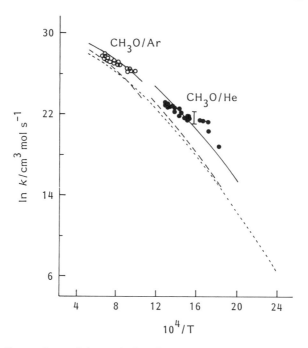

Figure 11.7 Comparison of theoretical and experimental values of the second-order rate constant for the decomposition of the methoxy radical. (Reaction 11.33)[115]. (○) M = Ar, (●) M = He experimental points. Solid and long dash curves are calculated results with and without tunnelling corrections respectively. The dotted curve is the recommended value of Tsang and Hampson, *J. Phys. Chem. Ref. Data*, **15**, 1087 (1986). Reprinted with permission from M. Page, M. C. Lin, Y. He and T. K. Choudhury, *J. Phys. Chem.*, **93**, 4404 (1989), Fig. 1. Copyright 1989 American Chemical Society.

Another recent and more direct study of radical decomposition employing time-resolved measurements is that of Gutman and co-workers[120] for the decomposition of the neopentyl radical

$$neo\text{-}C_5H_{11} \xrightarrow{k} CH_3 + i\text{-}C_4H_8 \tag{11.37}$$

Chlorine atoms (from the excimer laser photolysis of CCl_4) were used to generate the neopentyl radicals from neopentane, in the presence of various bath gases. Decay of the radicals was monitored by photoionisation mass spectrometry.

Fall-off curves were obtained at 10 temperatures between 560 and 650 K in the presence of helium bath gas.

Calculations of RRKM using a vibrational model and incorporating weak collisions gave good agreement with experiment and extrapolated to give the high-pressure rate constant expression

$$k = 10^{13.9 \pm 0.5} \exp\left(-[129.2 \pm 4.2]\,\text{kJ mol}^{-1}/RT\right)\text{s}^{-1} \qquad (11.38)$$

The step-size down for He used was $200\,\text{cm}^{-1}$. This result was combined with calculations of the equilibrium constant to obtain a rate constant expression for the reverse reaction, the non-terminal addition of methyl radical and isobutene which is difficult to study directly as the terminal addition is favoured.

The isomerisation of free radicals is of importance in the quantitative interpretation of hydrocarbon oxidation where it has been shown[121] that at temperatures around 770 K the isomerisation of larger alkyl radicals involving 4- or 5-membered transition states compete effectively with radical decomposition and oxidation. Peroxyalkyl radicals are involved in the oxidation of hydrocarbons. A well-established example is provided by the neopentyl peroxy radical in the oxidation of neopentane.

Baldwin, Hisham and Walker[122] have derived the Arrhenius parameters $\log(A/\text{s}^{-1}) = 13.08$ and $E/\text{kJ mol}^{-1} = 120$ for the 1,5 H-transfer reaction

$$\overset{\displaystyle \overset{\textstyle \dot{C}H_2}{|}}{(CH_3)_3CCH_2\dot{O}_2 \rightarrow (CH_3)_2C\text{—}CH_2OOH} \qquad (11.39)$$

Comparison with the earlier measured[123] values for the ethylperoxy radical isomerisation (11.40) of $\log(A/\text{s}^{-1}) = 13.3$ and $E/\text{kJ mol}^{-1} = (143.7 \pm 10)$ enabled the above-mentioned authors to estimate the changes in A-factors and activation energies expected for a number of 1,5- and 1,4-H transfer processes for alkylperoxy radicals with increasing ring size in the transition state.

$$C_2H_5OO \rightarrow C_2H_4OOH \qquad (11.40)$$

11.2 OTHER METHODS OF ENERGISATION

In the last 20 years, the enormous expansion of experimental data relating to unimolecular reactions precludes a comprehensive review in this book of all the methods of energisation which now exist other than thermal energisation. In this section a few examples of each of some of the more important methods are presented to illustrate the type of work carried out and the information obtained.

ENERGISATION BY SHOCK WAVES

The use of shock tubes either in conventional mode, where the incident shock wave produces chemical reaction in a gas, or using reflected shock waves, or by the single pulse technique is well established and experimental methods[124-126] and theoretical considerations[127] are covered by excellent texts.

Many studies of the kinetics of unimolecular decompositions have been made using the single pulse technique and this has enabled extension of the temperature range of conventionally thermally energised reactions to temperatures up to 10^3–10^4 K. Uncertainties in temperature measurement, typically ± 50 K are minimised by comparative studies using an internal standard of known rate parameters. Examples of such studies are provided by the early work of Tsang on alkyl halide decompositions[128,129], hydrocarbons[130] and alcohols[131].

A more recent example is the work of O'Neal and co-workers[132] on the decomposition of germane in a shock tube at temperatures between 950 and 1060 K and pressures around 4000 Torr. The reaction mixture contained 0.5–1% GeH_4 and small percentages of Xe and cyclopropane as internal mass spectrometric and kinetic standards, together with the major bath gas, argon. Under these conditions the initial reaction was found to be the unimolecular elimination (11.41) which was in its fall-off region.

$$GeH_4 \rightarrow GeH_2 + H_2 \qquad (11.41)$$

RRKM calculations enabled derivation of the high-pressure rate expression

$$\log_{10}(k_\infty/s^{-1}) = 15.5 - 227.2 \text{ kJ mol}^{-1}/2.303\mathbf{R}T$$

Extrapolation to the conditions (~ 600 K, 200 Torr) of previous experiments carried out under static thermal energisation[133,134] gave much lower rates and thus showed that these were probably predominantly heterogeneous.

The isomerisation of cyclopropene to allene and propyne has been studied in shock-tube experiments in the temperature range 800–1200 K by Lifshitz and co-workers[20]. The rate constants k_1–k_4 in the scheme

$$\text{allene} \underset{k_4}{\overset{k_3}{\rightleftharpoons}} \text{cyclopropene} \underset{k_1}{\overset{k_2}{\rightleftharpoons}} \text{propyne} \qquad (11.42)$$

were evaluated by RRKM and CTST calculations based on *ab initio* determinations of the energy levels and vibrational frequencies of the two transition states.

The high-pressure rate constants were used to predict propyne/allene ratios at various temperatures and pressures which were compared with experimental values and also the values obtained by Bailey and Walsh[135] at lower temperatures (466–516 K). The good agreement obtained between these calculated and experimental values over this wide temperature range is shown in Figure 11.8.

In a study of the shock-wave decomposition of dilute mixtures of toluene

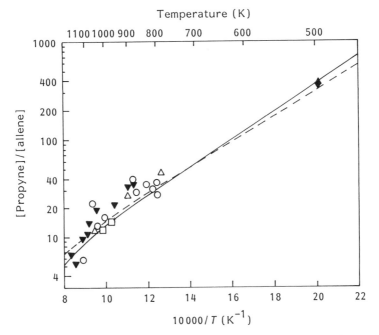

Figure 11.8 Calculated and experimental values of the ratio [propyne]/[allene] from the shock wave isomerisation of cyclopropene at 800–1200 K[20]. Total gas concentrations (experimental points): (▼), $(1-1.2) \times 10^{-5}$ mol cm^{-3}; (○), $(2-2.5) \times 10^{-5}$ mol cm^{-3}; (△), $(4-5.0) \times 10^{-5}$ mol cm^{-3}; (□), $(8-10) \times 10^{-5}$ mol cm^{-3}; (—) calculated (ref. 20); (---) extrapolated using Walsh's[135] values for rate constants; (◆) an experimental point from ref. 135. Reprinted with permission from M. Karni, I. Oref, S. Barzilac-Gilboa and A. Lifshitz, *J. Phys. Chem.*, **92** 6924 (1988), Fig. 1. Copyright 1988 American Chemical Society.

in argon at 1450–1900 K, Troe and co-workers[136] found the main reactions to be

$$C_6H_5CH_3 \rightarrow C_6H_5CH_2 + H \qquad (11.43)$$

$$C_6H_5CH_2 \xrightarrow{\text{fast}} \text{other products} \qquad (11.44)$$

Toluene, benzyl radicals and 'benzyl fragments' were analysed by UV absorption. A competing carbon–carbon fission (11.45) occurred only to the extent of 10%.

$$C_6H_5CH_3 \rightarrow C_6H_5 + CH_3 \qquad (11.45)$$

The high-pressure rate constant for reaction (11.43) was modelled by simplified statistical adiabatic channel theory which gave $k(E, J)$ values which were thermally averaged to yield k_∞. Experimental results obtained from laser activation experiments for $k(E, J)$ were also compared with these calculations[137,138]. The agreement shown in Fig. 11.9 is reasonable in view of the nature of the calculations and is within 30%. The use of experimental data derived from thermal (shock-wave) and laser activation experiments with different rotational energy distributions provides a stringent test of the theory.

In a related study, Müller-Markgraf and Troe[139] studied the shock-wave

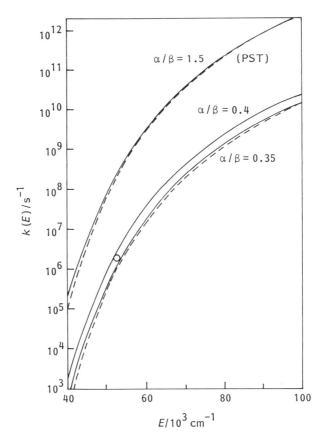

Figure 11.9 Comparison of the laser-activated and shock wave results for the decomposition of toluene from refs 136–8. The solid and dashed lines represent $k(E, J)$ values at $J = 0$ and 100 respectively obtained from the oberved k values for reaction (11.43) from shock-tube data by the use of SACM theory; experimental value derived from laser-activation experiments at room temperature; $k(E) = (1.9 \pm 0.1) \times 10^6 \, s^{-1}$ at average thermal energy plus photon energy and average angular momentum ($\langle J \rangle = 38$), i.e. $\langle E \rangle = 52.7 \, cm^{-1}$ for $\lambda = 193.3$ nm. Reprinted with permission from L. D. Brouwer, W. Muller-Markgraf and J. Troe, *J. Phys. Chem.*, **92**, 4905 (1988). Copyright 1988 American Chemical Society.

decompositions of ethyl benzene, styrene and bromophenyl ethane. In contrast to toluene, the principal mode of decomposition of ethyl benzene was the carbon–carbon fission

$$C_6H_5C_2H_5 \rightarrow CH_3 + C_6H_5CH_2 \qquad (11.46)$$

Other unimolecular processes reported in this work are the decomposition of styrene via a bicyclic intermediate

$$C_6H_5CH{=}CH_2 \rightarrow C_6H_6 + C_2H_2 \qquad (11.47)$$

and the production of styrene by HBr elimination from 1-bromo-1-phenyl ethane.

$$\underset{C_6H_5}{\overset{H}{\diagdown}}C{-}\underset{Br\ H}{\overset{}{CH_2}} \rightarrow C_6H_5CH{=}CH_2 + HBr \qquad (11.48)$$

The rapid fragmentation of benzyl radicals and isomerisation to dibenzyl which occurs in these systems under shock-wave conditions has also been observed in the shock wave decomposition of benzyl iodide[140].

Recent shock-wave decomposition data for the C—H or C—C fissions of a number of aromatic species have been compared by Brand et al.[138] with laser activation experiments similar to those referred to above for toluene. Theoretical calculations using simplified SACM theory have produced very satisfactory agreement within a factor of two in the value of derived rate constants for molecules activated in these two ways.

An example of a study of the unimolecular decomposition of a hetero-cyclic molecule is provided by the work of Lifshitz and co-workers on the single-pulse shock tube decomposition of tetrahydrofuran[141]. This molecule was decomposed in reflected shock waves at temperatures of 1070–1530 K and total concentrations from $(2 \rightarrow 8) \times 10^{-5}\ mol\,cm^{-3}$. The initial ring fission (11.49) is followed by subsequent decomposition of the biradical to yield ultimately CO. In contrast to previous suggestions from a low-temperature study[142], no acetaldehyde was found.

$$(11.49)$$

$$\dot{C}H_2{\diagdown}_{O}{\diagup}\dot{C}H_2 \longrightarrow CH_3 + \left[\dot{C}HO \longrightarrow CO + H \right]$$

$$(11.50)$$

Studies using tetrahydrofuran-d_8 showed a secondary hydrogen isotope effect of 1.65, indicating that ring cleavage is close to the high-pressure limit (see Chapter 10).

The rate expression for the high-pressure rate constant of (11.49) is

$$k(C_2H_4)/s^{-1} = 3.30 \times 10^{16} \exp\left(-347.2 \text{ kJ mol}^{-1}/RT\right)$$

The high pre-exponential factor is typical of this type of cleavage (cf. Table 11.1).

Many smaller molecules when decomposed at high temperatures and pressures in shock tubes are in their second-order decomposition regions and second-order rate constants may be directly derived. Quite frequently the purpose of these studies is to gain information on energy transfer with the bath gas molecules present. The mechanistic information, however, is also of interest and complements and sometimes elucidates problems in low-temperature work. An example is in the thermal decomposition of isocyanic acid in shock waves. In a study of the decomposition of isocyanic acid in argon behind incident shock waves at 2100–2500 K[143], light absorption of HNCO and NH($^3\Sigma^-$) at 206 and 336 nm was used to follow the bimolecular process

$$\text{HNCO} + \text{Ar} \rightarrow \text{NH}(^3\Sigma^-) + \text{CO} + \text{Ar} \qquad (11.51)$$

The rate constant for (11.51) was given by

$$k_{\text{bim}}/\text{cm}^3 \text{ mol}^{-1} \text{ s}^{-1} = 10^{17.23 \pm 0.36} \exp\left(-402 \pm 17\right) \text{ kJ mol}^{-1}/RT$$

In contrast, a low-temperature thermal decomposition study[144] was found to be so complex that the primary step could not be identified.

Many other second-order rate constants have been directly determined in shock-tube experiments for the unimolecular reactions of small molecules. Some examples are HN_3[145], NH_3[146], C_2N_2[147,148], H_2O_2[149], SO_3[150], HCO_2H[151], and the HO_2 radical[152].

PHOTOACTIVATION

Highly vibrational excited molecules prepared by thermal energisation are produced with a Boltzmann distribution of energies whose width depends on the temperature.

Photoactivation in principle is capable of producing a narrower energy distribution since the energy deposited in the molecule is the photon energy plus the thermal energy and the latter is smaller at the lower temperatures normally employed. Photoactivation is a term encompassing a number of techniques including single photon excitation, multiple photon excitation, overtone excitation and, in addition, radicals generated by photolysis are often involved in chemical activation.

Single Photon Activation

Single photon activation employing UV radiation raises the molecules to excited electronic states, and subsequent internal conversion then occurs rapidly by a radiationless process to overlapping high vibrational levels of the ground electronic state molecules. This process is illustrated schematically in Fig. 11.10.

An example of the study of a unimolecular reaction induced by single photon excitation is provided by the work of Atkinson and Thrush[153] who studied the photoisomerisation of cycloheptatriene to toluene

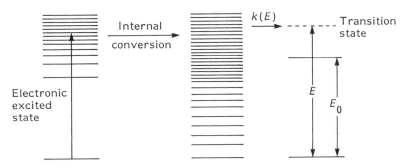

$$ \text{(11.52)} $$

This isomerisation was studied at various wavelengths from 228.8 to 313.0 nm. The initially produced electronically excited cycloheptatriene undergoes a rapid internal conversion to vibrationally excited ground state molecules which then either undergo unimolecular reaction or collisional deactivation. A major goal of this work was to study the relative efficiencies of various deactivating gases (Stern-Volmer quenching) as has been described in similar studies in Chapter 9. The proof that highly vibrationally excited ground state molecules, and not electronically excited singlet or triplet states, were involved was based on a number of lines of reasoning including the fact that specific triplet quenchers showed the same pattern of behaviour as other added gases, and that toluene was the major observed product and not bicyclo 3,2,0-heptadiene which would have been predicted

Figure 11.10 Schematic energy level diagram for single photon activation with internal conversion.

from the excited electronic state by the Woodward–Hofman symmetry rules.

Application of RRKM theory to such a steady-state photoisomerisation resembles the application to chemically activated systems discussed in Chapter 7. The quantum yield (ϕ) was calculated by the above authors[154] from the equation

$$\frac{1}{1 - \phi} = \prod_{i=m+1}^{n} \left(1 + \frac{1}{a_i[M]}\right)$$

where [M] is the effective bath gas concentration and a_i is given by $a_i = Z/k_i(E)$; Z being the collision frequency and k_i the rate constant for vibrationally excited molecules in energy state i.

The product in the equation above is taken over the range of energy states $m + 1$ to n, in increments ΔE, and state m corresponds to the critical energy for isomerisation E_0.

Agreement between experimental and calculated values of ϕ then gives ΔE values for various bath gases corresponding to a set of $k_i(E)$.

Experiments by Troe and Wieters[155] on the photoisomerisations of cycloheptatriene, 7-methylcycloheptatriene and 7-ethylcycloheptatriene with no added bath gases used a different approach. Stern–Volmer plots were calibrated using $k(E)$ values previously obtained from thermal isomerisation studies[156] which provided absolute values of $\langle \Delta E \rangle$ indicating that in self-quenching experiments these molecules were all strong colliders. The resulting derived $k(E)$ values could then be compared with RRKM-calculated values.

More recent experiments[157] on cycloheptatriene and substituted cycloheptatrienes have been used to measure $k(E)$ by a direct, time-resolved method. Excitation of the parent molecules by single photon UV laser and subsequent internal conversion has been used to prepare vibrationally excited ground state molecules. The isomerisation of these molecules has been followed directly using time-resolved UV absorption spectroscopy to follow the formation of product molecules. The results gave $k(E)$ values about a factor of 2 lower than those in ref. 155. However, correcting the latter by use of directly measured ΔE values[158] gave good agreement between $k(E)$ values measured by the steady-state photoisomerisation and laser time-resolved techniques, as is seen in Figure 11.11.

The method has been extended to the study of 1,3,5,7-cyclooctatetraene[159] which undergoes parallel reactions, one an isomerisation process and the other decomposition to benzene and acetylene. Once again despite the increased complexity, good agreement was obtained between experimental $k(E)$ values from stationary photolysis experiments with those calculated by RRKM theory by combining the directly measured $k(E)$ from UV laser experiments with thermal high-pressure data.

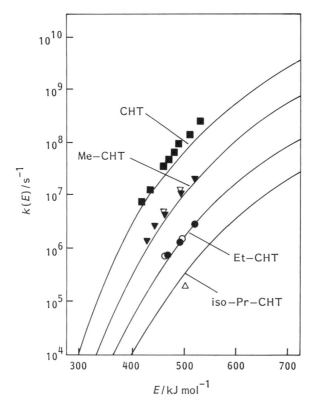

Figure 11.11 Rate constants $k(E)$ for the unimolecular isomerisations of cycloheptatrienes. Open symbols: direct measurements from ref. 157; filled symbols: steady-state photoisomerisation from ref. 155, Stern–Volmer constants reinterpreted with direct (ΔE) measurements from ref. 158; full lines RRKM fits to high-pressure thermal isomerisation lowered by a factor of 2.5. From H. Hippler, K. Luther, J. Troe and H. J. Wendelken, *J. Chem. Phys.*, **79**, 239 (1983), Fig. 11. Reproduced by permission of the American Institute of Physics.

Bond fission reactions have also been studied in this way such as the decompositions of toluene and various alkyl benzenes[138,160–162]. Some of these experiments have been referred to previously since the results have been combined with those obtained using shock-wave activation.

A detailed and elegant study of the dissociation of NCNO

$$NCNO \rightarrow CN + NO \tag{11.53}$$

has been carried out by Zewail and co-workers[163].

Two picosecond laser pulses of wavelengths ~ 580 nm and ~ 610 nm were used to pump and then probe the reactant molecules which were generated in a supersonic molecular beam.

In this method the first pulse is considered to raise the molecules to an excited electronic state, the vibrational levels of which overlap with those of vibrationally excited ground state molecules. Rapid intramolecular transfer to this ground state precedes dissociation. The second pulse (Probe 1), delayed in time is used either to pump the ground state product CN or to monitor absorption of the parent NCNO molecule (Probe 2).

In these experiments, which are illustrated by Figure 11.12, detailed state-to-state rate measurements were possible by varying the wavelengths of the pump and probe lasers and the delay time between pulses. In this way values of $k(E)$ were obtained by monitoring the appearance of the CN product as a function of excess energy (E) of the parent molecule. The experimental results are illustrated in Figure 11.13 compared with one of a number of RRKM curves calculated for various models of the transition state. Although the experimental curve is reasonably close to the calculated curve at energies near the threshold energy $(204.2\ \text{kJ}\,\text{mol}^{-1})$, at higher

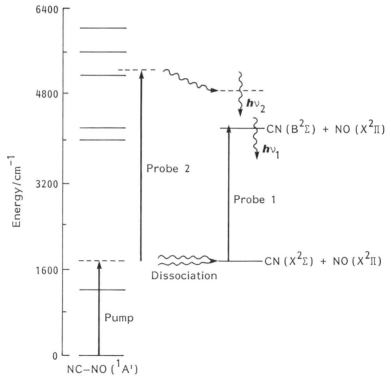

Figure 11.12 Experimental pump-probe scheme for the dissociation of NCNO[163]. From L. R. Khundar, J. L. Knee and A. H. Zewail, *J. Chem. Phys.*, **87**, 77 (1987), Fig. 6. Reproduced by permission of the American Institute of Physics.

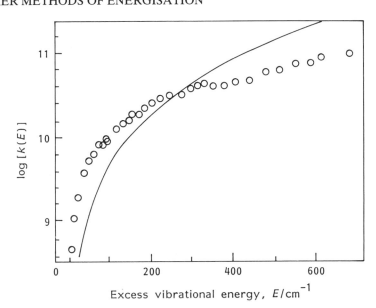

Figure 11.13 Calculated and experimental results for the dissociation of NCNO[163]. The solid curve is the calculated RRKM curve for one of several assumed models of the activated complex. From L. R. Khundar, J. L. Knee and A. H. Zewail, *J. Chem. Phys.*, **87**, 77 (1987), Fig. 15. Reproduced by permission of the American Institute of Physics.

energies it shows an inflection and some structure which was not easily explained in terms of either rigid RRKM or phase-space theories[164].

Subsequent application of the form of RRKM theory modified for flexible transition states[165] and also of modified phase-space theory has resolved some of these discrepancies.

Overtone Vibrational Excitation

An alternative single photon excitation method is the method of overtone vibrational excitation. It is well known that molecular vibrations may be represented by a superposition of *normal mode* vibrations with characteristic harmonic vibration frequencies v_1, v_2, etc. (see Appendix 3). On a purely harmonic model, direct observation of the overtone frequencies $2v_1$, $3v_1$, etc. is forbidden, but the existence of anharmonic coupling between the modes allows transitions from the ground vibrational state of the ground electronic state to various upper overtone states to be observed with low intensities[166]. Direct observations, therefore, require sensitive absorption measurements and photoacoustic methods have proved particularly suitable. Interpretation of the X—H overtone band frequencies is possible in terms of a *local mode* rather than a normal mode description. In this interpretation the stretching motions of X—H bonds (or other bonds to light atoms) are treated as independent anharmonic Morse oscillators only weakly coupled to other modes of vibration in the molecule. The intensities of the observed bands are proportional to the number of equivalent X—H bonds in the molecule, and bandwidths, which are generally narrow, reflect the low degree of coupling to other modes. The theoretical treatment of local modes has been the subject of several reviews[167,168], and Henry[169(a)] first suggested that the ideas implicit in the local mode description, i.e. of energy localised in specific oscillators might produce the consequence of 'non-RRKM' or 'non-statistical' energy randomisation subsequent to the initial excitation.

This aspect of the application of overtone vibrational excitation to the study of unimolecular reactions has been a principal motive for many investigations, but will not be over-elaborated here since it has been discussed in Chapter 9.

Suffice it to say for present purposes that early attempts to provide evidence for 'mode-specific chemistry' have largely been discredited. Such effects are now generally considered unlikely to be observed unless the coupling of local modes to other modes is exceptionally weak. As in Slater's purely harmonic theory (Chapter 2) it is also unlikely that only a few modes are involved in the reaction coordinate. There has, however, been some recent evidence that water molecules prepared by vibrational overtone excitation reacting with translationally excited chlorine atoms in the bimolecular reaction (11.53a) produce OH radicals with a vibrational distribution depending upon the degree of vibrational excitation of the water[169(b)]. This is most easily interpreted in terms of excitation into different overtones of local mode O—H vibrations. Bond selective chemistry has also been observed in the reaction of chlorine atoms with vibrationally excited HOD molecules involving the third overtone of the O–H stretching vibration. Here an eightfold excess of OD over OH is produced:

$$Cl + H_2O \rightarrow OH + HCl \tag{11.53a}$$

The advantage of single photon activation of high-energy overtone vibrational states over other methods of activation is that high state selectivity is possible. Disadvantages apart from the low observed absorptions already mentioned, include the restricted wavelengths which may be selected for excitation and the generally low degree of excitation (typically $E_{tot} \approx$ 200 kJ mol^{-1})[169(a)] which may be achieved. This last fact has restricted the examples of unimolecular processes studied, in general, to those with relatively low energy barriers.

An early application of this technique was that of Berry and co-workers[170,171] to the isomerisation of methyl isocyanide (11.54).

$$CH_3NC \rightarrow CH_3CN \tag{11.54}$$

This reaction has been extensively studied by Rabinovitch and co-workers using thermal energisation in the presence of a wide variety of added bath gases[172] and the kinetics conform with the predictions of RRKM theory. In the apparatus designed by Reddy and Berry[170], the sample cell containing methyl isocyanide at a pressure of a few Torr and also the photoacoustic detector were mounted inside the cavity of a CW dye laser, pumped by a krypton ion laser and tuned to the absorption at ≈ 730 nm in the visible region. The apparatus was used to record the photoacoustic spectrum shown in Figure 11.14(a) which shows two peaks tentatively assigned as the $0 \to 5$, i.e. fourth overtone transitions of two C—H stretching vibrations at 726.5 nm and 733.0 nm. Timed irradiations at 726.5 nm with fixed laser

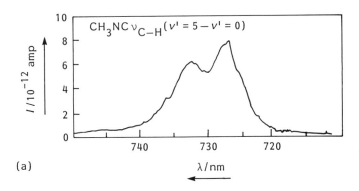

(a)

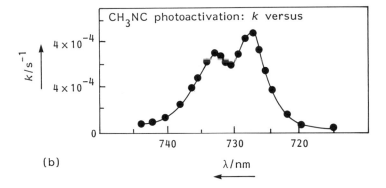

(b)

Figure 11.14 (a) Photoacoustic spectrum of CH_3NC near 730 nm; (b) Apparent isomerisation rate constant for CH_3NC as a function of wavelength. From K. V. Reddy and M. J. Berry, *Chem. Phys. Lett.*, **52**, 111 (1977), Fig. 2. For interpretation and details see text and ref. 170. Reproduced by permission of Elsevier Science Publishers BV.

power and various reactant pressures were analysed by gas–liquid chromato-graphy to obtain first-order rate constants. In the lower Figure 11.14(b) taken from ref. 170, the plot of rate constant versus irradiation wavelength is shown, which clearly indicates the correspondence between the number of photons absorbed and the reaction rate.

A simple analysis in terms of the following mechanism:

$$CH_3NC + h\nu \xrightarrow{k_a} CH_3NC^* \text{ photoactivation}$$

$$CH_3NC^* + M \xrightarrow{k_d} CH_3NC + M \text{ collisional deactivation}$$

$$CH_3NC^* \xrightarrow{k(E)} CH_3CN \text{ isomerisation}$$

and using the steady-state approximation leads to the expression for the reciprocal of the apparent first-order rate constant (k_{app}^{-1}) which is of the Stern–Volmer type and predicts a linear plot of k_{app}^{-1} against total pressure:

$$k_{app}^{-1} = \frac{1}{k_a(h\nu)} + \frac{k_d}{k(E)} \cdot \frac{1}{k_a(h\nu)}[M] \qquad (11.55)$$

This analysis, which resembles that applicable to other types of steady-state photoactivation, enables $k(E)$ to be calculated from the slope of the Stern–Volmer plot if the collisional deactivation rate constant is known. Reddy and Berry[170] assumed that strong collisions occurred under the conditions of their experiment since E_{tot} only just exceeded the critical energy for the reaction using irradiation at 726.5 nm. The derived $k(E)$ values agreed in this case quite well with RRKM predictions. Subsequent work, however, using the 0–6 overtone transition led to a curved Stern–Volmer plot and less good agreement with RRKM calculations[171]. The strong collision assumption is clearly less appropriate in the case of molecules which are more highly excited over the critical energy. Baggott[173] has carried out experiments on the overtone excitation of cyclobutene in the presence of a number of deactivating bath gases. It was concluded that failure to consider weak collisions gives overestimates of $k(E)$. Incorpora-tion of inefficient *intermolecular* energy transfer, for example by insertion of the collision efficiency γ_c of Troe[174] into equations such as (11.55) usually gives satisfactory agreement between rate constants $k(E)$ from this type of experiment and from RRKM calculations.

Allyl isocyanide is a related molecule chosen for study by overtone excitation because of the possibility that the excitatation of the three different types of C—H bond present might lead to different isomerisation rates if subsequent energy transfer to the critical coordinate were slow. The early evidence that this was so[175] has since been shown to be not necessarily the case and capable of other interpretation[176].

Another molecule which has been studied using a continuous dye laser to excite overtone vibrations is the molecule of tertiary butyl hydroperoxide.

This molecule possesses a number of advantages for this type of experiment including a low critical energy for dissociation and the possibility of exciting either the O—H or C—H stretching overtone vibrations. Chandler, Farneth and Zare[177] used excitation of the 0–5 overtone of the O—H stretching vibration at 619.0 nm and found that the rate constants for dissociation at low pressures below 40 Torr showed linear Stern–Volmer plots and were well described by statistical RRKM calculations. Some deviation above this pressure was attributed to non-statistical effects.

A different approach was used by Crim and co-workers[178,179], who measured product yield in a time-resolved method.

The initial dissociation

$$tBuOOH \rightarrow tBuO + OH \qquad (11.56)$$

is followed by the rapid hydrogen abstraction process

$$tBuO + RH \rightarrow tBuOH + R \qquad (11.57)$$

and the yield of the product OH radical can be measured by laser-induced fluorescence (LIF). Varying the time delay between the pump and probe laser pulses enables the time dependence of the OH concentration to be measured and explained in terms of the rate constants for its production by step (11.56) and subsequent pressure quenching. This technique also yields the LIF excitation of the OH products and hence information on the energy partitioning into particular vibrational–rotational levels.

Another time-resolved technique which has been used to study products from overtone excitation incorporates measurements of chemiluminescence, although this is less widely applicable than laser-induced fluorescence. One molecule which has proved particularly suitable is tetramethyl dioxetane which on overtone excitation produces two molecules of acetone, one electronically excited (principally to the triplet-excited state):

$$(11.58)$$

The electronically excited molecule emits chemiluminescence in the visible region of the spectrum and its time dependence can be used to derive the rate constant for step (11.58) if the molecules are assumed to carry with them the room-temperature thermal energy into the excited state upon excitation. For this molecule, excitation to the third or fourth overtone level has been used to derive rate constants which are fitted quite well by RRKM calculations[180].

The problem of broadening of the energy distribution of photoactivated molecules by their initial thermal energies is reduced in experiments involving picosecond techniques and free jet expansion. This work was pioneered by Zewail and co-workers[181,182] who monitored by laser-induced fluorescence OH radicals produced by picosecond pulses from a dye laser which caused overtone excitation and subsequent unimolecular decomposition of H_2O_2. In this technique, overtone excitation is produced in molecules which are vibrationally and rotationally cooled by expansion from a relatively high-pressure inert gas through a fine nozzle into a molecular beam. As a consequence, molecules are produced which have excitation energies very close to the photon energy.

The technique has been applied to the tetramethyl dioxetane molecule[183] and measurements for the rate constant of molecules raised to the $v = 5$ overtone level have been used to 'calibrate' RRKM calculations which can then be compared with room-temperature measurements referred to above. Good agreement has been obtained subject to the usual reservations concerning the insensitivity of these calculations to details of the transition states apart from some low-frequency vibrations and the threshold energy.

Numerous other applications of overtone excitation have been reported and further details may be found in published reviews[169,184].

Infrared Multiphoton Activation

It was first demonstrated in the early 1970s following the development of intense coherent infrared laser sources that these could be used to cause molecules to absorb sufficient photons to bring about unimolecular decomposition. The mechanism by which this occurs is commonly described in terms of a diagram showing the vibrational energy levels represented in terms of three regions: (I) distinct separated vibrational levels; (II) a quasi-continuum of closely spaced anharmonic vibrational levels occurring at higher energy; (III) a true continuum existing above the dissociation threshold. Such a diagram is shown in Figure 11.15. The details of the absorption process and their interpretation are the concern of photophysics and have been treated in a number of reviews[185-188].

The implication for unimolecular reactions is that this process enables molecules to acquire high levels of vibrational excitation within the ground electronic state; the high intensity of these laser sources coupled with relatively strong absorption at selected infrared wavelengths enables much higher levels of excitation to be achieved than are possible by overtone excitation. A disadvantage, however, is that unlike excitation by overtone vibrations or by internal conversion, the energy selectivity of the multipho-

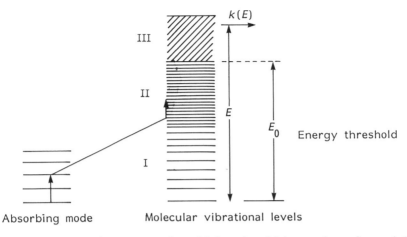

Figure 11.15 Schematic representation of infrared multiphoton absorption and decomposition.

ton infrared excitation is much worse[168]. The energy distribution of molecules excited by infrared multiphoton absorption is a consequence of the rates of absorption (or 'pumping'), unimolecular reaction, and also collisional activation–deactivation processes. These are incorporated into the master equation formulation based on statistical–mechanical concepts developed largely by Quack[189–192]. Such master equations when applied to infrared multiphoton processes contain terms for stimulated absorption and emission of radiation which depend upon the laser variables such as fluence (energy per unit area), intensity (fluence per unit time) and line shape and wavelength as well as upon the absorption cross-section and pressure of the reactant molecules.

In order to achieve the high levels of vibrational excitation (typically $\approx 240 \text{ kJ mol}^{-1}$) necessary to bring about many unimolecular gas phase processes, with a CO_2 gas laser operating at 10 μm wavelength for example, at least 20 photons of infrared radiation must be absorbed. If thermal energisation also participates via collisions at moderately high pressures, then this can be achieved by low-power CW lasers, sometimes with the aid of absorbing gases as sensitisers[193]. Although some interesting studies have been carried out using CW lasers[194] and they are in some respects easier to model theoretically[195], much more work has been done with pulsed infrared lasers. The high power available in these lasers enables direct excitation to highly vibrationally excited levels even under collisionless conditions during the pulse. As an example a transversely excited atmospheric (TEA) CO_2 gas laser with a typical incident energy of 120 mJ per pulse and an effective pulse length of 125 ns has a power of about 10^6 W.

Except at the highest excitation energies, the factors affecting the absorption process will affect the product yields.

It is necessary in experimental studies to measure both absorption and product yields as a function of all the variables such as pressure, fluence, intensity, line shape and wavelength. The most meaningful experiments are those carried out under pressure conditions where no collisions occur during the laser pulse. The effects of increased pressure and hence increased number of intermolecular collisions may be to promote enhanced reaction by increased 'thermal' activation, to reduce reaction by deactivating collisions or to enhance reaction by inducing increased absorption by 'rotational hole-filling'.

Danen, Rio and Setser[196] studied the infrared multiphoton decompositions of a number of organic esters and measured both absorption and product yields. With relatively high pressures of added bath gases, the principal effect was found to be collisional deactivation and a reduced yield. At pressures below 0.1 Torr, however, using a mixture of 97:3 ethyl acetate to isopropyl bromide, it was possible to use the isopropyl bromide decomposition to propene as a 'thermal monitor'. Increased propene yields were found at higher pressures; below 0.1 Torr, however, it was concluded that thermal contribution to the reaction rate was negligible.

Some molecules exhibit enhancement of the reaction rate by increased pressure of added inert gases which has been attributed to 'rotational hole-filling'. This effect which is due to repopulation of the absorbing levels of the reactant gas by collisions has been invoked by Jang and Setser[197] to explain the effect of inert gases upon the infrared multiphoton decomposition of CH_3CF_3 under high fluence and short pulse conditions. Jang, Setser and Ferrero[198] found a dependence of the absorption coefficient upon pressure for both CH_3CF_3 and C_2H_5F which were classified as 'small molecules' based upon their absorption characteristics in contrast to those molecules exhibiting Beer's law behaviour and pressure-independent absorption coefficients. The phenomenon of rotational hole-filling is, as expected, dependent upon the nature of the bath gas and also upon the energy-level density of the absorbing molecule.

The molecule 2,2-dimethyloxetane[199] was also found to exhibit a pressure-dependent absorption coefficient. An attempt was made using the infrared laser-induced decomposition of this molecule, to model the results using different assumed forms of the energy distribution. The product yields (of isobutene and ethene) were modelled using RRKM calculated rate constants and assuming that the absorbed energy was distributed among all the molecules according to either a Boltzmann or a Poisson distribution.

The results shown in Figure 11.16 show that neither distribution fits the observed results unless it is assumed that not all the molecules in the irradiated volume absorb the radiation. Distinction between different forms

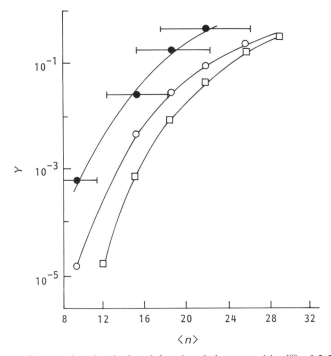

Figure 11.16 Observed and calculated fractional decomposition[199] of 2,2-dimethyl-oxetane (Y) as a function of the average number of photons absorbed per molecule $\langle n \rangle$ for 5 mTorr 2,2-dimethyloxetane plus 45 mTorr n-hexane. R20 (10 μm) ● observed, □ calculated—Poisson distribution, ○ calculated—Boltzmann distribution. Both calculations assumed a step size of 21 kJ. From K. A. Holbrook, C. J. Shaw, G. A. Oldershaw and P. J. Dyer, *Int. J. Chem. Kinet.*, **18**, 1215 (1986), Fig. 7. Reproduced by permission of John Wiley & Sons, Inc.

of energy distribution is, therefore, not possible on this basis. A direct measurement of the energy distribution of tetramethyl dioxetane molecules energised by infrared multiphoton absorption at various CO_2 laser energies has been made by Haas and co-workers[200]. These workers measured the chemiluminescence produced from the excited molecules of acetone produced at various times after the laser pulse and compared this with that similarly produced following overtone excitation of tetramethyl dioxetane[180]. By using three different overtone C—H stretch frequencies it was possible knowing the narrowly defined energy distributions in the overtone excitation experiments to derive coarsely grained three-point energy distributions applicable to infrared multiphoton excitation. This comparison depends on the assumption of rapid intramolecular energy distribution which ensures that decomposition of the energised molecule (in this case producing chemiluminescence) only depends upon its internal energy content and not upon its mode of production. While this assumption is reasonably sound, it was shown that in general the dependence of yield upon fluence in infrared multiphoton experiments is not uniquely determined by simple master equation models because of uncertainties in the dependence of absorption coefficients upon energy.

Doubts of this kind and lack of knowledge of the energy distributions continue to limit the usefulness of infrared multiphoton excitation experiments in their application to unimolecular processes. A large number of experimental studies has, however, now accumulated and many have been reported in the sources quoted earlier and especially in the excellent comprehensive review of Lupo and Quack[201].

Some features of infrared multiphoton excitation and reaction deserve special mention. One of these is the study of competing reaction channels. The study of molecules which absorb radiation to produce different products by genuinely independent channels is not unique to infrared multiphoton excitation. However, the use of different narrowly defined laser frequencies in an attempt to achieve reaction by one channel rather than another and hence to produce 'mode-selective' chemistry was the motivation for many early studies such as those on cyclopropane[202,203], ethylvinyl ether[204] and cyclobutanone[205,206]. It is now generally recognised that mode selective chemistry may only rarely be possible[207], nevertheless, the study of systems with competing channels offers in principle some advantages over that of single-channel systems. This is the principal conclusion of Krajnovich *et al.*[208] who have reviewed much of the early data and emphasised the need for the measurement of branching ratios under collision-free conditions. One technique by which this is achieved is the use of molecular beam infrared multiphoton experiments; a method also used by, among others, Lee and co-workers[208–212] to obtain information on product translational energy distributions. For example, the CF_2Cl_2 molecule is known to decompose by two channels of similar endothermicity:

$$CF_2Cl_2 \rightarrow CF_2Cl + Cl \qquad \Delta H = 334.7 \text{ kJ mol}^{-1} \qquad (11.59a)$$

$$\rightarrow CF_2 + Cl_2 \qquad \Delta H = 305.4 \text{ kJ mol}^{-1} \qquad (11.59b)$$

Using a crossed molecular beam of CF_2Cl_2 molecules and the output of a high repetition rate CO_2 TEA laser at low temperature and pressure it was possible using a rotatable mass-spectrometer detector to determine the time-of-flight product distributions from each channel. From these, angular and velocity distributions of the products were obtained and used to deduce the probability distribution for the centre of mass translational energy in the separating fragments.

For dissociation into fragments where there is little or no reverse activation energy (exit channel zero barrier) RRKM theory can be used to predict translational energy distributions[210]. In this case the excess energy in the transition state is statistically distributed between all the contributing modes including translation in the reaction coordinate which ends up as translation and vibration in the fragments. The initial translational energy distribution for the fragments produced by channel (11.59(a)) peaked close to zero as predicted by RRKM theory. The products of channel (11.59(b)), on the other hand, were formed with a mean translational energy of ~ 32 kJ mol^{-1} with a distribution peaking sharply about this value indicating an appreciable exit barrier. Many examples now exist of the statistical distribution of product translational energy and low exit barriers in simple bond fission processes, for example in halogenated alkanes[210], diethyl ether[211] and methyl and ethylacetates[212]. Evidence for exit barriers has, however, been found for ethyl vinyl ether[208] and recently for anisole and the secondary decomposition of its product phenoxy radical[213]. For some two-channel decompositions experimental measurements of translational energy distributions for a bond-rupture channel have been combined with RRKM calculated product ratios (using known A-factors) to find the excitation energy distribution in the parent molecule and hence the barrier height and activation energy of the second channel[214].

Further detailed work on the chemical reaction dynamics of these processes has involved the study of the kinetic energy of species formed in well-defined initial quantum states, for example the work of King and Stephenson[215] on the fragments produced by dissociating methyl nitrite in a molecular beam irradiated by an infrared laser of well-defined intensity. Time-resolved product studies also provide further information, such as the use of resonance-enhanced multiphoton ionisation (REMPI) to monitor the products from the infrared multiphoton decomposition of CF$_3$I and CF$_3$Br[216].

Despite the vast amount of work which has been done on infrared multiphoton absorption and decomposition in the last two decades, many fundamental aspects of the interpretation of these experiments remain obscure. The anomalous behaviour of methyl and ethyl isocyanides both of which above a certain pressure appear to react completely upon a single pulse is still lacking in complete interpretation[217]. An alternative explanation of pressure effects upon reaction yield is that of 'vibrational energy pooling'[218] involving collisional self-activation between excited reactant or excited reactant and product molecules. It has been suggested that this effect is responsible for the pressure dependence of a secondary channel to give radical products in the infrared laser-induced decomposition of t-butyl-d_9-bromide[219]. It is further suggested that the possible involvement of vibrational energy pooling could invalidate many of the assumptions made prior to 1990 in the interpretation of experimental data and require some reinterpretation to be made.

11.3 ION AND ION–MOLECULE REACTIONS

The study of unimolecular processes involving vibrationally excited ions has proceeded in parallel with that of comparable processes of neutral species. The interpretation has largely been in terms of the so-called 'quasi-equilibrium theory' (QET) proposed to explain the mass spectra of polyatomic molecules by Rosenstock *et al.*[220].

In the production of ions from neutral molecules by electron impact, photoionisation, field ionisation or charge-transfer processes involving other ions, a range of high-energy electronic states, in general, is produced. In QET it is assumed that rapid radiationless conversion to the ground electronic state of the ion occurs with a residual level of vibrational excitation which may or may not result in unimolecular reaction. Forst[221] has discussed this assumption in some detail which requires a close matching of vibrational levels of two or more electronic states producing a cascade terminating in the ground electronic state. While the high densities of states occurring in large polyatomic ions makes this a reasonable assumption in most cases, he has drawn attention to some theoretical and experimental evidence to suggest that such is not always the case.

It is a consequence of this assumption that energised ions with the same internal energy are then independent of the manner of their production. The subsequent intramolecular energy redistribution is treated by QET by application of statistical transition state theory in essentially the same manner as the RRKM theory for neutral molecules, with the result that the equations developed in the earlier chapters of this book may be applied to calculate the expected unimolecular rate constants.

The advantages of the study of unimolecular ion decompositions or isomerisations in comparison with those of neutral species have been given by Lifshitz[222] as (a) the ability to produce ions at very low pressures in the absence of molecular collisions, and (b) the ability to produce ions with well-defined internal energies.

Techniques for producing ions with well-defined internal energies have led to a big expansion in the study of gas-phase ion chemistry in recent years. Many such studies are described in review articles[223–227] to which the reader is referred for a more comprehensive coverage of this topic than is possible here. In the present section the discussion is confined to a few examples chosen to illustrate well-designed experiments yielding results of relevance to unimolecular theories.

One technique of particular importance in the production of state-selected ions is that of photoion–photoelectron coincidence (PIPECO) first developed by Brehm and von Puttkamer[228]. In this method molecules are ionised by photon impact, producing at the same time electrons of a certain kinetic energy.

$$M \overset{h\nu}{\rightarrow} M^+ + e$$

The energy of the ejected electron is governed by the thermal energy of the molecule M plus the photon energy minus the ionisation potential and minus the internal energy of the M^+ ion. For a given photon energy, therefore, ions of a particular internal energy are produced in coincidence with electrons of a particular kinetic energy. One kind of PIPECO experiment is to keep the photon energy fixed and to collect electrons of different energies which relate to production of M^+ ions with different internal energies. Another version is to keep the energy of the detected electrons fixed and to vary the photon energy. In a typical PIPECO experiment, a time-of-flight mass spectrometer is used to measure the times of arrival at the detector of the electrons and their corresponding ions. The basic principles of the method have been discussed by Eland[229] and more recently by Baer[230]. When a parent ion decomposes to give a fragment ion, the 'breakdown graphs' which are obtained show the dependence of the fractional abundance of both parent and fragment ions upon the internal energy of the parent ion. These results may be compared with the predictions of RRKM theory for the energy dependence of the fragmentation rate constant.

One reaction which has been much studied in this way because of its apparently simple transition state involving the extension of the C–Br bond is the fragmentation of the bromobenzene ion

$$C_6H_5Br^+ \rightarrow C_6H_5^+ + Br \tag{11.60}$$

This reaction was studied by Baer and co-workers[231] who obtained good agreement with statistical theory. It was later suggested, however, that the results obtained inferred an unreasonably tight transition state[232]. In the work of Rosenstock, Stockbauer and Parr[232], a slightly different experimental method was used which enabled parent/fragment ratios to be obtained as a function of parent ion residence time. Breakdown curves for the parent ($C_6H_5Br^+$) and fragment ($C_6H_5^+$) ions obtained by these workers are shown in Figure 11.17. This figure shows the experimental results obtained for two different residence times of the ions in the mass spectrometer before their ejection by application of a pulse to the ion source region. Breakdown curves were calculated according to QET by assuming sets of transition state frequencies and activation energies, i.e. reaction threshold energies. The best combination of these parameters was chosen by a sensitivity analysis

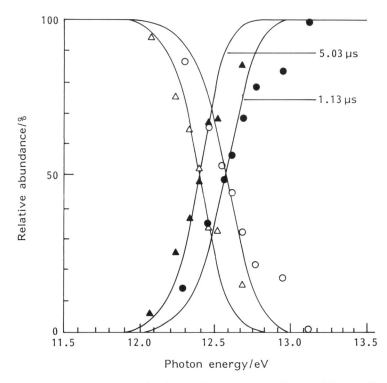

Figure 11.17 Breakdown curves for bromobenzene for 1.13 and 5.03 μs effective residence times[232]. Experimental points: ($\bigcirc$) ($\triangle$) $C_6H_5Br^+$ parent; ($\bullet$) ($\blacktriangle$) $C_6H_5^+$ fragment. ($-$) Best fit calculations. Arrows indicate the thermochemical thresholds. From H. M. Rosenstock, R. Stockbauer and A. C. Parr, *J. Chem. Phys.*, **89**, 2253 (1979), Fig. 3. Reproduced by permission of the American Institute of Physics.

and the resulting calculated breakdown curves are shown by the solid lines in Figure 11.17.

Some disagreement between the calculated curves and the experimental points occurs at both high and low energies, but agreement is better near the cross-over points (50% fragmentation).

From the results, an activation energy of 266 kJ mol^{-1} was derived and also an entropy of activation corresponding to a temperature of 1000 K. The value obtained for the latter was, however, later recalculated by Lifshitz and co-workers[233] who reinvestigated this decomposition by time-resolved photoionisation mass spectrometry (TPIMS). In this technique, ions created by a pulsed vacuum UV light source are trapped within an ion trap for an accurately known delay time before ejection by a draw-out pulse into a quadrupole mass filter. It is then possible to obtain photoionisation efficiency curves for both parent and fragment ions at delay times between 10^{-6}

and 10^{-3} s which can be compared with QET calculations. In these calculations the critical energy E_0 was assumed to be the activation energy of 266 kJ mol^{-1} found by Rosenstock, Stockbauer and Parr[232] and a similar set of transition state frequencies corresponded to $\Delta S^{\ddagger} = 8.07$ cal mol^{-1} K^{-1} and A_{∞} (1000 K) $= 3.29 \times 10^{15}$ s^{-1}. Very similar values were obtained by Pratt and Chupka[234] who interpreted their results on the basis of the Klots–Langevin[235] phase-space model. These results and also measurements of kinetic energy release[236] for reaction (11.60) support a loose transition state and a negligible reverse activation energy. Such models may well be valid for many ionic dissociations of this type[237] and require special methods similar to those discussed in Chapter 6 for location of the transition state.

As in the case of the study of the unimolecular reactions of neutral species, many unimolecular studies of ion decompositions have been concerned with tests of the assumption of energy randomisation. In the case of ions, several different types of randomisation may be distinguished, for example randomisation of electronic energy originally deposited in the molecule and hence the ion upon ionisation. This has been tested by examination of the breakdown curves of ions generated in different ways[238].

The assumption of rapid randomisation of vibrational energy in the propene oxide cation decomposition has been questioned by experiments of Lifshitz and co-workers[239]. This ion decomposes via a highly excited acetone ion which has been shown by isotopic labelling to lose its two symmetrical methyl groups at different rates (r_a, r_b, equation 11.61).

$$(11.61)$$

Calculations show that the lifetime of the excited acetone ion is of the order of 5×10^{-13} s which is commensurate with the rate of intramolecular energy transfer found from experiments on chemically activated species and other sources (see Chapter 9).

Finally, the randomisation of energy between translation in the reaction coordinate and the other degrees of freedom in the transition state is tested by measurements of the kinetic energy of product ions. It appears that many results cannot be simply explained by either the QET or the phase space treatment. The failure may be due to breakdown of some of the assumptions such as the one that excited electronic states rapidly relax into a single electronic state, usually assumed to be the ground electronic state prior to dissociation. Bimodal kinetic energy release distributions are obtained in some cases indicating incomplete randomisation[222].

Among many examples of the study of metastable ions (which have lifetimes of the order of 10^{-5}–10^{-6} s and therefore decay within a conventional mass spectrometer), one of particular relevance to the theory of unimolecular reactions is that of the H_3^+ ion. This ion has been studied theoretically by Pollak and Schlier[240] and experimentally by Carrington and McNab[241]. The experimental evidence from photodissociation following single-photon excitation showed that this ion decomposes much more slowly than expected from statistical calculations for a given excitation energy. This work has shown the relevance of rotational barriers and of the inclusion of angular momentum in the formulation of unimolecular decay rate constants as well as the importance of the conservation of total angular momentum.

An alternative to the PIPECO technique for the production of ions with specific energy content is that of multiphoton ionisation. Absorption of relatively few photons of visible or ultraviolet light are needed to bring about ionisation of many neutral species. This is geneally considered to occur via an intermediate absorbing level of the molecule. Once the ionisation threshold has been crossed, further absorption occurs in the parent ion which for low energies of incident radiation is then produced with a relatively narrow energy distribution. If the parent ion decomposes into fragment ions on the microsecond time-scale which defines it as a metastable ion, the absorption then occurs in the fragments produced. This is the so-called ladder-switching model illustrated in Figure 11.18 and first proposed by Boesl, Neusser and Schlag[242(a)].

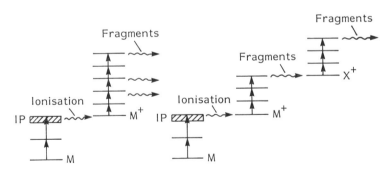

Figure 11.18 Ladder switching mechanism in multiphoton ionisation. (a) Multiphoton ionisation occurring by autoionising the first level of molecule M to cross the ionisation potential, thereafter absorption in the ion (M^+). (b) Same as (a) except that absorption in the ion (M^+) can lead to fragmentation to give secondary ions (eg. X^+) which can in turn absorb photons from the laser pulse. From E. W. Schlag and H. J. Neusser, *Acc. Chem. Res.*, **16**, 355 (1983), Fig. 2. Reproduced by permission of the American Chemical Society.

An example of the use of multiphoton ionisation to study decay of a metastable ion is given by the work of Neusser *et al.*[242(c)]. Benzene cations have been shown to decompose by two H-loss (11.62, 11.63) and two competing C-loss (11.64, 11.65) channels:

$$C_6H_6^+ \rightarrow C_6H_5^+ + H \qquad (11.62)$$

$$C_6H_5^+ \rightarrow C_6H_4^+ + H \qquad (11.63)$$

$$C_6H_6^+ \rightarrow C_4H_4^+ + C_2H_2 \qquad (11.64)$$

$$C_6H_6^+ \rightarrow C_3H_3^+ + C_3H_3 \qquad (11.65)$$

When benzene molecules are ionised in a two-photon absorption process using frequency-doubled light from a nitrogen-laser pumped dye laser, the resulting $C_6H_6^+$ ions are produced in their vibrational ground state. As they are accelerated under application of an electric field into a time-of-flight mass spectrometer, a second laser displaced spatially by about 1 mm from the first laser, produces their decomposition into the fragment ions. The spatial displacement of the second laser ensures that no further parent ions are produced and variation of its energy together with determination of the rates of production in the high-resolution reflectron time-of-flight mass spectrometer enables the rates of the decay processes to be measured as a function of the internal energy of the ions. The results are illustrated in Figure 11.19 which for the C-loss channel (11.64) are in good agreement with PIPECO measurements of Eland and Schulte[243]. The points corresponding to the H-loss channel (11.62) show a similar rate dependence for this competing channel. In general the small mass change for H-loss processes precludes their investigation by the PIPECO technique, and this example was the first demonstration of the viability of the multiphoton ionisation technique in the study of such processes.

ION–MOLECULE REACTIONS

Ion–molecule reactions have received much attention in recent years, and much of the work has derived its impetus from the desire to understand the synthesis of molecules in dense interstellar clouds[244,245]. In the earth's atmosphere most studies have been concerned with the region above 60 km, i.e. the ionosphere[246] or in the upper stratosphere[247,248]. Much less is known at present about the ion chemistry of the troposphere[248(b)].

Ion–molecule reactions are naturally related to the reverse process of ionic dissociation and therefore proceed through a common transition state. They are generally faster than neutral association reactions of atoms or radicals because they are influenced by long-range forces. For ions with neutral molecules which are non-polar, these involve ion-induced dipole

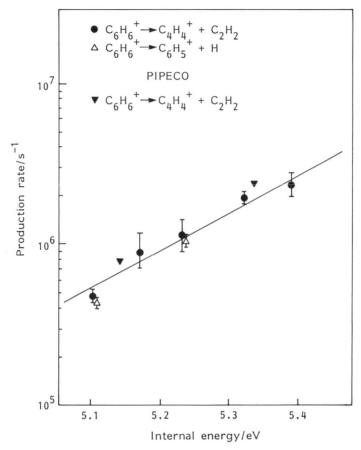

Figure 11.19 Two-laser multiphoton ionisation of benzene. Measured production rates for $C_4H_4^+$ (●) and $C_6H_5^+$ (△) as a function of the internal energy of $C_6H_6^+$ after two-laser excitation. For comparison, PIPECO results for the production rate of $C_4H_4^+$ (▼) are given. From H. J. Neusser, H. Kühlewind, U. Boesl and E. W. Schlag, *Ber. Bunsenges. Phys. Chem.*, **89**, 276 (1985), Fig. 8.

forces and their collision rates can be calculated from Langevin theory[249,250]. Collisions between ions and polar molecules may be treated by average dipole orientation theory (ADO theory)[251,252]. The consequence for transition states and stable intermediates in such processes are that these occur on potential energy surfaces showing one or more potential energy minima, i.e. Type I or Type II potential energy surfaces (see Figure 5.1) either with or without exit barriers to decomposition. Some attention has been given to the loose nature of ion-neutral transition states which undergo virtually free rotation and facile 1–2 hydrogen transfer processes. This may account for some observed ion isomerisations such as the conversion of 1-propyl cation to 2-propyl cation[253].

While the rate constants of ion–molecule processes at high pressures can be relatively easily explained, recent work has been mostly involved with the low-pressure rate constants and their temperature dependence. As predicted by unimolecular theory, third-order behaviour with rate constants exhibiting negative temperature dependence has been observed. Patrick and Golden[254] have calculated such rate constants for the three reactions (11.66), (11.67) and (11.68):

$$N_2^+ + 2N_2 \rightarrow N_4^+ + N_2 \qquad (11.66)$$

$$O_2^+ + 2O_2 \rightarrow O_4^+ + O_2 \qquad (11.67)$$

$$O_2^+ + 2N_2 \rightarrow N_2O_2^+ + N_2 \qquad (11.68)$$

Values of the high-pressure second-order (k_∞) and low pressure third-order (k_0) rate constants calculated from the parameterised equations of Troe[255,256] for these reactions are given in Table 11.4.

Agreement with experimental data for the rate constants at various temperatures in this range was found to be reasonable for reactions (11.67) and (11.68). The calculated rate constants of (11.66) at low temperatures were larger than experimental values[257] at 50 K by a factor of 3.3 which was attributed to this reaction being in its fall-off region at the pressures used rather than at the low-pressure third-order limit. The $(N_2)_2^+$ cluster which is the product of reaction (11.66) has also been studied because of its importance in stratospheric chemistry[258]. In this connection it is important to know whether the $(N_2)_2^+$ cluster decays on a microsecond time scale (metastable decay) or on a shorter time scale. Experiments have been carried out using a high-energy ion beam crossed with a dye laser beam at 458–514 nm[259]. Photodissociation of the $(N_2)_2^+$ cluster occurred and product kinetic energy distributions were measured using an electrostatic analyser. The results implied that dissociation occurred by direct transition to a repulsive potential energy surface.

A number of techniques exist for the determination of the rate constants of ion–molecule reactions in the gas phase. One technique referred to above

Table 11.4   Values of k_0 and k_∞ used in the range 20–500 K in the calculations of Patrick and Golden[254]

Reaction	$k_0/cm^6\ molecule^{-2}\ s^{-1}$	$k_\infty/cm^3\ molecule^{-1}\ s^{-1}$
(11.66)	$1.49 \times 10^{-23}\ T^{-1.99}$	7.19×10^{-10}
(11.67)	$2.53 \times 10^{-25}\ T^{-1.94}$	6.71×10^{-10}
(11.68)	$5.71 \times 10^{-26}\ T^{-1.89}$	7.11×10^{-10}

is the use of ion beams of high energy, others have been developed to study ion–molecule interactions at lower energies with less chance of complications arising from the presence of electronically or vibrationally excited ions. The flowing afterglow technique, for example, was developed by Ferguson, Fehsenfeld and Schmeltekopf[260] to study the kinetics of thermalised ion–molecule reactions of particular relevance to the ion chemistry of the stratosphere. In this technique, ions are created in a fast flow of carrier gas such as helium and are then carried downstream, producing a thermalised plasma of ions, metastable atoms and neutral species. Neutral gases can be added at this point and disappearance of the reactant ions and appearance of the product ions from their reactions with the neutrals or other species present can be monitored by a mass spectrometer as a function of carrier gas flow rate. The technique can be modified to allow the study of the reactions of a large number of both positive and negative ions with neutral species under thermalised conditions. Temperature variation over a wide range is possible, enabling the study of a number of weakly bound complexes. Compilations of the data for many ion–molecule reactions by this method exist[261,262]. One disadvantage of the flowing afterglow technique is that ions are created in the flow tube and can react with their source gas as well as with the desired reactant gas. This is avoided in the selected ion flow tube designed by Smith and Adams[263,264] since in this technique the ions are generated from a source gas remote from the flow tube and subsequently injected into the carrier gas flow and hence into the reactant. A further development is the use of supersonic expansion to study reactions at temperatures as low as 8 K by the so-called CRESU technique ('Cinétique de réactions en ecoulement supersonique uniforme'). This technique and other low-temperature techniques have been reviewed and the results for a number of radiative and termolecular association processes compare with theoretical predictions by Rowe[265].

Ion cyclotron techniques have also been developed and used to study proton transfer processes relating to the measurement of gas-phase acidities and basicities[266,267].

Unimolecular theory has been applied to many ion–molecule reactions occurring through an association complex. One example of such an application is that of Brauman and co-workers[268] who applied the standard RRKM theory to some experimental results of Meot-Ner and Field[269] on the proton-bound dimers of NH_3, CH_3NH_2 and $(CH_3)_2NH$.

In the case of the ammonia dimer, the system was represented by

$$NH_4^+ + NH_3 \underset{k_2}{\overset{k_1}{\rightleftharpoons}} [N_2H_7^+]^* \xrightarrow{k_3[CH_4]} N_2H_7^+ \tag{11.69}$$

The collision rate constants k_1 and k_3 were calculated from ADO[251,252] and Langevin[249,250] theory and the rate constant for decomposition of the complex k_2 was found from the experimental rate of formation of the product

$N_2H_7^+$ and assumed RRKM models for the transition state. The location of the transition state was based upon a Gorin-type model[270]. This early example of the application of RRKM theory was reasonably successful in showing evidence for intramolecular randomisation of energy in the transition state and predicting the temperature and pressure dependence of its formation. However, a number of assumptions were made, one of which was the strong-collision rate for the stabilisation of the transition state by the methane bath gas.

More sophisticated treatments of ion–molecule reactions have now been made, such as those of Gilbert and co-workers[271–273] which are based upon a master equation formulation and which incorporate many refinements. The reaction between CH_3^+ and HCN for example has been studied experimentally[274,275] and a RRKM master equation treatment incorporating angular momentum conservation and weak-collision effects produces excellent agreement with the experimental data at low pressures.

11.4   ASSOCIATION REACTIONS

Association reactions are the reverse of unimolecular dissociation reactions and include combination (e.g. $CH_3 + CH_3 + M$) and addition (e.g. $H + C_2H_4 + M$).

Under all conditions of pressure, the forward and reverse rate constants are related via the equilibrium constant at constant volume $K_c(T)$. Thus, the reversible dissociation of a molecule AB in the presence of bath gas M, may be written as (11.70), and the rate constants k_{diss} and k_{ass}, which are functions of [M] and T, are related to $K_c(T)$ by (11.71):

$$AB + M \underset{k_{ass}}{\overset{k_{diss}}{\rightleftharpoons}} A + B + M \qquad (11.70)$$

$$\frac{k_{diss}([M], T)}{k_{ass}([M], T)} = K_c(T) \qquad (11.71)$$

Unimolecular rate theory may be used directly to calculate k_{diss}, with k_{ass} being calculated indirectly via K_c. Alternatively, k_{ass} can be calculated directly, for example using the master equation approach developed in Chapter 7. Experimentally measured association rate constants can also be used to calculate dissociation rate constants, using the equilibrium constant and vice versa. The rate constant k_{ass} has a much weaker temperature dependence than k_{diss} and, since a particular experimental technique has a specific dynamic range, association rate constants can generally be measured over a wider temperature range than dissociation rate constants. Often, also, the temperature ranges studied are complementary, with k_{ass} being measured at lower temperatures and k_{diss} at higher ones. This complementarity permits testing of unimolecular rate theory over a wider range of temperatures.

Recent years have seen a considerable increase in activity in the measurement of rate constants for association reactions. This increase is partly due to the developments in experimental technique and particularly to the continuing improvements in discharge-flow and flash photolysis methods. The demands for rate data, especially for atmospheric chemistry but also for combustion, have probably played a greater part and have been central to the drive for improved techniques.

These applications have also been responsible for the widespread adoption of standardised forms of parameterisation of pressure-dependent rate constants. The Troe formulation is widely used (see Table 11.5, refs 1–3) with the general unimolecular rate constant k_{uni} given by

$$k_{\text{uni}} = \frac{k_\infty k_0[\text{M}]}{k_\infty + k_0[\text{M}]} \cdot F \qquad (11.72)$$

Table 11.5 Rate data for association reactions. (Ref. page 378)

Reaction	$k_\infty/\text{cm}^3\,\text{molecule}^{-1}\,\text{s}^{-1}$ $k_0/\text{cm}^6\,\text{molecule}^{-2}\,\text{s}^{-2}$ F_c	T/K	Evaluation (ref. number)
H + H + Ar	$k_0 = 1.8 \times 10^{-30}\,(\text{T/K})^{-1.0}$	300–2500	1
H + O$_2$ + N$_2$	$k_0 = 3.9 \times 10^{-30}\,(\text{T/K})^{-0.8}$ $k_0 = 5.7 \times 10^{-28}\,(\text{T/K})^{-1.6}$	300–2000 200–600	1 2
OH + OH + N$_2$	$k_0 = 6.1 \times 10^{-29}\,(\text{T/K})^{-0.76}$ $k_\infty = 1.2 \times 10^{-10}\,(\text{T/K})^{-0.37}$ $F_{\text{cent.}} = 0.5$ (equivalent expressions given by ref. 2 over a narrower range)	250–1400 200–1500	1
CH$_3$ + CH$_3$ + Ar	$k_0 = 3.5 \times 10^{-7}(\text{T/K})^{-7.0}\exp(-1390\,\text{K/T})$ $k_\infty = 6.0 \times 10^{-11}$ $F_{\text{cent.}} = 0.38\exp(-\text{T}/73\,\text{K})$ $\qquad + 0.62\exp(-\text{T}/1180\,\text{K})$	300–2000 300–2000 300–2000	1
H + C$_2$H$_4$ + He	$k_0 = 1.3 \times 10^{-29}\exp(-380\,\text{K/T})$ $k_\infty = 6.6 \times 10^{-15}\,(\text{T/K})^{1.28}\exp(-650\,\text{K/T})$ $F_{\text{cent.}} = 0.24\exp(-\text{T}/40\,\text{K})$ $\qquad + 0.76\exp(-\text{T}/1025\,\text{K})$	300–800 200–1100 300–800	3
ClO + NO$_2$ + N$_2$	$k_0 = 4.2 \times 10^{-23}\,(\text{T/K})^{-3.4}$ $k_\infty = 2 \times 10^{-11}$ $F_{\text{cent.}} = \exp(-\text{T}/430\,\text{K})$	200–300 200–300 200–300	2

where F is a broadening factor (for further explanation see Chapter 8, p. 226) given by

$$\log F = \log F_{\text{cent.}}/\{1 + [\log (P_{\text{r}})/N]^2\} \qquad (11.73)$$

and

$$N = (0.75 - 1.27 \log F_{\text{cent.}}) \quad \text{or} \quad N = 1 \qquad (11.74)$$

(Table 11.5, refs 1–3) $\qquad\qquad$ (Table 11.5, ref. 2)

As discussed in Chapter 8, F describes the broadening of the fall-off curve from the Lindemann form as a result of the dependence of the microcanonical dissociation rate constants on energy and of weak collisions. Thus the pressure and temperature dependence of both dissociation and association rate constants can be described using only three temperature-dependent parameters, k_0, k_∞ and $F_{\text{cent.}}$. Modified Arrhenius equations of the form (11.75) are used for k_0 and k_∞ while the form proposed for $F_{\text{cent.}}$ is (11.76). Not all of these parameters may be employed for specific reactions; in atmospheric chemistry where quite low temperatures (200–300 K) are significant, $F_{\text{cent.}}$ is often set to 0.6.

$$k = A T^n \exp (-B/T) \qquad (11.75)$$

$$F_{\text{cent.}} = (1 - a) \exp (-T/T^{***}) + a \exp (-T/T^*) - \exp (-T^{**}/T) \qquad (11.76)$$

where T^*, T^{**} and T^{***} are empirical parameters obtained from fitting the experimental data.

Specific rate constants are required for incorporation in computer models employed in both combustion and atmospheric chemistry. Because of the difficulties involved in measuring rate constants, and because there are often considerable discrepancies between different measurements, evaluation methods have been developed and compilations of evaluated rate data are frequently presented. Much of the methodology was developed by Baulch and co-workers and their initial evaluations were largely directed at high-temperature data[276].

The availability of these evaluations obviates the need to present a large table of rate data for association reactions. Data for a selected group of reactions are given in Table 11.5 and are taken from refs 1, 2 and 3 (of Table 11.5) which are from *Evaluated Kinetic Data for Combustion Modelling*, compiled by the EC Task Group on Combustion Kinetics (refs 1, 3) and *Evaluated Kinetics and Photochemical Data for Atmospheric Chemistry, Supplement IV*, compiled by the IUPAC subcommittee on Gas Kinetic Data Evaluation for Atmospheric Chemistry (ref. 2). Each of the association reactions listed in Table 11.5 is now briefly discussed and the relevant rate data are compared.

THE H + H REACTION

This reaction has been studied by a wide range of techniques, from discharge flow using catalytic detection on a tungsten filament[277] and ESR[278] at low temperatures, to numerical simulations of flames[279] and shock-tube studies with resonance absorption[280] at high temperatures. There is reasonably good agreement in the rate constants measured over the last 30 years and uncertainty in the rate constant of $\Delta \log k = \pm 0.5$ was proposed over the temperature range 300–2500 K. The reaction is at the low-pressure limit under all conditions studied and the rate constant shows the anticipated decrease with temperature which results from the generation of a wider distribution of energised molecules at higher temperatures.

THE H + O$_2$ REACTION

The H + O$_2$ reaction is of considerable importance in both atmospheric chemistry and in combustion. In the former it takes part in the HO$_x$ cycle and is thus involved in stratospheric ozone depletion. In the latter it acts as the terminating step at the second explosion limit in the H$_2$ + O$_2$ reaction. Interestingly, the competitive branching reaction,

$$H + O_2 \rightarrow OH + O \qquad (11.77)$$

occurs on the same potential energy surface as the combination reaction and Cobos, Hippler and Troe[281] have used the high-pressure limiting rate constant for the HO$_2$ forming reaction to determine microcanonical rate constants and hence deduce characteristics of the reverse of reaction (11.77).

The combination reaction has been extensively investigated. At room temperature, and at pressures normally investigated ($\leqslant 1$ atm), third-order kinetics prevail. The most recent set of measurements were performed by Hsu, Durant and Kaufmann[282] using a flow technique generating H by both microwave discharge and by decomposition on a tungsten filament. They detected H by resonance absorption and OH (from reaction 11.77) by resonance fluorescence, covering the pressure range 4.8–30 Torr. Pirraglia et al.[283] investigated the reaction under higher temperature conditions (746–987 K) using flash photolysis of NH$_3$ or H$_2$O in a reflected shock and detecting H by resonance absorption. Both of these sets of measurements were in agreement with the mass of earlier data, certainly within the rather wide uncertainty limits quoted in ref. 2 of Table 11.5 ($\Delta \log k = \pm 0.5$). The parameters proposed for the reaction differ widely in the two evaluations shown in Table 11.5. These differences reflect partly the difference in emphasis of the two. The room temperature estimate in ref. 1 is based on a wide body of work, while that in ref. 2 is based on the results of Hsu, Durant and Kaufmann[282] which were judged to be the most complete and accurate. The rate constant recommendations differ by only 50%. The difference in the temperature exponent derives from this discrepancy in room temperature recommendations and the differences in temperatures covered. At 600 K, the upper limit of the IUPAC range, the two expressions for k_0 in Table 11.5 agree to within 15%.

Cobos, Hippler and Troe[281] used laser flash photolysis to provide a distinctive set of room temperature measurements at pressures up to 200 atm. Under these high-pressure conditions, the reaction is no longer at the low-pressure limit, and the curvature in the fall-off plot allowed extrapolation to a high-pressure limiting value for the rate constant of 7.5×10^{-11} cm^3 molecule^{-1} s^{-1}. Analysis of the reaction using the canonical method discussed in Section 6.5 enabled the temperature dependence to be assessed theoretically, giving $k_\infty = 7.5 \times 10^{-11} (T/300 \text{ K})^{0.6}$ cm^3 molecule^{-1} s^{-1}.

THE OH + OH REACTION

The experimental basis of recommendations for this reaction is primarily the experiments of Zellner et al.[284]. They employed flash photolysis of water vapour to generate OH coupled with detection by laser-induced fluorescence over the temperature range 253–353 K and at pressures of 20–840 Torr. The extrapolation to the high-pressure limit was made with $F_c = 0.6$.

These data, data for the dissociation of H_2O_2 and state-resolved dissociation data were compared with statistical adiabatic channel calculations by Brouwer et al.[285]. Both detailed and simplified (Chapter 6) SACM calculations were employed with β calculated from spectroscopic data and α varied to obtain agreement with the high-pressure limiting recombination rate constant of Zellner et al.[284]. Comparisons were also made with measurements of the line widths in the fifth vibrational overtone of H_2O_2[286]. These measurements provide an upper limit to the microcanonical dissociation rate constant at low J (only levels with $J < 5$ are populated because of the jet cooling). Satisfactory agreement between theory and experiment was obtained. A comparison was also made with direct, time-resolved lifetime measurements in the fourth overtone region[287] with similarly satisfactory results.

The paper of Brouwer et al.[285] exemplifies a method of analysis which will be increasingly useful, in which experimental data from a variety of sources are employed to validate a theoretical model which can then be used, with increased confidence, to estimate both microcanonical and canonical rate constants.

THE $CH_3 + CH_3$ REACTION

Methyl recombination is one of the most widely studied association reactions occurring on a Type II potential energy surface. Earlier measurements, which employed rotating sector, molecular modulation and flash photolysis experiments, have been reviewed by Baulch and Duxbury[101]. The more recent measurements are based on flash photolysis and shock-tube experiments.

Hippler et al.[288] used the laser flash photolysis of azomethane at 193 nm to generate CH_3 which was detected, as a function of time, by absorption spectroscopy at 216 nm. Very high pressures, up to 200 bar, were employed in order to establish unequivocally the high-pressure limit. A value for the rate constant of 5.8×10^{-11} cm^3 molecule^{-1} s^{-1} was obtained at 296 K.

The most comprehensive set of data was obtained by the combined efforts of research groups at Oxford and Chicago[289,290]. Both employed laser flash photolysis of acetone or azomethane. The Chicago group detected CH_3 by photoionisation mass spectrometry (PIMS) which operates optimally at low pressures (0.5–10 Torr), while the Oxford group used absorption spectroscopy at higher pressures (10–800 Torr). The lower pressure limit of the latter experiments was determined by the need to relax vibrationally excited CH_3: photolysis of acetone leads to the generation of CH_3 with excess energy in the 'umbrella' vibrational mode. The collisional relaxation time varies inversely with pressure and must be significantly less than the time for recombination if interferences are to be minimised. The PIMS technique must be used at low pressures in order to optimise sampling from the reaction gas, but is more sensitive so that lower $[CH_3]$ and hence longer recombination times were employed, thus ensuring complete vibrational relaxation on the experimental time scale, even at the low pressures employed. These combined measurements were extended to even lower pressures by Walter et al.[291], who generated CH_3 in a discharge-flow system by the $F + CH_4$ reaction. They used mass spectrometry to detect CH_3.

There have been two determinations of the rate constant at higher temperatures using shock tubes, coupled with absorption spectroscopy, by Glänzer, Quack and Troe[292] and by Hwang, Wagner and Wolf[293]. These data, together with those obtained at lower temperatures, are shown in Figure 11.20.

The reaction has been modelled in a variety of ways. Wagner and Wardlaw[294] employed flexible transition state theory to calculate $k(E, J)$ values which were incorporated in a modified strong collision integral form for $k(P, T)$ (Chapter 3) and fitted to the experimental data. They suggested that the reaction has a temperature-dependent value of k_∞ given by $k_\infty = 1.5 \times 10^{-7} T^{-1.18} \exp(-329 \text{ K}/T)$ cm^3 molecule^{-1} s^{-1}; with k_∞ decreasing by a factor of 3.7 over the temperature range 300–2000 K. Stewart, Larson and Golden[295] reached a similar conclusion using a modified

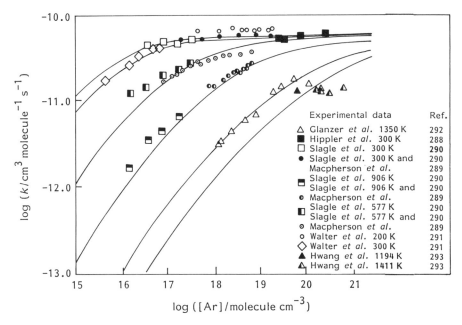

Figure 11.20 Experimental data over a range of pressures and temperatures for $CH_3 + CH_3 + M$. The curves correspond to the recommendations of ref. 3, Table 11.5. From D. L. Baulch, C. J. Cobos, R. A. Cox, P. Frank, G. Hayman, Th. Just, J. A. Kerr, T. Murrells, M. J. Pilling, J. Troe, R. W. Walker and J. Warnatz, *J. Phys. Chem. Ref. Data*, Suppl. 1, **123**, 847 (1994).

Gorin[270] model. Quack and Troe[296], on the other hand, using the statistical adiabatic channel model, have proposed that k_∞ is temperature independent, remaining at its room temperature value of 6.0×10^{-11} cm^3 molecule^{-1} s^{-1} over the experimental range. Figure 11.20 shows fits to the experimental data, based on this assumption, which are reasonable. Further experiments are needed at both low and high temperatures before this uncertainty in the temperature dependence of k_∞ can be resolved.

THE H + C₂H₄ ADDITION REACTION

This reaction has been studied by discharge flow[297], flash photolysis[298] and pulse radiolysis[299], using resonance fluorescence and resonance absorption as a hydrogen atom monitoring technique. At low temperatures the reaction is close to the high pressure limit at 1 atm pressure, and k_∞ is well defined from the experimental data. Above room temperature the reaction is in the fall-off region, even at 1 atm; k_∞ can only be obtained by extrapolation and is, in consequence, less well determined.

At temperatures around 700–800 K, the ethyl radical starts to decompose on the time scale of the addition reaction. Hanning-Lee et al.[300] and Brouard, Lightfoot and Pilling[119] employed laser flash photolysis to study both the addition and reverse H-atom dissociation reactions in a classical relaxation experiment in which H atoms are generated at zero time and relax exponentially to equilibrium with a time constant equal to the sum of the pseudo-first-order addition and first-order dissociation rate constants. Experiments of this type also enable the equilibrium constant and hence the reaction enthalpy and enthalpy of formation of C_2H_5 to be determined. In an equilibrating system, the distribution of energised C_2H_5 radicals differs from that found for either the forward or reverse reaction under irreversible conditions. In the equilibrating system, the distribution is time dependent and relaxes towards the Boltzmann distribution at infinite time for all pressures. As discussed in Chapter 7, the distribution for an irreversible reaction assumes a constant shape which is depleted relative to the Boltzmann distribution except at infinite pressure. Despite these qualitative differences, Hanning-Lee et al.[300] and Green et al.[301] showed that the rate constants measured under reversible and irreversible conditions are the same, provided some simple, and easily realised, conditions are fulfilled.

Gutman and co-workers[302] have measured rate constants for ethyl radical decomposition, using laser flash photolysis/photoionisation mass spectrometry, close to the low-pressure limit. Their work extended the temperature range over which the $H + C_2H_4/C_2H_5$ system has been studied using direct techniques. In particular they employed their data and the lower temperature data for the addition reaction to demonstrate that the average energy transferred downwards, $\langle \Delta E \rangle_d$, increases markedly with temperature. These data have already been referred to in Chapter 9.

The addition reaction of H atoms and C_2H_4 has a positive temperature dependence and an activation energy of $9.0 \, kJ \, mol^{-1}$. The reaction occurs on a Type I potential energy surface, with a constrained transition state. It has been modelled by Hase and Schlegel[303] who included a simple tunnelling correction. Their model predicts significant curvature in the Arrhenius plot and a non-Arrhenius expression has been proposed in the latest combustion evaluation (see ref. 3 of Table 11.5); the expression is based not only on the addition data, but on the dissociation data linked via the equilibrium constant. Over the temperature range 211–800 K, Hanning-Lee et al. found no evidence of curvature[300].

THE ClO + NO₂ REACTION

This reaction is important in stratospheric chemistry. The association reaction forms $ClONO_2$ which acts as an unreactive reservoir for chlorine. Its reaction with the HCl on ice and nitric acid surfaces in polar stratospheric clouds and the related release of photolabile Cl_2 (11.78) are central to the mechanism of ozone depletion in the Antarctic spring.

$$ClONO_2 + HCl \rightarrow HNO_3 + Cl_2 \qquad (11.78)$$

The reaction has been studied most recently by flash photolysis, detecting ClO by absorption spectroscopy[304,305], or by modulated photolysis coupled with detection of ClO by absorption[306] or by diode laser spectroscopy of $ClONO_2$[307]. The latter experiments demonstrated that only the isomer $ClONO_2$ is formed. Earlier experiments had suggested that other isomers were also formed because of an inconsistency between the association data, the reverse dissociation data and the equilibrium constant (see ref. 2 of Table 11.5). This problem has now been resolved following new and consistent measurements of the dissociation rate constants[308-310].

The reaction is well into the fall-off region at 1 atm pressure and extrapolation to evaluate k_∞ is unreliable. The recommended value is based on a theoretical fit to several experimental fall-off curves. An alternative expression (11.79) has been proposed by Smith and Golden[311], based on a modified Gorin[270] model.

$$k_\infty = 1.5 \times 10^{-11} \, (T/300 \text{ K})^{-1.9} \text{ cm}^3 \text{ molecule}^{-1} \text{s}^{-1} \qquad (11.79)$$

REFERENCES

1. P. J. Robinson in *Reaction Kinetics*, A Specialist Periodical Report, ed. P. G. Ashmore, The Chemical Society, London, Vol. 1, p. 93 (1975).
2. K. A. Holbrook, *Chem. Soc. Rev.*, 163 (1983).
3. D. L. Baulch, R. A. Cox, R. F. Hampson Jr, J. A. Kerr, J. Troe and R. T. Watson, *J. Phys. Chem., Ref. Data*, **9**, 295 (1980).
4. NIST Chemical Kinetics Database Version 5.0, National Institute of Standards and Technology, Gaithersburg MD 20899, USA (1993).
5. S. W. Benson, *Thermochemical Kinetics*, 2nd edn., Wiley, N. York and London, (1976); see also H. E. O'Neal and S. W. Benson, *J. Phys. Chem.*, **72**, 1866 (1968).
6. R. C. S. Grant and E. S. Swinbourne, *Chem. Commun.*, 620 (1966).
7. K. A. W. Parry and P. J. Robinson, *J. Chem. Soc. (B)*, 1019 (1970).
8. H. Heydtmann and B. Körbitzer, *Z. Phys. Chem (Frankfurt)*, **125**, 255 (1981); see also K. Eichler and H. Heydtmann, *Int. J. Chem. Kinet.*, **13**, 1107 (1981).
9. K. A. Holbrook and K. A. W. Parry, *J. Chem. Soc. (B)*, 1019 (1970).
10. M. C. Flowers and H. M. Frey, *J. Chem. Soc.*, 3547 (1961).
11. R. J. Ellis and H. M. Frey, *J. Chem. Soc.*, 5578 (1964).
12. V. Dalacker and H. Hopf, *Tetrahedron Lett.*, 15 (1974) see also ref. 13.
13. H. Hopf, G. Wachholz, R. Walsh, A. Demeijere and S. Teichmann, *Chem. Ber.*, **122**, 377 (1989).
14. R. T. Conlin and Y. W. Kwak, *Organometallics*, **5**, 1205 (1986).

15. J. P. Chesick, *J. Amer. Chem. Soc.*, **85**, 2720 (1963).
16. W. R. Dolbier, S. F. Sellers, B. H. Al-Sader and B. E. Smart, *J. Am. Chem. Soc.*, **102**, 5398 (1980).
17. K. A. W. Parry and P. J. Robinson, *Chem. Commun.*, 1377 (1968).
18. J. C. Ferrero and E. H. Staricco, *Int. J. Chem. Kinet.*, **11**, 1287 (1979).
19. R. Walsh, S. Untiedt, M. Stohlmeier and A. Demeijere, *Chem. Ber.*, **122**, 637 (1989).
20. M. Karni, I. Oref, S. Barzilai-Gilboa and A. Lifshitz, *J. Phys. Chem.*, **92**, 6924 (1988).
21. J. A. Berson in *Rearrangements in Ground and Excited States*, ed. P de Mayo, Academic Press, New York, p. 311 (1980).
22. P. B. Dervan and D. S. Santilli, *J. Am. Chem. Soc.*, **102**, 3863 (1980).
23. H. M. Frey and R. Pottinger, *J. Chem. Soc., Faraday Trans. I*, **74**, 1827 (1978).
24. T. C. Brown, K. D. King and T. T. Nguyen, *J. Phys. Chem.*, **90**, 419 (1986).
25. T. C. Brown and K. D. King, *Int. J. Chem. Kinet.*, **21**, 257 (1989).
26. H. M. Frey, *Advan. Phys. Org. Chem.*, **4**, 148 (1966).
27. J. E. Douglas, B. S. Rabinovitch and F. S. Looney, *J. Chem. Phys.*, **23**, 315 (1955).
28. K. E. Lewis and H. Steiner, *J. Chem. Soc.*, 3080 (1964).
29. H. M. Frey and R. J. Ellis, *J. Chem. Soc.*, 4770 (1965).
30. H. M. Frey and R. K. Solly, *Trans. Faraday Soc.*, **64**, 1858 (1968).
31. D. W. Setser, *J. Phys. Chem.*, **70**, 826 (1966).
32. M. Lenzi and A. Mele, *J. Chem. Phys.*, **43**, 1974 (1965).
33. M. C. Flowers and T. Öztürk, *J. Chem. Soc., Faraday Trans. I*, **71**, 1509 (1975).
34. M. C. Flowers and M. R. Honeyman, *J. Chem. Soc., Faraday Trans. I*, **79**, 2185 (1983).
35. K. A. Holbrook and R. A. Scott, *J. Chem. Soc., Faraday Trans. I*, **71**, 1849 (1975).
36. L. Zalotai, Zs Hunyadi-Zoltan, T. Bérces and F. Márta, *Int. J. Chem. Kinet.*, **15**, 505 (1983).
37. P. Hammonds and K. A. Holbrook, *J. Chem. Soc., Faraday Trans. I*, **78**, 2195 (1982).
38. M. C. Flowers and L. E. Gusel'nikov, *J. Chem. Soc (B)*, 428, 1396 (1968).
39. I. M. T. Davidson, A. Fenton, S. Ijadi-Maghsoodi, R. J. Scampton, N. Auner, J. Grobe, N. Tillman and T. J. Barton, *Organometallics*, **3**, 1593 (1984); **4**, 957 (1985).
40. R. T. Conlin, H. B. Huffaker and Y. W. Kwak, *J. Amer. Chem. Soc.*, **107**, 733 (1985).
41. I. M. T. Davidson, *Ann. Rep. Chem. Soc., C, Physical Chemistry*, **82**, 47 (1985).
42. R. Walsh, *Acc. Chem. Res.*, **14**, 246 (1981).
43. H. M. Frey and M. T. H. Liu, *J. Chem. Soc., A,* 1916 (1970) and further refs quoted in ref. 44.
44. M. T. H. Liu, *Chemical Society Reviews*, **11**, 127 (1982).
45. A. Maccoll and P. J. Thomas, *Nature*, **176**, 392 (1955).
46. S. W. Benson and A. N. Bose, *J. Chem. Phys.*, **39**, 3463 (1963).
47. S. W. Benson and G. R. Haugen, *J. Amer. Chem. Soc.*, **87**, 4036 (1965).
48. A. Maccoll in *Theoretical Organic Chemistry* (Kekulé Symposium of the Chemical Society 1958), Butterworths, London (1959).
49. K. A. Holbrook, *Proc. Chem. Soc.*, 418 (1964), see also K. A. Holbrook and

J. J. Rooney, *J. Chem. Soc.*, 247 (1965).

50. G. Chuchani, A. Rotinov and R. M. Dominguez, *Int. J. Chem. Kinet.*, **14**, 381 (1982) and refs cited therein.
51. D. H. R. Barton and P. F. Onyon, *Trans. Faraday Soc.*, **45**, 725 (1949).
52. K. E. Howlett, *Trans. Faraday Soc.*, **48**, 25 (1952).
53. P. J. Agius and A. Maccoll, *J. Chem. Soc.*, 973 (1955).
54. K. A. Holbrook, G. A. Oldershaw and M. Matthews, *Int. J. Chem. Kinet.*, **17**, 1275 (1985).
55. P. Kubát and J. Pola, *Z. Phys. Chem., Leipzig*, **268**, 849 (1987).
56. G. Huybrechts and Y. Hubin, *Int. J. Chem. Kinet.*, **17**, 157 (1985).
57. G. Huybrechts, Y. Hubin and B van Mele, *Int. J. Chem. Kinet.*, **21**, 575 (1989).
58. S. W. Benson, *J. Chem. Phys.*, **38**, 1945 (1963).
59. J. L. Holmes and A. Maccoll, *J. Chem. Soc.*, 5919 (1963).
60. D. J. Rakestraw and B. E. Holmes, *J. Phys. Chem.*, **95**, 3968 (1991).
61. G. G. Smith and F. W. Kelly, *Prog. Phys. Org. Chem.*, **8**, 75 (1971) and refs cited therein.
62. R. Taylor in *Suppl. B. The Chemistry of Acid Derivatives*, ed. S. Patai, Wiley, Chichester, p. 860 (1979).
63. G. Chuchani and R. M. Dominguez, *Int. J. Chem. Kinet.*, **13**, 377 (1981) and earlier papers in this series.
64. C. D. Hurd and F. H. Blunck, *J. Amer. Chem. Soc.*, **60**, 2419 (1938).
65. K. A. Holbrook, *Suppl. B2. The Chemistry of Acid Derivatives*, ed. S. Patai, Wiley, Chichester, Vol. 2, Ch. 14. (1993).
66. E. Dyer and G. C. Wright, *J. Amer. Chem. Soc.*, **81**, 2138 (1959).
67. R. Taylor, *J. Chem. Soc., Perkin Trans.*, **2**, 291 (1983).
68. E. S. Lewis and W. C. Herndon, *J. Amer. Chem. Soc.*, **83**, 1955 (1961).
69. N. Barroeta, V de Santis and M. Rincon, *J. Chem. Soc., Perkin Trans.*, **2**, 911 (1974).
70. See K. H. Leavell and E. S. Lewis, *Tetrahedron*, **28**, 1167 (1972) and refs cited therein.
71. G. Chuchani, R. M. Dominguez and A. Rotinov, *Int. J. Chem. Kinet.*, **18**, 203 (1986).
72. G. Chuchani and A. Rotinov, *Int. J. Chem. Kinet.*, **19**, 789 (1987).
73. V. S. Rao, K. Takeda and G. B. Skinner, *Int. J. Chem. Kinet.*, **20**, 153 (1988); see also S. W. Benson, *Int. J. Chem. Kinet.*, **21**, 233 (1989).
74. M. Rossi and D. M. Golden, *Int. J. Chem. Kinet.*, **11**, 715 (1979).
75. P. G. Blake and A. Speis, *J. Chem. Soc (B)*, 1877 (1971).
76. R. Taylor, *J. Chem. Soc., Perk. Trans. II*, 89 (1983).
77. N. Al-Awadi and R. Taylor, *J. Chem. Soc. Perk Trans. II*, 1255 (1986).
78. D. W. Placzek, B. S. Rabinovitch, G. Z. Whitten and E. Tschuikow-Roux, *J. Chem. Phys.*, **43**, 4071 (1965).
79. K. M. Maloney and B. S. Rabinovitch, *J. Phys. Chem.*, **72**, 4483 (1968).
80. D. F. Kelley, B. D. Barton, L. Zalotai and B. S. Rabinovitch, *J. Chem. Phys.*, **71**, 538 (1979).
81. K. D. King, D. M. Golden, G. N. Spokes and S. W. Benson, *Int. J. Chem. Kinet.*, **3**, 411 (1971).
82. K. D. King and R. D. Goddard, *Int. J. Chem. Kinet.*, **7**, 109 (1975).
83. J. Troe, *J. Phys. Chem.*, **83**, 114 (1979); see also J. Troe, *Ber. Bunsenges. Phys. Chem.*, **87**, 161 (1983), and R. G. Gilbert, K. Luther, J. Troe, *Ber. Bunsenges. Phys. Chem.*, **87**, 169 (1983).
84. G. M. Wieder and R. A. Marcus, *J. Chem. Phys.*, **37**, 1835 (1962).

85. M. C. Lin and K. J. Laidler, *Trans. Faraday Soc.*, **64**, 927 (1968).
86. D. C. Tardy and B. S. Rabinovitch, *J. Chem. Phys.*, **48**, 1282 (1968).
87. J. Troe, *J. Chem. Phys.*, **66**, 4745 (1977).
88. J. Troe, *J. Chem. Phys.*, **66**, 4758 (1977).
89. L. Zalotai, Zs. Hunyadi-Zoltán, T. Bérces and F. Márta, *Int. J. Chem. Kinet.*, **15**, 505 (1983).
90. I. M. Bailey and H. M. Frey, *J. Chem. Soc., Faraday Trans. I*, **77**, 709 (1981).
91. N. Chow and D. J. Wilson, *J. Phys. Chem.*, **66**, 342 (1962).
92. E. V. Waage and B. S. Rabinovitch, *J. Phys. Chem.*, **76**, 1695 (1972).
93. See for example P. J. Papagiannakopoulos and S. W. Benson, *Int. J. Chem. Kinet.*, **14**, 63, (1982).
94. K-H Jung, S. H. Kang, C. U. Ro and E. Tschuikow-Roux, *J. Phys. Chem.*, **91**, 2354 (1987).
95. Th. Just and J. Troe, *J. Phys. Chem.*, **84**, 3068 (1980).
96. J. P. Burton and C. J. S. M. Simpson, *Chem. Phys.*, **105**, 307 (1986).
97. S. H. Bauer and K. I. Lazaar, *J. Chem. Phys.*, **79**, 2808 (1983).
98. P. S. Chu and N. S. True, *J. Phys. Chem.*, **89**, 5613 (1985); J. P. Chauvel Jr, C. B. Conboy and N. S. True, *J. Chem. Phys.*, **80**, 3561 (1984); C. B. Conboy, J. P. Chauvel and N. S. True, *J. Phys. Chem.*, **90**, 4388 (1986).
99. M. C. Lin and K. J. Laidler, *Trans. Faraday Soc.*, **64**, 94 (1968).
100. S. H. Bauer, *Int. J. Chem. Kinet.*, **17**, 367 (1985).
101. D. L. Baulch and J. Duxbury, *Comb. and Flame*, **37**, 313 (1980).
102. E. V. Waage and B. S. Rabinovitch, *Int. J. Chem. Kinet.*, **3**, 105 (1971).
103. A. B. Trenwith, *J. Chem. Soc., Faraday Trans. I*, **75**, 614 (1979).
104. G. B. Skinner, D. Rogers and K. B. Patel, *Int. J. Chem. Kinet.*, **13**, 481 (1981).
105. D. L. Baulch, C. J. Cobos, R. A. Cox, C. Esser, P. Frank, Th. Just, J. A. Kerr, M. J. Pilling, J. Troe, R. W. Walker and J. Warnatz, *J. Phys. Chem. Ref. Data*, **21**, 411 (1992).
106. M. C. Lin and M. H. Back, *Can. J. Chem.*, **44**, 505 (1966).
107. H. M. Frey and I. C. Vinall, *Int. J. Chem. Kinet.*, **5**, 523 (1973).
108. L. Batt, *Int. J. Chem. Kinet.*, **11**, 977 (1979).
109. K. Y. Choo and S. W. Benson, *Int. J. Chem. Kinet.*, **13**, 833 (1981).
110. L. Batt and G. N. Robinson, *Int. J. Chem. Kinet.*, **19**, 391 (1987).
111. L. Batt and G. N. Robinson, *Int. J. Chem. Kinet.*, **14**, 1053 (1982).
112. P. G. Ashmore, A. J. Owen and P. J. Robinson, *J. Chem. Soc., Faraday Trans. I*, **78**, 677 (1982).
113. P. G. Ashmore, J. W. Gardner, A. J. Owen, B. Smith and P. R. Sutton, *J. Chem. Soc., Faraday Trans. I*, **78**, 657 (1982).
114. G. Huybrechts, J. Kathabwa, G. Martens, M. Nejszaten and J. Olbregts, *Bull. Soc. Chim. Belg.*, **81**, 65 (1972).
115. M. Page, M. C. Lin, Y. He and T. K. Choudhury, *J. Phys. Chem.*, **93**, 4404 (1989).
116. B. C. Garrett and D. G. Truhlar, *J. Phys. Chem.*, **83**, 1079 (1979).
117. W. H. Miller, *J. Amer. Chem. Soc.*, **101**, 6810 (1979).
118. B. C. Garrett, D. G. Truhlar, *J. Phys. Chem.*, 2921 (1979).
119. M. Brouard, P. D. Lightfoot and M. J. Pilling, *J. Phys. Chem.*, **90**, 445 (1986).
120. I. Slagle, L. Batt, G. W. Gmurczyk, D. Gutman and W. Tsang, *J. Phys. Chem.*, **95**, 7732 (1991).
121. R. W. Walker in *Reaction Kinetics*, Specialist Periodical Report, ed. P. G. Ashmore, The Chemical Society, London, Vol. 1, p. 161 (1975).
122. R. R. Baldwin, M. W. M. Hisham and R. W. Walker, *J. Chem. Soc., Faraday*

Trans. I, **78**, 1615 (1982).

123. R. R. Baldwin, I. A. Pickering and R. W. Walker, *J. Chem. Soc., Faraday Trans. I*, **74**, 2229 (1978).

124. J. N. Bradley, *Shock Waves in Chemistry and Physics*, Methuen/Wiley (1962).

125. J. N. Bradley, *Fast Reactions*, Oxford Chem. Series, Clarendon Press, Oxford, (1975).

126. A. Lifshitz, ed. *Shock Waves in Chemistry*, M. Dekker, New York, pp. 59–129 (1981).

127. E. F. Greene and J. P. Toennies, *Chemical Reactions in Shock Waves*, Arnold, London (1964).

128. W. Tsang, *J. Chem. Phys.*, **41**, 2487 (1964).

129. W. Tsang, *J. Chem. Phys.*, **40**, 1171 (1964).

130. W. Tsang, *Int. J. Chem. Kinet.*, **2**, 311 (1970) and refs cited therein.

131. W. Tsang, *Int. J. Chem. Kinet.*, **8**, 173 (1976).

132. C. G. Newman, J. Dzarnaski, M. A. Ring and H. E. O'Neal, *Int. J. Chem. Kinet.*, **12**, 661 (1980).

133. K. Tamaru, M. Boudart and H. Taylor, *J. Phys. Chem.*, **59**, 806 (1955).

134. G. G. Devyatykh and I. A. Frolov, *Russ. J. Inorg. Chem.*, **11**, 385 (1966).

135. I. M. Bailey and R. Walsh, *J. Chem. Soc., Faraday Trans. I*, **74**, 1146 (1978).

136. L. D. Brouwer, W. Müller-Markgraf and J. Troe, *J. Phys. Chem.*, **92**, 4905 (1988).

137. H. Hippler, V. Schubert, J. Troe and H. J. Wendelken, *Chem. Phys. Lett.*, **84**, 253 (1981);

138. U. Brand, H. Hippler and L. Lindemann, J. Troe, *J. Phys. Chem.*, **94**, 6305 (1990).

139. W. Müller-Markgraf and J. Troe, *J. Phys. Chem.*, **92**, 4914 (1988).

140. W. Müller-Markgraf and J. Troe, *J. Phys. Chem.*, **92**, 4899 (1988).

141. A. Lifshitz, M. Bidani and S. Bidani, *J. Phys. Chem.*, **90**, 3422 (1986).

142. C. H. Klute and W. D. Walters, *J. Amer. Chem. Soc.*, **68**, 506 (1946).

143. O. Kajimoto, O. Kondo, K. Okada, J. Fujikane, T. Fueno, *Bull. Chem. Soc., Jpn.*, **58**, 3469 (1985).

144. R. A. Back, J. Childs, *Can. J. Chem.*, **46**, 1023 (1968).

145. O. Kajimoto, T. Yamamoto, T. Fueno, *J. Phys. Chem.*, **83**, 429 (1979).

146. K. W. Michael, H. G. Wagner, *Proc. 10th Int. Sym. Comb.*, **1964**, 353 (1965).

147. W. Tsang, S. H. Bauer, M. Cowperthwaite, *J. Chem. Phys.*, **36**, 1768 (1962).

148. T. Fueno, K. Tabayashi, O. Kojimoto, *J. Phys. Chem.*, **77**, 575 (1973).

149. H. Kijewski, J. Troe, *Int. J. Chem. Kinet.*, **3**, 223 (1971).

150. D. C. Astholtz, K. Glänzer, J. Troe, *J. Chem. Phys.*, **70**, 2409 (1979).

151. K. Saito, T. Kakumoto, H. Kuroda, S. Torii, A. Imamura, *J. Chem. Phys.*, **80**, 4989 (1984).

152. H. Hippler, J. Troe, J. Willner, *J. Chem. Phys.*, **93**, 1755 (1990).

153. R. Atkinson and B. A. Thrush, *Proc. Roy. Soc. (London)*, **A316**, 123 (1970).

154. R. Atkinson and B. A. Thrush, *Proc. Roy. Soc. (London)*, **A316**, 131 (1970).

155. J. Troe and W. Wieters, *J. Chem. Phys.*, **71**, 3931 (1979).

156. D. Astholtz, J. Troe and W. Wieters, *J. Chem. Phys.*, **70**, 5107 (1979).

157. H. Hippler, K. Luther, J. Troe and H. J. Wendelken, *J. Chem. Phys.*, **79**, 239 (1983).

158. H. Hippler, J. Troe and H. J. Wendelken, *J. Chem. Phys.*, **78**, 6709, 6718 (1983).

159. D. Dudek, K. Glänzer and J. Troe, *Ber. Bunsenges. Phys. Chem.*, **83**, 788 (1979).

160. H. Hippler, V. Schubert, J. Troe and H. J. Wendelken, *Chem. Phys. Lett.*, **84**, 253 (1981).
161. K. Tsukiyama and R. Bersohn, *J. Chem. Phys.*, **86**, 745 (1987).
162. K. Luther, J. Troe and K.-M. Weitzel, *J. Phys. Chem.*, **94**, 6316 (1990).
163. L. R. Khundar, J. L. Knee and A. H. Zewail, *J. Chem. Phys.*, **87**, 77 (1987).
164. (a) P. Pechukas and J. C. Light, *J. Chem. Phys.*, **42**, 3281 (1965); (b) P. Pechukas, R. Rankin and J. C. Light, *J. Chem. Phys.*, **44**, 794 (1966).
165. S. Klippenstein, L. R. Khundar, A. H. Zewail and R. A. Marcus, *J. Chem. Phys.*, **89**, 4761 (1988).
166. G. Herzberg, *Infrared and Raman Spectra*, Van Nostrand, New York (1945).
167. M. L. Sage and J. Jortner, *Adv. Chem. Phys.*, **47**, 293 (1981).
168. F. F. Crim, *Ann. Rev. Phys. Chem.*, **35**, 657 (1984).
169. (a) B. R. Henry, *Acc. Chem. Res.*, **10**, 207 (1977); (b) A. Sinha, J. D. Thoemke and F. F. Crim, *J. Chem. Phys.*, **96**, 372 (1992).
170. K. V. Reddy and M. J. Berry, *Chem. Phys. Lett.*, **52**, 111 (1977).
171. K. V. Reddy, R. G. Bray and M. J. Berry in *Advances in Laser Chemistry*, ed. A. H. Zewail, Springer-Verlag, Berlin, p. 48 (1978).
172. S. C. Chan, B. S. Rabinovitch, J. T. Bryant, L. D. Spicer, T. Fujimoto, Y. N. Lin and S. P. Pavlou, *J. Phys. Chem.*, **74**, 3160 (1970).
173. J. E. Baggott, *Chem. Phys. Lett.*, **119**, 47 (1985).
174. J. Troe, *J. Phys. Chem.*, **87**, 1800 (1983).
175. K. V. Reddy and M. J. Berry, *Chem. Phys. Lett.*, **66**, 223 (1979).
176. J. Segal and R. N. Zare, *J. Chem. Phys.*, **89**, 5704 (1988).
177. D. W. Chandler, W. E. Farneth and R. N. Zare, *J. Chem. Phys.*, **77**, 4447 (1982).
178. T. R. Rizzo and F. F. Crim, *J. Chem. Phys.*, **76**, 2754 (1982).
179. T. R. Rizzo, C. C. Hayden and F. F. Crim, *Disc. Farad. Soc.*, **75**, 223 (1983).
180. B. D. Cannon and F. F. Crim, *J. Chem. Phys.*, **75**, 1752 (1981).
181. N. F. Scherer, F. E. Doany, A. H. Zewail and J. W. Perry, *J. Chem. Phys.*, **84**, 1932 (1986).
182. N. F. Scherer and A. H. Zewail, *J. Chem. Phys.*, **87**, 97 (1987).
183. E. S. McGinley and F. F. Crim, *J. Chem. Phys.*, **85**, 5748 (1986).
184. H. Reisler and C. Wittig, *Ann. Rev. Phys. Chem.*, **37**, 307 (1986).
185. R. G. Harrison and S. R. Butcher, *Contemp. Phys.*, **21**, 19 (1980).
186. D. M. Golden, M. J. Rossi, A. C. Baldwin and J. R. Barker, *Acc. Chem. Res.*, **14**, 56 (1981).
187. M. N. R. Ashfold and G. Hancock in *Gas Kinetics and Energy Transfer*, The Chemical Society, London and New York, Vol. 4, p. 73 (1981).
188. W. C. Danen and J. C. Jang in *Laser-induced Chemical Processes*, ed. J. I. Steinfeld, Plenum, New York (1981).
189. M. Quack, *J. Chem. Phys.*, **69**, 1282 (1978).
190. M. Quack, *Ber. Bunsenges. Phys. Chem.*, **83**, 757 (1979).
191. M. Quack, *Ber. Bunsenges. Phys. Chem.*, **83**, 1287 (1979).
192. M. Quack, *Ber. Bunsenges. Phys. Chem.*, **85**, 318 (1981).
193. K. A. Holbrook, G. A. Oldershaw and M. Matthews, *Int. J. Chem. Kinet.*, **17**, 1275 (1985).
194. W. M. Shaub and S. H. Bauer, *Int. J. Chem. Kinet.*, **7**, 509 (1975).
195. J. Zhu and E. S. Yeung, *J. Phys. Chem.*, **92**, 2184 (1988).
196. W. C. Danen, V. C. Rio and D. W. Setser, *J. Am. Chem. Soc.*, **104**, 5431 (1982).
197. J. C. Jang and D. W. Setser, *J. Phys. Chem.*, **83**, 2809 (1979).

198. J. C. Jang-Wren, D. W. Setser and J. C. Ferrero, *J. Phys. Chem.*, **89**, 414 (1985).
199. K. A. Holbrook, C. J. Shaw, G. A. Oldershaw and P. J. Dyer, *Int. J. Chem. Kinet.*, **18**, 1215 (1986).
200. S. Ruhman, O. Anner and Y. Haas, *J. Phys. Chem.*, **88**, 6397 (1984).
201. D. W. Lupo and M. Quack, *Chem. Rev.*, **87**, 181 (1987).
202. R. B. Hall and A. Kaldor, *J. Chem. Phys.*, **70**, 4027 (1979).
203. R. L. Woodin, R. B. Hall, C. F. Meyer and A. Kaldor in *Advances in Laser Spectroscopy*, ed. B. A. Garetz and J. R. Lombardi, Heydon and Son, London, Vol. 1, 1982.
204. D. M. Brenner, *Chem. Phys. Lett.*, **57**, 357 (1978).
205. R. A. Back, *Can. J. Chem.*, **60**, 2542 (1982).
206. P. John, M. R. Humphries, R. G. Harrison and P. G. Harper, *J. Chem. Phys.*, **79**, 1353 (1983).
207. I. Oref and B. S. Rabinovitch, *Acc. Chem. Res.*, **12**, 166 (1979).
208. D. Krajnovich, F. Huisken, Z. Zhang, Y. R. Shen and Y. T. Lee, *J. Chem. Phys.*, **77**, 5977 (1982).
209. F. Huisken, D. Krajnovich, Z. Zhang, Y. R. Shen and Y. T. Lee, *J. Chem. Phys.*, **78**, 3806 (1983).
210. Aa. S. Sudbø, P. A. Schulz, E. R. Grant, Y. R. Shen and Y. T. Lee, *J. Chem. Phys.*, **70**, 912 (1975).
211. L. J. Butler, R. J. Buss, R. J. Brudzynski and Y. T. Lee, *J. Phys. Chem.*, **87**, 5106 (1983).
212. E. J. Hintsa, A. M. Wodtke and Y. T. Lee, *J. Phys. Chem.*, **92**, 5379 (1988).
213. A. M. Schmoltner, D. S. Anex and Y. T. Lee, *J. Phys. Chem.*, **96**, 1236 (1992).
214. A. M. Wodtke and Y. T. Lee in *Advances in Gas-Phase Photochemistry and Kinetics, Molecular Photodissociation Dynamics*, ed. M. N. R. Ashfold and J. E. Baggott, p. 31 (1987).
215. D. S. King and J. C. Stephenson, *Chem. Phys. Lett.*, **114**, 461 (1985).
216. R. M. Robertson, D. M. Golden and M. J. Rossi, *J. Chem. Phys.*, **89**, 2925 (1988).
217. M. J. Shultz, L. M. Yam and R. E. Tricca, *J. Phys. Chem.*, **93**, 1884 (1989).
218. I. Oref, *J. Chem. Phys.*, **75**, 131 (1981).
219. B. M. Toselli, G. A. McRae, M. Ivanco and R. D. McAlpine, *J. Phys. Chem.*, **96**, 4912 (1992).
220. H. N. Rosenstock, M. B. Wallenstein, A. L. Wahrhaftig and H. Eyring, *Proc. Natl. Acad. Sci.*, **38**, 667 (1952); see also M. L. Vestal in *Fundamental Processes in Radiation Chemistry*, ed, P. Ausloos, Interscience, New York, p. 59 (1968).
221. W. Forst, *Theory of Unimolecular Reactions*, Academic Press, New York (1973).
222. C. Lifshitz, *J. Phys. Chem.*, **87**, 2304 (1983).
223. P.B. Comita and J. I. Brauman, *Science*, **227**, 863 (1985).
224. A. G. Brenton, R. P. Morgan and J. H. Beynon, *Ann. Rev. Phys. Chem.*, **30**, 51 (1979).
225. *Gas Phase Ion Chemistry*, ed. M. T. Bowers, Academic Press, New York and London, Vols I–III (1979) and (1984).
226. C. Lifshitz, *Int. Rev. Phys. Chem.*, **6**, 35 (1987).
227. H. J. Neusser, *Int. J. Mass. Spec., Ion. Proc.*, **79**, 141 (1987).
228. B. Brehm and E. von Puttkamer, *Z. Naturforsch. A*, **22**, 8 (1967).
229. J. H. D. Eland, *Int. J. Mass Spec., Ion. Phys.*, **8**, 143 (1972).
230. T. Baer in *Gas Phase Ion Chemistry*, ed. M. T. Bowers, Academic Press, New

York and London, Vol. 1, p. 153 (1979).

231. T. Baer, B. P. Tsai, D. Smith and P. T. Murray, *J. Chem. Phys.*, **64**, 2460 (1976).

232. H. M. Rosenstock, R. Stockbauer and A. C. Parr, *J. Chem. Phys.*, **73**, 773 (1980).

233. Y. Malinovitch, R. Akawa, G. Haase and C. Lifshitz, *J. Phys. Chem.*, **89**, 2253 (1985).

234. S. T. Pratt and W. A. Chupka, *Chem. Phys.*, **62**, 153 (1981).

235. C. E. Klots, *J. Phys. Chem.*, **75**, 1526 (1971).

236. J. L. Franklin in *Gas Phase Ion Chemistry*, ed. M. T. Bowers, Academic Press, New York and London, Vol. 1, p. 273 (1979).

237. C. Lifshitz, *Adv. Mass Spectrom.*, **7A**, 3 (1978).

238. See the examples quoted by W. Forst *Theory of Unimolecular Reactions*, Academic Press, New York, p. 303 (1979).

239. C. Lifshitz, T. Peres, N. Ohmichi, I. Pri-bar and L. Radom, Paper D7, Int. Conf. on Chem. Kinetics, Bordeaux (1986); see also *J. Phys. Chem.*, **87**, 2304 (1983).

240. E. Pollak and C. Schlier, *Acc. Chem. Res.*, **22**, 223 (1989).

241. A. Carrington and J. R. McNab, *Acc. Chem. Res.*, **22**, 218 (1989).

242. (a) U. Boesl, H. J. Neusser and E. W. Schlag, *J. Chem. Phys.*, **72**, 4327 (1980); (b) E. W. Schlag and H. J. Neusser, *Acc. Chem. Res.*, **16**, 355 (1983); (c) H. J. Neusser, H. Kühlewind, U. Boesl and E. W. Schlag, *Ber. Bunsenges. Phys. Chem.*, **89**, 276 (1985).

243. J. H. D. Eland and H. Schulte, *J. Chem. Phys.*, **62**, 3835 (1975).

244. W. T. Huntress, *Chem. Soc. Rev.*, **6**, 295 (1977).

245. Chemistry in the interstellar medium, *Spec. issue J. Chem. Soc., Farad. Trans.*, **2** (10) (1989).

246. R. P. Wayne *Chemistry of Atmospheres*, 2nd edn, Oxford Univ. Press, Ch. 6 (1991).

247. *Rate Coefficients in Astrochemistry*, ed. T. J. Millar and D. A. Williams, Kluwer, Dordrecht (1988).

248. (a) E. E. Ferguson, F. C. Fehsenfeld and D. L. Albritton in *Gas Phase Ion Chemistry*, ed. M. T. Bowers, Academic Press, New York and London, Vol. 1, p. 45 (1979); (b) E. E. Ferguson and F. Arnold, *Acc. Chem. Res.*, **14**, 327 (1981).

249. M. P. Langevin, *Am. Chim. Phys.*, **5**, 245 (1905).

250. G. Gioumousis and D. P. Stevenson, *J. Chem. Phys.*, **29**, 294 (1958).

251. T. Su and M. T. Bowers, *J. Chem. Phys.*, **58**, 3027 (1973).

252. T. Su and M. T. Bowers, *Int. J. Mass Spectrum. Ion. Phys.*, **12**, 347 (1973).

253. R. D. Bowen, *Acc. Chem. Res.*, **24**, 364 (1991).

254. R. Patrick and D. M. Golden, *J. Chem. Phys.*, **82**, 75 (1985).

255. J. Troe, *J. Chem. Phys.*, **66**, 4758 (1977).

256. J. Troe, *Ber. Bunsenges. Phys. Chem.*, **87**, 161 (1983).

257. H. Bohringer and F. Arnold, *J. Chem. Phys.*, **77**, 5534 (1982).

258. W. Lindinger, I. Dotan, D. L. Albritton and F. C. Fehsenfeld, *J. Chem. Phys.*, **68**, 2607 (1968).

259. M. F. Jarrold, A. J. Illies and M. T. Bowers, *J. Chem. Phys.*, **81**, 214, 222 (1984).

260. E. E. Ferguson, F. C. Fehsenfeld and A. L. Schmeltekopf, *Adv. in Atomic and Mol. Phys.*, **5**, 1 (1969).

261. D. L. Albritton, *Atom Data Nucl. Data Tables*, **22**, 1 (1978).

262. V. G. Anicich and W. T. Huntress Jr., *Astrophys. J. Suppl.*, **62**, 553 (1986).

263. D. Smith and N. G. Adams in *Gas Phase Ion Chemistry*, ed. M. T. Bowers, Academic Press, New York and London, Vol 1, p. 1 (1979).
264. D. Smith and N. G. Adams, *Adv. in Atomic and Mol. Phys.*, **24**, 1 (1988).
265. B. R. Rowe, *Atom Data Nucl. Data Tables*, **22**, 1 (1978).
266. D. H. Aue and M. T. Bowers in *Gas Phase Ion Chemistry*, ed. M. T. Bowers, Academic Press, New York and London, Vol. 2, p. 2 (1979).
267. J. E. Bartmess and R. T. McIver in *Gas Phase Ion Chemistry*, ed. M. T. Bowers, Academic Press, New York and London, Vol. 2, p. 88 (1979).
268. W. N. Olmstead, M. Lev-on, D. M. Golden and J. I. Brauman, *J. Amer. Chem. Soc.*, **99**, 992 (1977).
269. M. Meot-Ner and F. H. Field, *J. Am. Chem. Soc.*, **97**, 5339 (1975).
270. E. Gorin, *Acta. Physicochim URSS*, **9**, 691 (1938).
271. R. G. Gilbert and M. J. McEwan, *Aust. J. Chem.*, **38**, 231 (1985).
272. S. C. Smith, M. J. McEwan and R. G. Gilbert, *J. Chem. Phys.*, **90**, 1630 (1989).
273. S. C. Smith, M. J. McEwan and R. G. Gilbert, *J. Chem. Phys.*, **90**, 4265 (1989).
274. P. R. Kemper, L. M. Bass and M. T. Bowers, *J. Phys. Chem.*, **89**, 1105 (1985).
275. J. S. Knight, C. G. Freeman and M. J. McEwan, *J. Amer. Chem. Soc.*, **108**, 1404 (1986).
276. D. L. Baulch, D. D. Drysdale, D. G. Horne and A. C. Lloyd, Evaluated Kinetic Data for High Temperature Reactions, Vol. 1: '*Homogeneous Gas Phase Reactions of the H_2-O_2 System*', Butterworths, London (1972).
277. D. W. Trainor, D. O. Ham and F. Kaufman, *J. Chem. Phys.*, **58**, 4599 (1973).
278. V. V. Azatyan, R. R. Borodulin and E. I. Intezarova, *Dokl. Akad. Nauk SSSR*, **213**, 856 (1973).
279. G. Dixon-Lewis, J. B. Greenberg and F. A. Goldsworthy, *15th Symp. (Int.) Comb.*, 717 (1975).
280. W. G. Mallard and J. H. Owen, *Int. J. Chem. Kinet.*, **6**, 753 (1974).
281. C. J. Cobos, J. Hippler and J. Troe, *J. Phys. Chem.*, **89**, 342 (1985).
282. K.-J. Hsu, J. L. Durant and F. Kaufman, *J. Phys. Chem.*, **91**, 1895 (1987).
283. A. N. Pirraglia, J. V. Michael, J. W. Sutherland and R. B. Klemm, *J. Phys. Chem.*, **93**, 282 (1982).
284. R. Zellner, F. Ewig, R. Paschke and G. Wagner, *J. Phys. Chem.*, **92**, 4184 (1988).
285. L. Brouwer, C. J. Cobos, J. Troe, H-R. Dübal and F. F. Crim, *J. Chem. Phys.*, **86**, 6171 (1987).
286. L. J. Butler, T. M. Ticich, M. D. Likar and F. F. Crim, *J. Chem. Phys.*, **85**, 2331 (1986).
287. N. F. Scherer, F. E. Duany, A. H. Zewail and J. W. Perry, *J. Chem. Phys.*, **84**, 1932 (1986).
288. H. Hippler, K. Luther, A. R. Ravishankara and J. Troe, *Z. Phys. Chem. NF*, **142**, 1 (1984).
289. M. T. Macpherson, M. J. Pilling and M. J. C. Smith, *J. Phys. Chem.*, **89**, 2268 (1985).
290. I. R. Slagle, D. Gutman, J. W. Davies and M. J. Pilling, *J. Phys. Chem.*, **92**, 2455 (1988).
291. D. Walter, H-H. Grotheer, J. W. Davies, M. J. Pilling and A. F. Wagner, *23rd Symp. (Int.) Combust.*, 107 (1990).
292. K. Glänzer, M. Quack and J. Troe, *16th Symp. (Int.) Combust.*, 949 (1977).
293. S. M. Hwang, H. Gg. Wagner and Th. Wolff, *23rd Symp. (Int.) Combust.*, 99 (1990).

294. A. F. Wagner and D. M. Wardlaw, *J. Phys. Chem.*, **92**, 2462 (1988).
295. P. H. Stewart, C. W. Larson and D. M. Golden, *Comb. and Flame*, **75**, 25 (1989).
296. M. Quack and J. Troe, *Ber. Bunsenges. Phys. Chem.*, **81**, 329 (1977).
297. J. H. Lee, J. V. Michael, W. A. Payne and L. J. Stief, *J. Chem. Phys.*, **68**, 1817 (1978).
298. P. D. Lightfoot and M. J. Pilling, *J. Phys. Chem.*, **91**, 3373 (1987).
299. K. Sugawara, K. Okazaki and S. Sato, *Bull. Chem. Soc. Jpn.*, **54**, 2872 (1981).
300. M. A. Hanning-Lee, N. J. B. Green, M. J. Pilling and S. H. Robertson, *J. Phys. Chem.*, **97**, 860 (1993).
301. N. J. B. Green, P. J. Marchant, M. J. Perona, M. J. Pilling and S. H. Robertson, *J. Chem. Phys.*, **96**, 5896 (1992).
302. Y. Feng, J. T. Niiranen, A. Bencsura, V. D. Knyazev, D. Gutman and W. Tsang, *J. Phys. Chem.*, **97**, 871 (1993).
303. W. L. Hase and H. B. Schlegel, *J. Phys. Chem.*, **86**, 3901 (1982).
304. W. Dasch, K. H. Sternberg and R. N. Schindler, *Ber. Bunsenges. Phys. Chem.*, **85**, 611 (1981).
305. V. Handwerk and R. Zellner, *Ber. Bunsenges. Phys. Chem.*, **88**, 405 (1984).
306. T. J. Wallington and R. A. Cox, *J. Chem. Soc. Faraday Trans. 2*, **82**, 275 (1986).
307. R. A. Cox, J. P. Burrows and G. B. Coker, *Int. J. Chem. Kinet.*, **16**, 445 (1984).
308. H. D. Knauth, *Ber. Bunsenges. P,ys. Chem.*, **82**, 212 (1978).
309. G. Schoenle, H. D. Knauth anc R. N. Schindler, *J. Phys. Chem.*, **83**, 3297 (1979).
310. L. C. Anderson and D. W. Fahey, *J. Phys. Chem.*, **94**, 644 (1990).
311. G. P. Smith and D. M. Golden, *Int. J. Chem. Kinet.*, **10**, 489 (1978).

REFERENCES FOR TABLE 11.1

1. W. E. Falconer, T. F. Hunter and A. F. Trotman-Dickenson, *J. Chem. Soc.*, 609 (1961).
2. E. W. Schlag and B. S. Rabinovitch, *J. Amer. Chem. Soc.*, **82**, 5996 (1960), note that the observed A-factor of $10^{16.4}$ refers to $k = k_f + k_r$.
3. Parameters calculated by Placzek and Rabinovitch (*J. Phys. Chem.*, **69**, 2141 (1965)) from data of D. W. Setser and B. S. Rabinovitch, *J. Amer. Chem. Soc.*, **86**, 564 (1964).
4. M. C. Flowers and H. M. Frey, *Proc. Roy. Soc.* **A260**, 424 (1961).
5. F. Casas, J. A. Kerr and A. F. Trotman-Dickenson, *J. Chem. Soc.*, 3655 (1964).
6. R. C. S. Grant and E. S. Swinbourne, *Chem. Commun.*, 620 (1966).
7. K. A. W. Parry and P. J. Robinson, *J. Chem. Soc. (B)*, 49 (1969).
8. H. Heydtmann and B. Körbitzer, *Z. Phys. Chem (Frankfurt)*, **125**, 255 (1981); see also K. Eichler and H. Heydtmann, *Int. J. Chem. Kinet.*, **13**, 1107 (1981).
9. K. A. Holbrook and K. A. W. Parry, *J. Chem. Soc. (B)*, 1019 (1970).
10. P. J. Robinson and M. J. Waller, *Int. J. Chem. Kinet.*, **11**, 937 (1979).
11. R. T. Conlin and Y. W. Kwak, *Organometallics*, **5**, 1205 (1986).
12. M. C. Flowers and H. M. Frey, *J. Chem. Soc.*, 3547 (1961).
13. R. J. Ellis and H. M. Frey, *J. Chem. Soc.*, 5578 (1964).
14. H. Hopf, G. Wachholz, R. Walsh, A. Demeijere and S. Teichmann, *Chem. Ber.*, **122**, 377 (1989).

15. B. Atkinson and D. McKeagan, *Chem. Commun.*, 189 (1966).
16. R. A. Mitsch and E. W. Neuvar, *J. Phys. Chem.*, **70**, 546 (1966).
17. C. T. Genaux, F. Kern and W. D. Walters, *J. Amer. Chem. Soc.*, **75**, 6196 (1953); R. W. Carr and W. D. Walters, *J. Phys. Chem.*, **67**, 1370 (1963).
18. M. N. Das and W. D. Walters, *Zeit. Phys. Chem (Frankfurt)*, **15**, 22 (1958).
19. H. R. Gerberich and W. D. Walters, *J. Amer. Chem. Soc.*, **83**, 4884 (1961).
20. H. R. Gerberich and W. D. Walters, *J. Amer. Chem. Soc.*, **83**, 3935 (1961); see also P. J. Conn, Ph.D Thesis, University of Oregon, 1966 [cf. *Diss. Abs. (B)*, **27**, 2311 (1966–7)].
21. K. D. King, B. J. Gaynor and R. G. Gilbert, *Int. J. Chem. Kinet.*, **11**, 11 (1979).
22. K. D. King and R. G. Gilbert, *Int. J. Chem. Kinet.*, **12**, 339 (1980).
23. H. M. Frey and R. T. Conlin, *J. Chem. Soc., Faraday Trans. I*, **75**, 2556 (1979).
24. M. N. Das, F. Kern, T. D. Coyle and W. D. Walters, *J. Amer. Chem. Soc.*, **76**, 6271 (1954).
25. A. T. Blades, *Can. J. Chem.*, **47**, 615 (1969).
26. P. M. Stacy, Ph.D. Thesis, University of Rochester, 1968 (cf. *Diss. Abs. (B)*, **29**, 1632 (1968–69)).
27. N. K. Dirjal and K. A. Holbrook, *J. Chem. Soc., Faraday Trans.*, **87**, 691 (1991).
28. H. M. Frey and R. Pottinger, *J. Chem. Soc., Faraday Trans. I*, **74**, 1827 (1978).
29. J. P. Chesick, *J. Phys. Chem.*, **65**, 2170 (1961); see also R. L. Brandauer, B. Short and S. M. E. Kellner, *J. Phys. Chem.*, **65**, 2269 (1961) and W. J. Engelbrecht and M. J. DeVries, *J. S. Afr. Chem. Inst.*, **23**, 163, 172 (1970).
30. T. C. Brown, K. D. King and T. T. Nguyen, *J. Phys. Chem.*, **90**, 419 (1986).
31. T. C. Brown and K. D. King, *Int. J. Chem. Kinet.*, **21**, 251 (1989).
32. I. M. Bailey and R. Walsh, *J. Chem. Soc., Faraday Trans. I*, **74**, 1146 (1978).
33. R. Srinavasan, *J. Chem. Soc., Chem. Commun.*, 1041 (1971).
34. R. Walsh, S. Untiedt, M. Stohlmeier and A. Demeijere, *Chem. Ber.*, **122**, 637 (1989).
35. R. W. Carr and W. D. Walters, *J. Phys. Chem.*, **69**, 1073 (1965).
36. H. M. Frey, *Trans. Faraday Soc.*, **58**, 957 (1962).
37. H. M. Frey, *Trans. Faraday Soc.*, **60**, 83 (1964).
38. R. Srinavasan, *J. Amer. Chem. Soc.*, **91**, 7557 (1969).
39. K. D. King, *Int. J. Chem. Kinet.*, **10**, 117 (1978).
40. D. C. Tardy, R. Ireton and A. S. Gordon, *J. Amer. Chem. Soc.*, **101**, 1508 (1979).
41. W. Tsang, *Int. J. Chem. Kinet.*, **5**, 651 (1973).
42. J. A. Barnard and T. K. Parrott, *J. Chem. Soc., Faraday I*, **72**, 2404 (1976).
43. W. Tsang, *Int. J. Chem. Kinet.*, **2**, 311 (1970).
44. G. Huybrechts, D. Rigaux, J. Vankeerberghen and B. van der Mele, *Int. J. Chem. Kinet.*, **12**, 253 (1980).
45. S. McLean, C. J. Webster and R. J. D. Rutherford, *Can. J. Chem.*, **47**, 1555 (1969).
46. H. M. Frey and D. H. Lister, *J. Chem. Soc. (A)*, 1800 (1967).
47. J. P. Chesick, *J. Amer. Chem. Soc.*, **84**, 3250 (1962).
48. C. Steel, R. Zand, P. Hurwitz and S. G. Cohen, *J. Amer. Chem. Soc.*, **86**, 679 (1964).
49. E. Ratajczak and A. F. Trotman-Dickenson, *J. Chem. Soc. (A)*, 509 (1968).
50. I. Haller, *J. Phys. Chem.*, **72**, 2882 (1968).
51. H. M. Frey and I. D. R. Stevens, *Trans. Faraday Soc.*, **61**, 90 (1965).
52. J. E. Douglas, B. S. Rabinovitch and F. S. Looney, *J. Chem. Phys.*, **23**, 315 (1955).

53. M. C. Flowers and N. Jonathan, *J. Chem. Phys.*, **50**, 2805 (1968); **52**, 1623 (1970).
54. B. S. Rabinovitch and K. W. Michel, *J. Amer. Chem. Soc.*, **81**, 5065 (1959).
55. K. E. Lewis and H. Steiner, *J. Chem. Soc.*, 3080 (1964).
56. W. R. Roth and J. König, *Ann.*, **699**, 24 (1966).
57. H. M. Frey and B. M. Pope, *J. Chem. Soc. (A)*, 1701 (1966).
58. V. Toscano and W. von E. Doering, quoted by W. von Doering and J. Gilbert, *Tetrahedron Supplement*, **7**, 397 (1966).
59. H. M. Frey and R. K. Solly, *Trans. Faraday Soc.*, **64**, 1858 (1968).
60. M. Lenzi and A. Mele, *J. Chem. Phys.*, **43**, 1974 (1965).
61. M. C. Flowers and M. R. Honeyman, *J. Chem. Soc., Faraday Trans. I*, **76**, 2290 (1980).
62. M. C. Flowers and M. R. Honeyman, *J. Chem. Soc., Faraday Trans. I*, **77**, 1923 (1981).
63. H. M. Frey and I. D. R. Stevens, *J. Chem. Soc.*, 3865 (1962).
64. M. R. Bridge, H. M. Frey and M. T. H. Liu, *J. Chem. Soc. (A)*, 91 (1969).
65. K. A. Holbrook and R. A. Scott, *J. Chem. Soc. Faraday Trans. I*, **71**, 1849 (1975).
66. L. Zalotai, Zs. Hunyadi-Zoltan, T. Bérces and F. Mártà, *Int. J. Chem. Kinet.*, **15**, 505 (1983).
67. L. Zalotai, T. Bérces and F. Mártà, *J. Chem. Soc., Faraday Trans.*, **86**, 21 (1990).
68. P. Hammonds and K. A. Holbrook, *J. Chem. Soc., Faraday Trans. I*, **78**, 2195 (1982).
69. M. C. Flowers and L. E. Gusel'nikov, *J. Chem. Soc. (B)*, 428, 1396 (1968).
70. C. A. Wellington and W. D. Walters, *J. Amer. Chem. Soc.*, **83**, 4888 (1961).
71. A. C. Thomas and C. A. Wellington, *J. Chem. Soc. (A)*, 2895 (1969).
72. C. A. Wellington, *J. Chem. Soc. (A)*, 2897 (1969).
73. H. M. Frey and S. P. Lodge, *J. Chem. Soc., Perkin Trans. 2*, 1463 (1979).
74. K. Okada, E. Tschuikow-Roux and P. J. Evans, *J. Phys. Chem.*, **84**, 467 (1980).
75. K. A. Holbrook and A. R. W. Marsh, *Trans. Faraday Soc.*, **63**, 643 (1967).
76. P. J. Evans, T. Ichimura and E. Tschuikow-Roux, *Int. J. Chem. Kinet.*, **10**, 855 (1978).
77. P. J. Thomas, *J. Chem. Soc.*, 1192 (1959).
78. T. J. Park and K. H. Jung, *Bull. Korean Chem. Soc.*, **1**, 30 (1980).
79. J. H. Yang and D. C. Conway, *J. Chem. Phys.*, **43**, 1296 (1965).
80. S. W. Benson and A. N. Bose, *J. Chem. Phys.*, **37**, 2935 (1962).
81. W. Tsang, *J. Chem. Phys.*, **41**, 2487 (1964).
82. W. Tsang, *J. Chem. Phys.*, **40**, 1498 (1964).
83. R. L. Failes, Y. Mollah and J. S. Shapiro, *Int. J. Chem. Kinet.*, **11**, 1271 (1979).
84. A. T. Blades, *Can. J. Chem.*, **32**, 366 (1954).
85. E. U. Emovon and A. Maccoll, *J. Chem. Soc.*, 335 (1962).
86. P. G. Blake and B. F. Shraydeh, *Int. J. Chem. Kinet.*, **13**, 463 (1981).
87. G. Chuchani, R. M. Dominguez and A. Rotinov, *Int. J. Chem. Kinet.*, **18**, 203 (1986).

REFERENCES FOR TABLE 11.2

1. H. O. Pritchard, R. G. Sowden and A. F. Trotman-Dickenson, *Proc. Roy. Soc.*, **A217**, 563 (1953); see also W. G. Falconer, T. F. Hunter and A. F. Trotman-Dickenson, *J. Chem. Soc.*, 609 (1961).

2. G. M. Wieder and R. A. Marcus, *J. Chem. Phys.*, **37**, 1835 (1962).
3. M. C. Lin and K. J. Laidler, *Trans. Faraday Soc.*, **64**, 927 (1968).
4. J. P. Chesick, *J. Amer. Chem. Soc.*, **82**, 3277 (1960).
5. D. W. Setser and B. S. Rabinovitch, *J. Amer. Chem. Soc.*, **86**, 564 (1964).
6. M. L. Halberstadt and J. P. Chesick, *J. Phys. Chem.*, **69**, 429 (1965).
7. K. A. Holbrook, J. S. Palmer, K. A. W. Parry and P. J. Robinson, *Trans. Faraday Soc.*, **66**, 869 (1970); see also K. Eichler and H. Heydtmann, *Int. J. Chem. Kinet.*, **13**, 1107 (1981).
8. J. N. Butler and R. B. Ogawa, *J. Amer. Chem. Soc.*, **85**, 3346 (1963).
9. R. W. Vreeland and D. F. Swinehart, *J. Amer. Chem. Soc.*, **85**, 3349 (1963).
10. A. F. Patarrachia and W. D. Walters, *J. Phys. Chem.*, **68**, 3894 (1964).
11. T. F. Thomas, P. J. Conn and D. F. Swinehart, *J. Amer. Chem. Soc.*, **91**, 7611 (1969).
12. W. P. Hauser and W. D. Walters, *J. Phys. Chem.*, **67**, 1328 (1963).
13. C. S. Elliott and H. M. Frey, *Trans. Faraday Soc.*, **62**, 895 (1966).
14. M. C. Lin and K. J. Laidler, *Trans. Faraday Soc.*, **64**, 94 (1968).
15. M. K. Knecht, *J. Amer. Chem. Soc.*, **91**, 7667 (1969).
16. D. C. Marshall and H. M. Frey, *Trans. Faraday Soc.*, **61**, 1715 (1965).
17. K. A. Holbrook and A. R. W. Marsh, *Trans. Faraday Soc.*, **63**, 643 (1967).
18. F. W. Schneider and B. S. Rabinovitch, *J. Amer. Chem. Soc.*, **84**, 4215 (1962).
19. F. J. Fletcher, B. S. Rabinovitch, K. W. Watkins and D. J. Locker, *J. Phys. Chem.*, **70**, 2823 (1966).
20. I. M. Bailey and H. M. Frey, *J. Chem. Soc. Faraday Trans. 1*, **77**, 709 (1981).
21. R. T. Conlin and H. M. Frey, *J. Chem. Soc., Faraday Trans. 1*, **76**, 322 (1980).
22. L. Zalotai, Zs. Hunyadi-Zoltán, T. Bérces and F. Mártà, *Int. J. Chem. Kinet.*, **15**, 505 (1983).
23. K. A. Holbrook and R. A. Scott, *J. Chem. Soc., Faraday Trans. 1*, **71**, 1849 (1975).
24. H. M. Frey and I. M. Pidgeon, *J. Chem. Soc., Faraday Trans. 1*, **84**, 1087 (1985).
25. H. M. Frey and H. P. Watts, *J. Chem. Soc., Faraday Trans. 1*, **79**, 1659 (1983).
26. J. P. Buxton and C. J. S. M. Simpson, *Chem. Phys.*, **105**, 307 (1986).
27. J. P. Chauvel Jr., C. B. Conboy and N. S. True, *J. Chem. Phys.*, **80**, 3561 (1984).
28. C. B. Conboy, J. P. Chauvel and N. S. True, *J. Phys. Chem.*, **90**, 4388 (1986).
29. B. D. Ross and N. S. True, *J. Amer. Chem. Soc.*, **105**, 1382 (1983).
30. P. S. Chu and N. S. True, *J. Phys. Chem.*, **89**, 2625 (1985).
31. P. S. Chu and N. S. True, *J. Phys. Chem.*, **89**, 5613 (1985).
32. C. B. Lemaster, C. L. Lemaster, C. Suarez, M. Tafazzd and N. S. True, *J. Phys. Chem.*, **93**, 3995 (1989).

REFERENCES FOR TABLE 11.3

1. K. Tabayahi and S. H. Bauer, *Comb. and Flame*, **34**, 63 (1979).
2. C. C. Chiang, J. A. Baker and G. B. Skinner, *J. Phys. Chem.*, **84**, 939 (1980).
3. D. B. Olson, T. Tanzawa and W. C. Gardiner, *Int. J. Chem. Kinet.*, **11**, 23 (1979).
4. O. Kondo, K. Saito and I. Murakami, *Bull. Chem. Soc. Jpn.*, **53**, 2133 (1980).
5. K. Saito, H. Tahara, O. Kondo, T. Yokubo, T. Higashihara, and I. Murakani, *Bull. Chem. Soc. Jpn.*, **53**, 1335 (1980).
6. J. E. Dove and W. S. Nip, *Can. J. Chem.*, **57**, 689 (1979).

REFERENCES FOR TABLE 11.5

1. D. L. Baulch, C. J. Cobos, R. A. Cox, C. Esser, P. Frank, Th. Just. J. A. Kerr, M. J. Pilling, J. Troe, R. W. Walker and J. Warnatz, *J. Phys. Chem Ref. Data*, **21**, 411 (1992).
2. R. Atkinson, D. L. Baulch, R. A. Cox, R. F. Hampson Jr., J. A. Kerr and J. Troe, *J. Phys. Chem. Ref. Data*, **21**, 1125 (1992).
3. D. L. Baulch, C. J. Cobos, R. A. Cox, P. Frank, G. Hayman, Th. Just, J. A. Kerr, T. Murrells, M. J. Pilling, J. Troe, R. W. Walker and J. Warnatz, *J. Phys. Chem Ref. Data*, Suppl **1**, **123**, 847 (1994).

Appendix 1   Nomenclature

The object of this appendix is to list and define the principal symbols used in the present work. Terms which are used only locally are mostly omitted, and this omission includes most of the special nomenclature used in the chapters on Slater theory, isotope effects and collisional energy transfer. On the other hand, the terms used in the basic RRKM theory are listed fairly exhaustively. A few symbols are used with more than one meaning, but the appropriate interpretation in these cases will always be clear from the context.

A1.1   GENERAL NOMENCLATURE

$\langle x \rangle$	Average value of a quantity x for molecules of interest (these being defined locally)
$\displaystyle\sum_{i=1}^{n} x_i$	$= x_1 + x_2 + \ldots + x_n$
$\displaystyle\prod_{i=1}^{n} x_i$	$= x_1 \times x_2 \times \ldots \times x_n$
$\Gamma(n)$	The gamma function of argument n (see Appendix 4)
$n!$	Factorial $n = \Gamma(n + 1)$
$(X)_i$	The ith element of a vector X
h, k, N_A, R	Planck, Boltzmann, Avogadro and gas constants
p, T	Pressure and absolute temperature
Q, Q_v, etc.	Molecular partition functions
σ	Symmetry number
$S(A)$	Entropy of a species A
$S_v(A), S_r(A)$	Vibrational and rotational contributions to $S(A)$
s	Number of vibrational degrees of freedom of a molecule (or Kassel's parameter in Chapter 2)
ν, ν_i	Vibration frequencies
v, v_i	Vibrational quantum numbers
p	Number of rotations of a molecule, i.e. the number of different moments of inertia
d_i	Degeneracy of ith rotation
r	Number of rotational degrees of freedom, $$r = \sum_{i=1}^{p} d_i$$
I, I_A, etc.	Moments of inertia

J, J_i	Rotational quantum numbers
E, E_v, etc.	Energy per molecule (unless it is stated otherwise or clear from the context that molar energies $N_A E$ are intended)
E_z	Zero-point energy of a molecule
$P(E_n)$	Number of quantum states of a given system at the quantised energy level E_n

$$W(E) \equiv \sum_{E_n=0}^{E} P(E_n)$$

Number of quantum states of a system at all energies up to and including E (see section 3.11 and Chapter 4)

$$\left[\equiv \sum P(E_n) \right]$$

GENERAL RATE PARAMETERS

k	Rate constants in general
k_{uni}	First-order rate constant, $-(1/[\text{A}]) \, \text{d}[\text{A}]/\text{d}t$
k_∞	Limiting high-pressure value of k_{uni}
k_{bim}, k_0	Second-order rate constant, $-(1/[\text{A}]^2) \, \text{d}[\text{A}]/\text{d}t = k_{\text{uni}}/[\text{A}]$
k_1, k_2	Rate constants (second order) for energisation and de-energisation respectively
k_3	Rate constant for conversion of energised molecules to products when this is independent of energy (HL theories)
$k_a(E)$	Same when energy dependent as in RRK and RRKM theories
$L(E_1, E_2, \ldots, E_n)$	Slater's specific dissociation probability for molecules having energies $E_1, \ldots, E_n$ in normal modes 1 to n
Z, σ_{d}	Second-order collision rate constant and collision diameter
ω	$= Zp = $ first-order rate constant for collisions
λ, β_c	Collisional energy transfer efficiencies (section 3.12 and Chapter 9)
E_0	Critical energy for reaction
E_{Arr}, A	Arrhenius activation energy (equation (1.3)) and A-factor $A = k \exp(E_{\text{Arr}}/kT)$
$A_\infty, E_\infty, E_{\text{bim}}$	Limiting high-pressure values of A and E_{Arr}, and low-pressure value of E_{Arr}
$p_{\frac{1}{2}}$	Pressure at which $k_{\text{uni}} = \frac{1}{2} k_\infty$
$L^{\ddagger}$	Statistical factor or reaction path degeneracy
$\Delta S^{\ddagger}$	Entropy of activation: $\Delta S^{\ddagger} = S(A^+) - S(A)$
D	Rate of production of decomposition or isomerisation products in a chemical activation experiment
S	Rate of production of stabilised molecules in a chemical activation experiment
$f(E)\delta E$	Boltzmann distribution function
$P_{i,j}$	Conditional probability of energy transfer from state j to state i
$\langle k_a \rangle$	Average value of $k_a(E)$ for the reacting molecules in a chemical activation experiment
$\langle k_a \rangle_0, \langle k_a \rangle_\infty$	Low- and high-pressure values of $\langle k_a \rangle$

NOMENCLATURE USED PARTICULARLY IN CONNECTION WITH RRKM THEORY

A^*, A^+	Energised molecule and activated complex
$E^*(\equiv E_{vr}^*)$	Total non-fixed energy in the active degrees of freedom of a given energised molecule $A^*(E^* = E_v^* + E_r^*$; called E_{active}^* in section 3.10)
E_v^*, E_r^*	Vibrational and rotational parts of E^*
E^+	Total non-fixed energy in the active degrees of freedom of a given transition state $A^+(E^+ = E_v^+ + E_r^+ + x)$
E_v^+, E_r^+	Vibrational and rotational contributions to E^+
E_{vr}^+	$(E_v^+ + E_r^+)$
x	Translational energy of A^+ in the reaction coordinate
E_J, E_J^+	Energy of adiabatic rotations in their Jth energy level in A^* and A^+ respectively; $\Delta E_J = E_J^+ - E_J$
$\rho(E^*)$	Density of quantum states of A^* at energy E^* (see section 3.11)
$\rho(E_{vr}^+, x)$	Density of quantum states of A^+ having energy E_{vr}^+ in the active degrees of freedom and energy x in the reaction coordinate
$\rho_{rc}^+(x)$	Density of quantum states for the translational motion of A^+ in the reaction coordinate with energy x
$P(E_{vr}^+)$	Number of vibrational–rotational quantum states of A^+ at the quantised energy level E_{vr}^+
$W(E) = \sum\limits_{E_{vr}^+=0}^{E^+} P(E_{vr}^+)$ or $\sum P(E_{vr}^+)$	Number of quantum states of A^+ at all energies up to and including E^+ (see section 3.11)
δ	Arbitrary length of region at top of potential energy barrier which is taken to define the activated complex
μ	Characteristic mass for motion in the reaction coordinate
$k^+(x)$	Rate constant with which complexes of energy x cross the barrier
Q_1, Q_1^+	Partition functions for adiabatic rotations in A and A^+ respectively
Q_2, Q_2^+	Partition functions for active degrees of freedom of A and A^+ respectively (A^+ has one fewer degree of freedom)
Q, Q^+	Complete vibrational–rotational partition functions for A and A^+ respectively; $Q = Q_1 Q_2$ and $Q^+ = Q_1^+ Q_2^+$
σ, σ^+	Symmetry numbers of A and A^+
Q_2^*, $Q_2^{*\prime}$	See section 3.8 and Appendix 2
$k_{EJ}(E)$	Symbol for $k_a(E)$ when it depends on the rotational quantum number J as well as the total energy E
E'	Reduced energy E/E_z
α, β_R, $w(E')$	Whitten–Rabinovitch functions; see section 4.4

Appendix 2 Statistical Mechanics

In this appendix we outline some parts of statistical mechanics which are important in rate theories. Particular emphasis is placed on some features which are important in the RRKM theory, but which are not dealt with prominently in the books on statistical mechanics. The treatment is mainly in terms of quantum statistical systems, but section A2.7 refers briefly to the classical treatment with its related concept of phase space.

Fuller details and more of the background can be found in the standard works on statistical mechanics[1].

A2.1 PARTITION FUNCTIONS: DEFINITION

If a molecular system is capable of existing in a series of quantised energy levels with total energy E_0, E_1, E_2, ..., then the *partition function* Q for the molecule is defined as

$$Q = g_0 \exp(-E_0/kT) + g_1 \exp(-E_1/kT) + g_2 \exp(-E_2/kT) + \dots$$
$$= \sum_{i=0}^{\infty} g_i \exp(-E_i/kT)$$

(A2.1)

where g_i is the *degeneracy* or *statistical weight* of the energy level E_i, defined as the number of distinct (i.e. physically distinguishable) quantum states of that energy. The degeneracy is alternatively the number of different independent wave functions of the system with total energy E_i, or the number of physically distinct ways this energy can be distributed in the molecule. These factors are introduced into equation (A2.1) so that the sum can be taken over all energy levels rather than evaluating $\Sigma \exp(-E_i/kT)$ over all quantum states (some of the same energy), which is mathematically more complicated. For a given system (i.e. for a given set of energy levels E_i) the numerical value of Q depends only on the temperature and on the zero chosen for the energy scale; if the energy zero is shifted down by an amount ΔE all the energies are increased by ΔE so that each exponential term in (A2.1) is multiplied by $\exp(-\Delta E/kT)$ and Q is simply multiplied by the same factor.

It is also possible to consider the partition function not for a whole molecule but for certain degrees of freedom of the molecule, the definition being as before, but with E_i being the energy specifically in these degrees of freedom, and g_i the number of different ways this energy can be distributed in these degrees of freedom. For example, it is common to consider separately the electronic, vibrational, rotational and translational partition functions, Q_e, Q_v, Q_r and Q_t, and it is not hard to show that if the energies are simply additive, i.e. if the total energy can be written as

$E_{tot} = E_e + E_v + E_r + E_t$, then the molecular partition function is the product of the individual partition functions:

$$Q = Q_e \cdot Q_v \cdot Q_r \cdot Q_t$$

In the same way it is not essential to consider together all the degrees of freedom of one type; for example Q_v can be factorised into contributions for different vibrations, provided that these are independent so that their energies are additive. Such an approach may be useful if an approximate treatment of Q_v is valid for some vibrations but not for others. Expressions for the partition functions of some specific systems are given in section A2.6.

A2.2   SIGNIFICANCE OF Q: POPULATION OF STATES

The basic usefulness of a partition function derives from the fact that each term of the series (A2.1) is proportional to the number of molecules existing in the corresponding energy level at thermal equilibrium. Since the sum of the terms (Q) therefore represents the total number of molecules (N) the fraction of molecules in the ith energy level is given by

$$\frac{N_i}{N} = \frac{g_i \exp\left(-E_i/kT\right)}{Q} \tag{A2.2}$$

It is interesting to note that for molecules of the same (or nearly the same) energy E_i the population is governed merely by the statistical weights g_i (the number of quantum states corresponding to each energy level). This is an expression of the fundamental postulate of statistical mechanics, *that all quantum states of a system are a priori equally probable* and will be equally populated in the absence of constraints imposed (for example) by energetic considerations. This point will be further illustrated in section A2.5.

It may be useful by way of background information to discuss briefly the significance of numerical values of Q in terms of the distribution of molecules between the energy levels. The result obtained by summing (A2.1) depends very much on the relative values of the energies E_i and the quantity kT which is the yardstick of thermal energy—the amount of energy which is readily available in a system. If E_0 is taken as zero, as is common, then for $E_i \gg kT$ there will be few molecules in the excited state since energy E_i will seldom be available to activate the molecules. If, on the other hand, $E_i \ll kT$, there will be almost as many molecules with energy E_i as with energy E_0. Thus only at high energies (relative to kT) will there be a significant decrease in the population of the energy levels.

These ideas are well illustrated by the cases of a typical vibration and a translation at room temperature. Consider, for example, the vibration of a bromine molecule Br_2; this has the fairly low vibration frequency of $v = 9.7 \times 10^{12}$ s^{-1} and hence energy levels (relative to the ground state) of $vhv = 0, 6.4 \times 10^{-21}, 12.8 \times 10^{-21}$, etc. J (per molecule). At 25 °C kT is 4.1×10^{-21} J so the series for the partition function is (to three decimal places):

$$Q_v(Br_2) = 1.00 + 0.210 + 0.044 + 0.009 + 0.002 + 0.000 + \ldots$$

$$= 1.265$$

$$(\equiv [1 - \exp\left(-hv/kT\right)] - 1 \text{ by algebraic summation})$$

It will be seen that this situation, in which kT is only of the same order of magnitude as the spacing of the energy levels, gives a partition function not much greater than unity. This in turn shows that the population of the ground state is high; from equation (A2.2),

$$N_0/N = \exp(-E_0/kT)/Q = 1.000/1.265 = 0.79 \text{ or } 79\%$$

Correspondingly, the population falls off rapidly in the excited energy levels, the figures being 17%, 3.5%, 0.7%, etc. in successive levels.

In contrast, the energy levels for translation are very closely spaced relative to kT for most molecules even at very low temperatures. For example, for the one-dimensional translation of a bromine molecule (mass $m = 2.7 \times 10^{-22}$ g) in a container of length $b = 10$ cm, the energy levels are given (relative to the ground state) by

$$(h^2/8mb^2)(n^2 - 1) = 0, 6.2 \times 10^{-41}, 1.6 \times 10^{-40}, 3.1 \times 10^{-40}, \ldots \text{ (for } n = 10^9)$$

$$2.1 \times 10^{-23}, \ldots \text{ (for } n = 10^{10}) 2.1 \times 10^{-21}, \ldots \text{ (for } n = 10^{11}) 2.1 \times 10^{-19} \text{ J}, \ldots$$

compared with $kT = 4.1 \times 10^{-21}$ J at 25 °C. There are clearly a very large number of energy levels easily accessible to molecules with the typical energy kT, and this is reflected in the series for the partition function, again given to three decimal places:

$$Q_t(Br_2) = 1.000 + 1.000 + 1.000 + 1.00 + \ldots$$

$$+ \text{ (for } n = 10^9) 0.995 + \ldots + \text{ (for } n = 10^{10}) 0.603 + \ldots$$

$$+ \text{ (for } n = 10^{11}) 0.007 + \ldots$$

$$\approx 6.2 \times 10^{10}$$

$$[\equiv (2\pi mb^2 kT/h^2)^{1/2} \text{ by integration}]$$

Thus when the thermal energy kT is large compared with the spacing of the energy levels, the partition function has a very large value and there is a correspondingly low fractional population of all energy levels. In the above example all the levels up to about the 10^9th have $N_i/N = 1/(6.2 \times 10^{10})$ to within $\frac{1}{2}\%$, and only at around the 10^{10}th level is the number of molecules decreasing significantly. As before, all quantum states are equally probable when the availability of energy imposes no restraint, i.e. at levels for which $E_i \ll kT$. It is under these circumstances, where the appropriate degrees of freedom are highly excited, that the effects of the quantisation of the energy disappear and a classical treatment of the motion in terms of a continuously variable energy becomes adequate.

A2.3 *Q* FOR MOLECULES RESTRICTED TO A SPECIFIED ENERGY RANGE

This aspect does not figure very prominently in the books on statistical mechanics but is fundamental to the RRKM theory and is therefore dealt with separately here. It is quite in order to consider the partition function for molecules restricted to a certain range of energies in the relevant degrees of freedom, but in this case only the relevant terms in equation (A2.1) must be included. The partition function for a molecule A in the energy range E to E' is therefore given by

$$Q(A_{(E \to E')}) = \sum_{E < E_i < E'} g_i \exp(-E_i/kT) \tag{A2.3}$$

For molecules with energy in a small range E to $E + \delta E$, the exponential terms in (A2.3) will be nearly the same for all the states of interest and can be replaced throughout the range by $\exp(-E/kT)$. The partition function then becomes

$$Q(A_{(E \to E+\delta E)}) = \left(\sum_{E < E_i < E+\delta E} g_i \right) \exp(-E/kT) \tag{A2.4}$$

in which the summation is now simply the total number of quantum states with energies in the relevant range $E \to E + \delta E$. If further energy levels are very closely spaced so that there are many levels in this small range, Σg_i can well be replaced by $\rho(E)\delta E$ where $\rho(E)$ is a continuous distribution function, the density of quantum states or the number of quantum states per unit energy range at energy E (see also sections 3.4 and 3.11). In this case

$$Q(A_{(E \to E+\delta E)}) = \rho(E) \exp(-E/kT)\delta E \tag{A2.5}$$

A2.4   ENERGY AND ENTROPY

If the partition function for a system is known as a function of temperature, the standard statistical mechanical results (A2.6) and (A2.7) can be used to calculate the average thermal energy per molecule, $\langle E \rangle$, and the molar entropy S of the system.

$$\langle E \rangle = kT^2 \, d \ln Q / dT \tag{A2.6}$$

$$S = R \frac{dT \ln Q}{dT} = R \ln Q + RT \frac{d \ln Q}{dT}$$
$$= R \ln Q + N_A \langle E \rangle / T \tag{A2.7}$$

The average energy $\langle E \rangle$ is given relative to the chosen ground state and is identical with the internal energy U of the system relative to the same energy zero. Equations (A2.6) and (A2.7) can also be applied to individual degrees of freedom of a system if the partition function is factorised into individual contributions as discussed in section A2.1.

A2.5   EQUILIBRIUM CONSIDERATIONS

One of the main uses of partition functions, and the use which is especially important in rate theories, is their use to calculate equilibrium constants. It is shown in the works on statistical mechanics that the equilibrium constant K for a reaction

$$A \rightleftharpoons B$$

is given by

$$K = \frac{Q(B)}{Q(A)} \exp(-\Delta E^\circ/kT)$$

where $Q(A)$ and $Q(B)$ are the partition functions for A and B respectively and ΔE° is the difference between the energy zeros of A and B used for the evaluation of the partition functions; ΔE° is closely related to the heat of the reaction.

It happens in the RRKM theory that the equilibria under consideration involve molecules for which the partition functions are calculated using the same energy zero, and hence $\Delta E° = 0$ and the equilibrium constant becomes

$$K = Q(B)/Q(A)$$

In one case both partition functions refer to excited molecules with energies in the same small range E to $E + \delta E$, in which case the partition functions can be expressed as in (A2.4) and we obtain

$$K = \sum g_i(B)/\sum g_i(A)$$

where the $\sum g_i$ are the number of quantum states of A and B in the small energy range. This gives the simple result, already mentioned in section A2.2, that if there are two conditions X and Y of a system, corresponding to n_X and n_Y quantum states respectively, the molecules will distribute themselves between these conditions in the ratio n_X/n_Y provided there are no constraints on the system.

Finally, if the energy levels of A and B in the range E to $E + \delta E$ are close enough together for a continuous distribution function to be used, (A2.5) gives

$$K = \rho_B(E)/\rho_A(E)$$

where $\rho_B(E)$ and $\rho_A(E)$ are the densities of quantum states of B and A respectively at energy E.

A2.6 EXPRESSIONS FOR SPECIFIC SYSTEMS

In the numerical application of reaction rate theories, specific expressions are required for the partition functions Q, average energies $\langle E \rangle$ and entropies S of the various degrees of freedom of the species involved. Translational contributions are of no concern in unimolecular reactions, and electronic contributions, which are only occasionally significant, may be obtained by direct summation of (A2.1). The equations required most are those for vibrational and rotational degrees of freedom, and some of the commonly used expressions for Q and $\langle E \rangle/kT$ are given in Table A2.1. The corresponding results for S are obtained simply from (A2.7).

The most widely used expressions are those for the simple harmonic oscillator, and the results here are conveniently expressed in terms of the dimensionless quantity $x = h\nu/kT - 1.4388(\bar{\nu}/cm^{-1})/(T/K)$. The oscillator is sometimes alternatively characterised by the quantum energy $E_q = N_A hc\bar{\nu}$, where $N_A hc = 0.01196$ kJ mol^{-1} cm, in which case $x = (E_q/kJ\,mol^{-1})/(8.3144T/K)$.

The expressions for molecular rotations given in Table A2.1 are all based on the classical treatment, as is clear from the $\langle E \rangle$ values which are all $\frac{1}{2}kT$ per rotational degree of freedom. Some care may be needed in the selection of the appropriate expressions for Q for a given calculation, and this problem is discussed in section 4.2. If moments of inertia are given in g cm^2 the appropriate value of $(8\pi^2 k/h^2)$ is 2.481×10^{45} kg^{-1} m^{2}K^{-1}. Moments of inertia are sometimes alternatively given in amu Å$^2 = 1.6604 \times 10^{-47}$ kg m^2. The case of hindered internal rotation is again more complicated[2,3], but the thermodynamic functions can be readily evaluated from tables given by Pitzer and Brewer[2]. State-counting for such systems is complicated and is generally avoided by substituting in the model a harmonic vibration or a free rotation which has the same entropy at the temperature of interest as the actual hindered rotation.

Table A2.1 Commonly used expressions for Q and $\langle E \rangle / kT$ for vibrational and rotational degrees of freedom; the corresponding molar entropies are given by $S = R \ln Q + R(\langle E \rangle / kT)$

Type of motion	Degrees of freedom	Q	$\langle E \rangle / kT$
Simple harmonic oscillator	1	$[1 - \exp(-h\nu/kT)]^{-1}$	$\dfrac{(h\nu/kT)}{[\exp(h\nu/kT) - 1]}$
Free internal rotation	1	$\pi^{1/2}(8\pi^2 kT/h^2)^{1/2}I^{1/2}$	$\frac{1}{2}$
Overall rotation of linear molecule	2	$(8\pi^2 kT/h^2)I$	1
Overall rotation of non-linear molecule[a]	3	$\pi^{1/2}(8\pi^2 kT/h^2)^{3/2}(I_A I_B I_C)^{1/2}$	$\frac{3}{2}$
p independent rotations (Marcus expression)[b]	$r = \displaystyle\sum_{i=1}^{p} d_i$	$\left\{\displaystyle\prod_{i=1}^{p}\Gamma(\tfrac{1}{2}d_i)\right\}(8\pi^2 kT/h^2)^{(\frac{1}{2})r}\displaystyle\prod_{i=1}^{p}I_{i,\frac{1}{2}d_i}$	$\frac{1}{2}r$

[a] Expression for Q applies strictly to spherical tops ($I_A = I_B = I_C$) and symmetric tops ($I_A \neq I_B = I_C$) and approximately to asymmetric tops ($I_A \neq I_B \neq I_C$).
[b] See Appendix 4 and section 4.5.

A2.7 PHASE SPACE AND CLASSICAL STATISTICAL MECHANICS

The concept of phase space is not used extensively in this book, and only a brief indication of some salient features will be given. A system of N particles behaving classically can be completely described by $3N$ position coordinates, together with $3N$ momenta, i.e. by the positions and momenta of each particle in three independent directions. The state of the whole system can thus be specified by the position of a point (the *representative point*) in a $6N$-dimensional space known as the phase space of the system. Any change in the position or velocity of one or more of the particles is described by a movement of the representative point in the phase space. Since changes in the system are subject to certain requirements such as the conservation of total momentum, the representative point can only move on certain defined lines in phase space, and these paths are known as *trajectories*. The concept is somewhat similar to the well-known mechanical analogy[4] of a ball rolling on a potential energy surface; once the ball is given a certain initial position and velocity its subsequent path is predetermined. For a system capable of chemical reaction some regions of the phase-space will be recognisable as reactants and some as products; these regions may be separated by a $(6N\text{-}1)$-dimensional hypersurface, and reaction corresponds to the crossing of this hypersurface by the representative point[5]. Crossings will occur most frequently at the lowest energy regions of the hypersurface, and these critical molecular configurations correspond loosely to the transition states of CTST.

The application of statistical mechanics to classical systems depends on the replacement of summations over discrete energy levels by integrations over regions of phase space. For example, the classical partition function is given by (A2.8) instead of (A2.1)

$$Q_{\text{class}} = \frac{1}{h^{3N}} \int\!\!\int \ldots \int\!\!\int \exp\left(-E/kT\right) \mathrm{d}x_1 \, \mathrm{d}x_2 \ldots \mathrm{d}x_{3N} \, \mathrm{d}p_1 \ldots \mathrm{d}p_{3N} \quad \text{(A2.8)}$$

In (A2.8) $x_1, \ldots, x_{3N}$ are the $3N$ position coordinates and $p_1, \ldots, p_{3N}$ the $3N$ momenta. The energy E is expressed as a function of these $6N$ variables, and the integration limits are such that the integration covers all relevant regions of phase space. The factor $(1/h^{3N})$ has been introduced so that for comparable systems (A2.8) gives the same result as (A2.1), although the multiplication of Q by a constant factor makes no difference for many purposes, including rate theory applications. The classical treatment can obviously be derived as a limiting case of the quantum version for cases where energy differences become small compared with kT, and treatments of this sort have been used in sections 4.4 and 4.5.

REFERENCES

1. See, for example, R. C. Tolman, *Foundations of Statistical Mechanics*, Oxford University Press (1938); R. H. Fowler and E. A. Guggenheim, *Statistical Thermodynamics*, Cambridge University Press (1939); N. Davidson, *Statistical Mechanics*, McGraw-Hill, New York (1962); G. S. Rushbrooke, *Introduction to Statistical Mechanics*, Oxford University Press (1949); D. A. McQuarrie, *Statistical Mechanics*, Harper/Collins (1990).
2. See, for example, G. N. Lewis and M. Randall, revised by D. S. Pitzer and L. Brewer, *Thermodynamics*, 2nd edn, McGraw-Hill, New York, Ch. 27 (1961).

3. G. Herzberg, *Molecular Spectra and Molecular Structure*, Vol. 2, *Infrared and Raman Spectra of Polyatomic Molecules*, sections IV, 5 and V, 1, Van Nostrand, Princeton (1945).
4. H. Eyring and M. Polanyi, *Zeit. Phys. Chem. (B)*, **12**, 279 (1931).
5. D. L. Bunker, *Theory of Elementary Gas Reaction Rates*, Pergamon, Oxford, p. 52 (1966).

Appendix 3 The Vibrational Analysis of Polyatomic Molecules

A3.1 DEVELOPMENT OF THE SECULAR EQUATION

Polyatomic molecules at normal temperatures undergo complicated vibrational motions. For a given non-linear molecule with N atoms it can be shown that, for small displacements of the atoms from their mean positions, the motion can be resolved into $3N - 6$ constituent *normal modes of vibration*. In a normal mode of vibration the atoms all execute simple harmonic motion about their mean positions with the same phase and with the same frequency.

Vibrational analysis is the treatment of the vibrations of a polyatomic molecule by the methods of classical mechanics. The motions of N atoms of masses $m_1, \ldots, m_N$ can be described by $3N$ Cartesian displacement coordinates $x_1, \ldots, x_{3N}$. These might be given the symbols $x_1, y_1, z_1, x_2, y_2, z_2, \ldots, z_N$, but the present terminology leads to simpler equations. The potential energy V is then given in general form, for small displacements, by

$$2V = k_{11}x_1^2 + k_{12}x_1x_2 + \ldots + k_{3N,3N}x_{3N}^2 \tag{A3.1}$$

where the k_{ij} are constants, and the kinetic energy T is given by

$$2T = m_1(\dot{x}_1^2 + \dot{x}_2^2 + \dot{x}_3^2) + m_2(\dot{x}_4^2 + \dot{x}_5^2 + \dot{x}_6^2) + \ldots + m_N(\ldots \dot{x}_{3N}^2) \tag{A3.2}$$

Equations of motion can then be written for each coordinate x_i. In Lagrangian form these are

$$\mathrm{d}/\mathrm{d}t(\partial \mathscr{L}/\partial \dot{x}_i) - \partial \mathscr{L}/\partial x_i = 0 \tag{A3.3}$$

where the Lagrange function $\mathscr{L}$ is related to the kinetic energy T and the potential energy V by

$$\mathscr{L} = T - V \tag{A3.4}$$

Hence

$$\mathrm{d}/\mathrm{d}t(\partial T/\partial \dot{x}_i) - (\partial T/\partial x_i) + (\partial V/\partial x_i) = 0$$

Since, from (A3.1) and (A3.2),

$$\partial V/\partial x_i = k_{11}x_1 + k_{12}x_2 + k_{13}x_3 + \ldots + k_{1,3N}x_{3N}$$
$$\partial T/\partial \dot{x}_1 = m_1\dot{x}_1 \quad \text{and} \quad \partial T/\partial x_1 = 0$$

the equation of motion involving x_1 is then

$$m_1\ddot{x}_1 + k_{11}x_1 + k_{12}x_2 + \ldots + k_{1,3N}x_{3N} = 0 \qquad (A3.5)$$

which is simply an extension of the Newtonian equation, *mass × acceleration = force*.

There will be $3N$ equations of this kind, applicable to the $3N$ coordinates $x_1, \ldots, x_{3N}$. In a normal mode of vibration (see above), each atom executes simple harmonic motion about its mean position with the same frequency v, and so it is necessary to apply the conditions defining simple harmonic motion in each coordinate, namely

$$\ddot{x}_i = -\lambda x_i \quad (\text{where } \lambda = 4\pi^2 v^2)$$

This leads to a new set of simultaneous equations

$$\left.\begin{array}{l} (k_{11} - m_1\lambda)x_1 + k_{12}x_2 + k_{13}x_3 + \ldots + k_{1,3N}x_{3N} = 0 \\ \cdot \\ k_{3N,1}x_1 + k_{3N,2}x_2 + \ldots + (k_{3N,3N} - m_N\lambda)x_{3N} = 0 \end{array}\right\} \qquad (A3.6)$$

which are linear in the x_i. The condition for the solution of these equations is that the *secular determinant* is equal to zero:

$$\begin{vmatrix} k_{11} - m_1\lambda & k_{12} & k_{13} & \ldots & k_{1,3N} \\ \vdots & \vdots & \vdots & \vdots & \vdots \\ k_{3N,1} & k_{3N,2} & k_{3N,3} & \ldots & k_{3N,3N} - m_N\lambda \end{vmatrix} = 0$$

This secular equation is a polynomial equation in λ^{3N}, and thus leads to $3N$ values of the roots λ. It can be shown, however, that for a non-linear polyatomic molecule, six of the roots are zero, corresponding to the three translations and three overall rotations of the molecule. The equation thus reduces to one of order $3N - 6$ and there are $3N - 6$ roots λ_k corresponding to $3N - 6$ normal mode frequencies v_k. For a linear polyatomic molecule there are $3N - 5$ non-zero roots and hence $3N - 5$ normal mode frequencies.

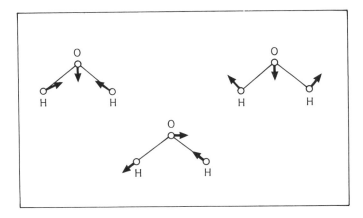

Figure A3.1 Normal modes of vibration of the water molecule. The directions of the arrows represent the directions of the relative motions of the nuclei. The magnitudes of the displacements are not to scale, however; those of the oxygen atom have been exaggerated for clarity.

A frequent use of vibrational analysis is the calculation of the normal mode frequencies for a molecule. In order to do this a potential energy expression of the form (A3.1) is assumed using reasonable values of the force constants k_{ij}. For the purposes of Slater theory the secular equation is used to obtain the form of the normal mode vibrations which may be described in terms of the coordinates x_i.

Since the equations (A3.6) are homogeneous, it is not possible to obtain absolute values of the coordinates x_i but only their ratios. Substitution of a particular value of a non-zero root λ_k into the secular equation enables the ratios $x_1 : x_2 : \ldots : x_{3N}$ to be determined. These ratios determine the form of the normal mode vibration of frequency v_k and indicate the relative displacements and directions of displacement of the individual atoms when the molecule is vibrating in this particular normal mode.

As a simple example, the water molecule is a non-linear triatomic molecule with $3 \times 3 - 6 = 3$ modes of vibration. Vibrational analysis enables the ratios of the displacements in each normal mode to be calculated. These determine the form of the normal mode vibrations which are illustrated in Figure A3.1.

A3.2   NORMAL COORDINATES

The solutions of the secular equation which yield $3N - 6$ non-zero values of λ_k can be described in terms of $3N - 6$ normal coordinates Q_k. If normal coordinates Q_k can be found which result in potential and kinetic energy expressions of the form (A3.7) and (A3.8), where $\lambda_k = 4\pi^2 v_k^2$, then the equation of motion becomes $3N - 6$

$$V = \sum_{k=1}^{3N-6} Q_k^2 \qquad (A3.7)$$

$$T = \sum_{k=1}^{3N-6} \dot{Q}_k^2 / \lambda_k \qquad (A3.8)$$

separable equations of the form

$$\ddot{Q}_k + \lambda_k Q_k = 0 \qquad (A3.9)$$

for which there are $3N - 6$ independent solutions of the form

$$Q_k = A_k \cos(2\pi v_k t + \psi_k)$$

Each solution corresponds to a normal mode of vibration with the corresponding normal mode frequency v_k, amplitude, A_k and phase ψ_k. For use in Slater theory the Q_k are normalised so that the coefficients of Q_k^2 in the potential energy expression (A3.7) are unity. This has the useful corollary that, as may readily be verified, $A_k = \sqrt{E_k}$, where E_k is the energy in the kth mode.

A3.3   INTERNAL AND SYMMETRY COORDINATES

In the preceding discussion the secular equation (A3.6) is expressed in terms of Cartesian coordinates $x_1 \ldots x_{3N}$ etc. It will be apparent from the examples of vibrational analysis given below that simplification of the secular determinant may be achieved by the use of *internal coordinates* (r) relating to bond distances within the molecule. For complicated molecules the relationship between internal and normal coordinates is not usually obvious, and instead knowledge of the symmetry of the molecules is utilised to construct *symmetry coordinates*[1] (s). These produce even greater simplification and may be used in the application of Slater theory in order to calculate the relevant contribution from each normal coordinate to the critical coordinate chosen for a particular reaction. This calculation and the detailed relationships between normal and symmetry coordinates are dealt with in section 2.7.

A3.5   EXAMPLES OF THE APPLICATION OF VIBRATIONAL ANALYSIS

APPLICATION TO TWO INDEPENDENT OSCILLATORS

Suppose we have two masses m_1 and m_2 at the end of two separate springs independently fixed to a firm support. If they are given displacements x_1 and x_2 from their mean positions, the total Lagrangian function (A3.4) for the system can be written as

$$\mathscr{L} = \tfrac{1}{2}m_1\dot{x}_1^2 + \tfrac{1}{2}m_2\dot{x}_2^2 - \tfrac{1}{2}k_1x_2^2 - \tfrac{1}{2}k_2x_2^2 \qquad (A3.10)$$

Application of (A3.3) then gives two separate equations of motion (A3.11) and (A3.12):

$$m_1\ddot{x}_1 + k_1x_1 = 0 \qquad (A3.11)$$

$$m_2\ddot{x}_2 + k_2x_2 = 0 \qquad (A3.12)$$

Since the oscillators are independent, no cross-terms occur in (A3.10), and hence (A3.11) and (A3.12) are completely separable. The solutions to (A3.11) and (A3.12) are each of the form $x_i = A_1 \cos(2\pi v_i t)$; thus the coordinates x_1 and x_2 in this simple case are themselves normal coordinates, and each oscillator executes simple harmonic motion independently of the other oscillator. For consistency with the forms (A3.7) and (A3.8) the renormalised coordinates

$$Q_1 = x_1\sqrt{(k_1/2)} \quad \text{and} \quad Q_2 = x_2\sqrt{(k_2/2)}$$

would be used.

APPLICATION TO A DIATOMIC MOLECULE

Suppose we have a diatomic molecule with atomic masses m_1 and m_2, and that the atoms are subjected to displacements x_1 and x_2 from their mean positions, as shown in Figure A3.2. The kinetic energy is given by

$$T = \tfrac{1}{2}m_1\dot{x}_1^2 + \tfrac{1}{2}m_2\dot{x}_2^2 \qquad (A3.13)$$

and the potential energy can be expressed in terms of the net extension of the internuclear distance ($x_2 - x_1$) by

$$V = \tfrac{1}{2}k(x_2 - x_1)^2 = \tfrac{1}{2}kx_1^2 + \tfrac{1}{2}kx_2^2 - kx_1x_2 \qquad (A3.14)$$

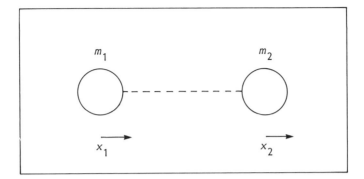

Figure A3.2 Coordinates for the diatomic molecule.

From (A3.13) and (A3.14),

$$\mathscr{L} = T - V = \tfrac{1}{2}m_1\dot{x}_1^2 + \tfrac{1}{2}m_2\dot{x}_2^2 - \tfrac{1}{2}k(x_1^2 + x_2^2 - 2x_1x_2)$$

Hence from (A3.3) the equations of motion are

$$m_1\ddot{x}_1 + kx_1 - kx_2 = 0 \qquad (A3.15)$$

$$m_2\ddot{x}_2 + kx_2 - kx_1 = 0 \qquad (A3.16)$$

Unlike (A3.11) and (A3.12) these equations are not completely separable since (A3.15) involves x_2 and (A3.16) involves x_1. It is possible, however, by suitable choice of coordinates, to write equations from which cross-terms are eliminated and which then become separable. Using new coordinates defined by

$$r = x_2 - x_1 \quad \text{and} \quad R = (m_1x_1 + m_2x_2)/M$$

where

$$M = m_1 + m_2$$

then the kinetic and potential energies become

$$T = \tfrac{1}{2}M\dot{R}^2 + \tfrac{1}{2}\mu\dot{r}^2 \quad \text{and} \quad V = \tfrac{1}{2}kr^2$$

where

$$\mu = m_1m_2/M$$

The Lagrangian function (A3.4) is then

$$\mathscr{L} = T - V = \tfrac{1}{2}M\dot{R}^2 + \tfrac{1}{2}\mu\dot{r}^2 - \tfrac{1}{2}kr^2$$

which leads to two equations of motion. The first involves only R:

$$d(M\dot{R})/dt = 0 \quad \text{or} \quad m_1\dot{x}_1 + m_2\dot{x}_2 = \text{constant} \qquad (A3.17)$$

This is simply a statement of the law of conservation of momentum, and if the centre of mass of the molecule is taken to be at rest the constant in (A3.17) is in fact zero. The second equation involves only r:

$$d/dt(\partial\mathscr{L}/\partial\dot{r}) - \partial\mathscr{L}/\partial r = 0$$

whence

$$\mu \ddot{r} + kr = 0$$

for which the solution is

$$r = A \cos\left(\sqrt{\frac{k}{\mu}}t + \psi\right)$$

which corresponds to the normal mode of vibration of a diatomic molecule. Thus the coordinate R and r in this case are both normal coordinates although only one (r) represents a genuine vibration. Again, normalization to fit (A3.7) and (A3.8) requires definition of a coordinate $Q = r\sqrt{(k/2)}$, in which case

$$Q = A' \cos\left(\sqrt{\frac{k}{\mu}}t + \psi\right)$$

$$= \sqrt{E_{\text{vib}}} \cos\left(\sqrt{\frac{k}{\mu}}t + \psi\right)$$

(A3.18)

APPLICATION TO A LINEAR TRIATOMIC MOLECULE

We consider in this section an extension of the above method to the stretching vibrations of a linear symmetric triatomic molecule ABA, i.e. the modes which involve displacements along the internuclear axis only. In restricting motion to the internuclear axis, we are ignoring bending vibrations and also rotations. If the masses and displacements of the atoms are defined as in Figure A3.3, then

$$T = \tfrac{1}{2}m_1\dot{x}_1^2 + \tfrac{1}{2}m_2\dot{x}_2^2 + \tfrac{1}{2}m_1\dot{x}_3^2$$

Defining new coordinates by

$$r_1 = x_2 - x_1 \qquad r_2 = x_3 - x_2$$

$$R = (m_1x_1 + m_2x_2 + m_1x_3)/M$$

where

$$M = 2m_1 + m_2$$

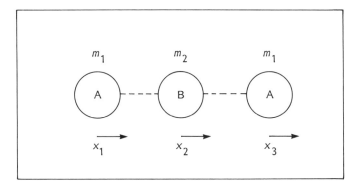

Figure A3.3 Coordinates for the linear triatomic molecule ABA.

then

$$V = \tfrac{1}{2}kr_1^2 + \tfrac{1}{2}kr_2^2 + ar_1r_2$$

where a is a constant, and

$$T = \tfrac{1}{2}M\dot{R}^2 + \frac{\tfrac{1}{2}m_1(m_1 + m_2)}{M}(\dot{r}_1^2 + \dot{r}_2^2) + \frac{m_1^2}{M}\dot{r}_1\dot{r}_2$$

Ignoring the coordinate R which, as before, leads to an equation referring to translation of the molecule as a whole, the Lagrangian in terms of coordinates r_1 and r_2 is given by

$$\mathscr{L} = \frac{\tfrac{1}{2}m_1(m_1 + m_2)}{M}(\dot{r}_1^2 + \dot{r}_2^2) + \frac{m_1^2}{M}\dot{r}_1\dot{r}_2 - \tfrac{1}{2}kr_1^2 - \tfrac{1}{2}kr_2^2 - ar_1r_2 \quad (\text{A3.19})$$

This equation is not directly separable in terms of the coordinates r_1 and r_2 since cross-products appear. It is, however, possible to introduce new coordinates to eliminate the cross-products; suitable expressions are the 'symmetry coordinates' (A3.20) and (A3.21)

$$Y_1 = (r_1 + r_2)/\sqrt{2} \qquad\qquad (\text{A3.20})$$

$$Y_2 = (r_1 - r_2)/\sqrt{2} \qquad\qquad (\text{A3.21})$$

Rotation of the molecule by 180° about the y or z axes (see Figure A3.4) changes r_1 into r_2 and r_2 into r_1. Under these conditions Y_1 remains unchanged, but Y_2 changes sign. Since the potential and kinetic energies must remain invariant towards such 'symmetry operations', terms involving the product Y_1Y_2 cannot occur in the expressions for T and V in terms of Y_1 and Y_2.

Further, since

$$r_1^2 + r_2^2 = Y_1^2 + Y_2^2$$

and

$$Y_1^2 - Y_2^2 = 2r_1r_2$$

the Lagrangian (A3.19) becomes

$$\mathscr{L} = \tfrac{1}{2}m_1\dot{Y}_1^2 + \tfrac{1}{2}\mu\dot{Y}_2^2 - \tfrac{1}{2}(k + a)Y_1^2 - \tfrac{1}{2}(k - a)Y_2^2$$

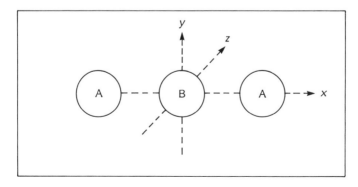

Figure A3.4 Rotation axes for the linear triatomic molecule.

where

$$\mu = m_1 m_2 / M$$

This is now separable and application of (A3.3) leads to

$$m_1 \ddot{Y}_1 - (k + a) Y_1 = 0 \tag{A3.22}$$

$$\mu \ddot{Y}_2 - (k - a) Y_2 = 0 \tag{A3.23}$$

Hence Y_1 and Y_2 are normal coordinates and describe two different stretching modes of vibration. For use in Slater theory we require coordinates normalised to fit the simple forms (A3.7) and (A2.8), and thus we write:

$$Q_1 = [(k + a)/2]^{1/2} Y_1 = [(k + a)^{1/2}/2](r_1 + r_2) \tag{A3.24}$$

$$Q_2 = [(k - a)/2]^{1/2} Y_2 = [(k - a)^{1/2}/2](r_1 - r_2) \tag{A3.25}$$

Then (A3.22) and (A3.23) become

$$\ddot{Q}_1 - [(k + a)/m_1] Q_1 = 0$$

$$\ddot{Q}_2 - [(k - a)/m_2] Q_2 = 0$$

whence

$$\lambda_1 = 4\pi^2 v_1^2 = (k + a)/m_1$$

$$\lambda_2 = 4\pi^2 v_2^2 = (k - a)/\mu$$

The variation of Q_1 with time in the normal vibration of frequency v_1 is given by

$$Q_1 = \sqrt{E_1} \cos(2\pi v_1 t + \psi_1) \tag{A3.26}$$

Since $Q_1 = \text{constant} \times (r_1 + r_2)$ and r_1 and r_2 both represent increases in the interatomic distances, it can be seen from (A3.26) that Q_1 is the normal coordinate describing the symmetrical mode of vibration. Similarly, Q_2 represents the asymmetric mode in which $(r_1 - r_2)$ increases and decreases with time according to

$$Q_2 = \sqrt{E_2} \cos(2\pi v_2 t + \psi_2) \tag{A3.27}$$

From (A3.24) and (A3.25) we can see how the internal coordinates r_1 and r_2 vary with time when the modes represented by Q_1 and Q_2 are both present; for r_1, for example:

$$r_1 = (k + a)^{-1/2} Q_1 + (k - a)^{-1/2} Q_2 \tag{A3.28}$$

$$= [E_1^{1/2}/(k + a)^{1/2}] \cos(2\pi v_1 t + \psi_1) + [E_2^{1/2}/(k - a)^{1/2}] \cos(2\pi v_2 t + \psi_2) \tag{A3.29}$$

Equation (A3.29) represents a complicated behaviour of the coordinate r_1 with time. Such behaviour for a typical internal coordinate q_1 of any polyatomic molecule is schematically represented by Figure 2.4 in Chapter 2. It can be seen that the coefficients multiplying the amplitude factors ($\sqrt{E_1}$ and $\sqrt{E_2}$ in (A3.29)) are the coefficients of the transformation from internal to normal coordinates in (A3.28).

REFERENCES

1. E. B. Wilson, J. C. Decius and P. C. Cross, *Molecular Vibrations*, McGraw-Hill, New York, p. 21 (1955).

Appendix 4 The Gamma Function $\Gamma(n)$

The gamma function $\Gamma(n)$ is defined, for $n > 0$, by the integral

$$\Gamma(n) = \int_{x=0}^{\infty} x^{n-1} \exp(-x)\,dx \quad (n > 0)$$

The function has many interesting and useful properties, but in the present work its argument n is restricted to positive integral or half-integral values, and only three points need to be noted:

$$\Gamma(n + 1) = n\Gamma(n)$$

$$\Gamma(1) = 1$$

$$\Gamma(\tfrac{1}{2}) = \sqrt{\pi}$$

Thus, if n is integral we have

$$\Gamma(n) = (n - 1)\Gamma(n - 1) = (n - 1)\Gamma(n - 2) = \ldots$$

$$= (n - 1)(n - 2) \ldots 3.2.1.\Gamma(1) = (n - 1)!$$

e.g.

$$\Gamma(\tfrac{7}{2}) = \tfrac{5}{2} \cdot \tfrac{3}{2} \cdot \tfrac{1}{2} \cdot \sqrt{\pi}$$

Table A4.1 gives values $\Gamma(n)$ for small values of n. Values for larger n will generally be evaluated as required in computer programs, or may occur as ratios of gamma functions which are better simplified by cancellation of corresponding terms.

Table A4.1 Gamma functions $\Gamma(n)$ for integral and half-integral n

n	$\Gamma(n)$	n	$\Gamma(n)$
0.5	$\sqrt{\pi} = 1.7725$	3.5	$\frac{15}{8}\sqrt{\pi} = 3.3234$
1.0	$0! = 1.0000$	4.0	$3! = 6.0000$
1.5	$\frac{1}{2}\sqrt{\pi} = 0.8862$	4.5	$\frac{105}{16}\sqrt{\pi} = 11.6317$
2.0	$1! = 1.0000$	5.0	$4! = 24.0000$
2.5	$\frac{3}{4}\sqrt{\pi} = 1.3293$	5.5	$\frac{945}{32}\sqrt{\pi} = 52.3428$
3.0	$2! = 2.0000$	6.0	$5! = 120.000$

Appendix 5   Monte Carlo Algorithm for the Evaluation of Vibrational State Sums

In this appendix a brief outline of Barker's[1] Monte Carlo algorithm for the calculation of state sums is given. Attention is confined exclusively to the case of a system of s coupled anharmonic oscillators the total energy, $E + E_z$, of which is given by equation (4.7). Only energies, E, in excess of the zero-point energy, E_z, are of interest and so equation (4.7) can be rewritten as

$$E/hc = \sum_{i=1}^{s} v_i v_i + \sum_{i=1}^{s} \sum_{j \leqslant 1}^{s} v_i v_j x_{ij} \qquad (A5.1)$$

where v_i is the quantum number associated with the ith mode, v_i is the corresponding harmonic frequency and x_{ij} are the anharmonicity and coupling coefficients.

Before looking in detail at Barker's algorithm, it is useful to examine the basic idea of the Monte Carlo approach as outlined in section 4.3. The approach can be illustrated by the simple example of finding the area of a circle as follows:

1. Consider a circle of radius r, in the Cartesian (x, y) plane it is described by

$$r^2 = x^2 + y^2 \qquad (A5.2)$$

   The circle is totally enclosed in a square centred at the origin and having an area $V_0 = (2r)^2$.
2. Two registers are initialised to zero, N, the total number of trials and n, the number of successful trials.
3. Values of x and y are now chosen at random from the interval $-r \leqslant x, y \leqslant r$ with uniform distribution and the function $f(x, y) = x^2 + y^2$ calculated.
4. If $f(x, y) \leqslant r^2$ then n is incremented by one, otherwise it is left unchanged. In either case N is always incremented by one.
5. Steps 3 and 4 are repeated many times until the fraction n/N converges to a prescribed precision. The area of the circle, S, is then obtained as

$$S = V_0 n/N = 4r^2 n/N \qquad (A5.3)$$

For the circle, of course, we have an analytic solution to the problem, i.e. $S = \pi r^2$

and so the fraction n/N is clearly an estimate of $\pi/4$. In section 4.3 the efficiency of the Monte Carlo procedure, e, was defined as

$$\varepsilon = n/N \tag{A5.4}$$

and it is desirable to keep ε as close to unity as possible, as this minimises the number of trials required for a given precision. For the above example the efficiency is thus $\pi/4$ or about 75%.

The same ideas can be applied to a sphere of radius r totally enclosed in a cube of side $2r$. Again, analytic solutions exist for the volume of these objects and it is easy to show the ε for the proposed Monte Carlo method for determining the volume of a sphere is given by

$$\varepsilon = \tfrac{4}{3}\pi r^3/(2r)^3 = \pi/6 \tag{A5.5}$$

or about 50%. As the number of dimensions increases the efficiency decreases rapidly. For example the efficiency for a 12-dimensional hypersphere of radius r enclosed in an equivalent hypercube is about 0.03%. This basic form of the Monte Carlo method thus becomes hopelessly inefficient for determining volumes enclosed by multidimensional functions such as given by equation (A5.1).

Barker[1] addressed this difficulty by finding ways of increasing the number of successful trials. The basis of the approach is to find a sampling domain that is as close to the shape of the desired function as possible, and so reduces the number of unsuccessful trials. In general this new domain will be of a more complex shape than the cuboids selected in the above examples, and, because of this, the range from which a given variable is chosen will be conditional on the values chosen for the previously selected variables. This has the side effect of producing uneven distributions throughout the sampling volume and so weighting factors are required to account for this. The procedure is best illustrated by example and the following is a description of Barker's algorithm as applied to the example of a system of coupled anharmonic oscillators:

1. A sampling domain is chosen which completely encloses the integration region. It is defined by an $(s - 1)$ dimensional boundary surface, B:

$$B = f(v_1, v_2, \ldots, v_s) \tag{A5.6}$$

   In the present case the most obvious surface to choose is that given by equation (A5.1), thus $B = E$. For more complex functions such a surface assignment is not always appropriate. Also note that in this example the v_i take only integer values.
2. All the values of v_i are set to zero and the maximum value of v_1, consistent with equation (A5.6), is calculated. There is a restriction on the maximum value of v_1, in that $v_{1,\text{max}}$ should be less than or equal to the dissociation threshold of that particular mode. The range of v_1, R_1 is set equal to $v_{1,\text{max}} + 1$.
3. A random value for v_1 is selected such by truncating UR_1, where U is a uniformly distributed random number ($0 \leqslant U < 1$).
4. Steps 2 and 3 are repeated for each v_i, but the values already selected for the preceding variables are used in equation (A5.6). In this way the boundary B is being refined throughout the selection procedure and because of this the value of R_i, the range allowed to the ith mode, is conditional on the values selected for the previous $i - 1$ modes.
5. The weighting factor, w_g for the gth point,

$$w_g = \prod_{i=1}^{s} R_i^{-1} \tag{A5.7}$$

is calculated. With a little thought it can be seen that this weighting factor corresponds to the reciprocal of the volume from which the sample point is drawn.

6. The following sum:

$$S = \frac{V_0}{GN} \sum_{g=1}^{N} w_g^{-1}$$ (A5.8)

is updated. The sum S will converge to the sum of states for the system described by equation (A5.1). The remaining terms are: w_g is the weight factor (A5.7), N is the number of trials, V_0 is the volume of a hypercuboid that totally encloses the domain of integration and G is the number of states contained with in that hypercuboid. In effect the ratio G/V_0 is the density of states points within the hypercuboid.

7. A statistical test is made to ascertain if the required precision has been reached; if so the calculation is terminated, otherwise the algorithm returns to step 1 and a new trial is initiated.

Equation (A5.8) requires a little more explanation: in step 5 it was seen that w_g^{-1} is related to the volume from which the sample point is drawn. The average that is defined in equation (A5.8) is, in a sense, an average volume from which sample points are drawn and it will eventually converge to the volume that is enclosed by equation (A5.1).

REFERENCES

1. J. R. Barker, *J. Phys. Chem.*, **91**, 3849 (1987).

Appendix 6 Matrix Solution of the Energy-grained Master Equation

In Chapter 7, the energy-grained master equation (equation 7.16) which needs to be solved is written as

$$\frac{d}{dt} n = \mathbf{M} n \qquad (A6.1)$$

where $\mathbf{M}$ is a matrix and n a column vector. A characteristic property of the matrix $\mathbf{M}$ is that there is a set of vectors $\{u_i\}$ such that

$$\mathbf{M} u_i = \lambda_i u_i \qquad (A6.2)$$

where λ_i is a constant associated with the vector u_i. Such a set of vectors are referred to as eigenvectors and the λ_i are the corresponding eigenvalues. The eigenvectors, which are column vectors, can be collected together to form a square matrix of the same dimensions as $\mathbf{M}$, and this allows (A6.2) to be rewritten in the form

$$\mathbf{M} \mathbf{U} = \mathbf{U} \mathbf{\Lambda} \qquad (A6.3)$$

where $\mathbf{\Lambda}$ is a diagonal matrix whose diagonal elements are the eigenvalues. Note the order of matrix multiplication is not arbitrary and because $\mathbf{U}$ is to the right of $\mathbf{M}$ the eigenvectors are referred to as the right eigenvectors. Equation (A6.3) allows $\mathbf{M}$ to be rewritten as

$$\mathbf{M} = \mathbf{U} \mathbf{\Lambda} \mathbf{U}^{-1} \qquad (A6.4)$$

Substitution of (A6.4) into (A6.1) gives

$$\frac{d}{dt} n = \mathbf{U} \mathbf{\Lambda} \mathbf{U}^{-1} n \qquad (A6.5)$$

Multiplying through by $\mathbf{U}^{-1}$ and interchanging $\mathbf{U}^{-1}$ and d/dt on the left-hand side gives

$$\frac{d}{dt} \mathbf{U}^{-1} n = \mathbf{\Lambda} \mathbf{U}^{-1} n \qquad (A6.6)$$

Defining

$$r = \mathbf{U}^{-1} n(t) \qquad (A6.7)$$

$$r_0 = \mathbf{U}^{-1} n(0) \qquad (A6.8)$$

where $n(0)$ is the initial population vector, followed by substitution gives

$$\frac{\mathrm{d}}{\mathrm{d}t}r = \Lambda r \tag{A6.9}$$

Equating elements of both sides gives,

$$\frac{\mathrm{d}r_i}{\mathrm{d}t} = \lambda_i r_i \tag{A6.10}$$

which is easily solved to give

$$r_i = \exp(\lambda_i t) r_{i,0} \tag{A6.11}$$

or in matrix notation

$$r = \exp(\Lambda t) r_0 \tag{A6.12}$$

Finally, substitution for r and r_0 using (A6.7) and (A6.8) gives

$$n(t) = \mathbf{U} \exp(\Lambda t) \mathbf{U}^{-1} n(0) \tag{A6.13}$$

This equation is identical to (7.17) used subsequently in Chapter 7.

Index